OUR FIRST FOREIGN WAR

AF086029

Newtown Park camp in Wellington, which housed the Second Contingent and elements of other contingents prior to their departure for South Africa. After rain, the camp became a muddy quagmire and in 1901 Volunteer corps members infuriated Premier Seddon by parading down Lambton Quay in a protest at the quality of camp rations. ALEXANDER TURNBULL LIBRARY, 1/1-006666-G

OUR FIRST FOREIGN WAR

THE IMPACT OF THE SOUTH AFRICAN WAR 1899–1902 ON NEW ZEALAND

NIGEL ROBSON

MASSEY UNIVERSITY PRESS

This book is dedicated to my wife, Cho Young-hae,
whose unwavering support, encouragement and patience
over many years has made it possible.

Contingent members ride onto Queen's Wharf in Wellington to board their troopship for South Africa. ALEXANDER TURNBULL LIBRARY, 1/2-110815-F

CONTENTS

Preface 9

1. 'The flag that floats over us'
   Patriotism and South Africa 17

2. 'Rally to the call of home and country'
   Domestic reaction to the war 33

3. 'An especially fine lot of fighting-men'
   The performance of New Zealand soldiers
   during the South African War 87

4. 'Loyalty to the British Empire'
   Māori responses to the South African War 139

5. 'Yelling yahoos in yellow'
   The behaviour of New Zealand soldiers
   during the South African War 167

6. 'Maimed, crippled and completely broken'
   The human cost of the war 219

7. 'These wars will always be popular'
   The economic impact of the South African War ... 275

Epilogue 335
Notes 344
Glossary 398
Acknowledgements 400
About the author 401
Index 402

PREFACE

On a busy street in the Wellington suburb of Johnsonville, a wrought-iron street lamp stands incongruously as a reminder of a largely forgotten war. Awkwardly wedged between a medical clinic and a real-estate agent, its design is in stark contrast to the architecture of its surroundings. Although its concrete base is chipped, its marble tablet discoloured and its three original glass globes long ago replaced by a single four-sided lantern, the lamp nonetheless hints at its former grandeur. Its unveiling on an autumn day in 1905 presented a very different spectacle.

Arriving from the city by special train, Sir Joseph Ward, a senior Cabinet minister and member of the House of Representatives (MHR) for the Southland electorate of Awarua, addressed the crowd that had gathered for the occasion. While New Zealand's governor, Lord Plunket, and Premier Richard Seddon forwarded their apologies for not attending, among those present were Defence Department officials, William Field, the MHR for Ōtaki electorate, the chairman and members of the Johnsonville Town Board, school cadets, and members of the public, including the parents and brother of Leonard Retter, the local blacksmith in whose honour the 'very handsome' acetylene lamp paid for by public subscription had been erected.[1]

Five years earlier and 11,000 kilometres away, war had broken out in South Africa. The conflict, which continued until 1902, pitted the combined military forces of the United Kingdom and contingents from other nations of the British Empire against those of the two Boer republics — the South African Republic and Orange Free State. One of many young New Zealanders eager to take part, Retter enlisted in the Seventh Contingent in April 1901. Nine months later, he was among 23 Seventh Contingent men who lost their lives during a desperate Boer night attack on New Zealand positions on a hillside in Orange Free State.

It was the biggest single loss of life by New Zealand troops during the war, and its significance was reflected in the Johnsonville unveiling ceremony. Yet, today, comparatively few New Zealanders are familiar with what occurred at Langverwacht (or the Battle of Bothasberg, as it was then known), and most have little more than a cursory knowledge of the war in which it took place. In my own case, I have no clear recollection of when I first heard of the 'Boer War', as the conflict was commonly known during my childhood, but it was sometime in the late 1960s or early 1970s.

Once a week, my grandmother would visit our home and, if I was lucky, she would ask my mother to retrieve a battered leather suitcase from its place in a cupboard beyond my reach. Inside were the remaining possessions of her husband, a First World War veteran who had survived the horrors of the Western Front only to die in a car accident in 1941. Lifting the suitcase's lid was like being transported back in time. Neatly arranged within were the three service medals of the grandfather I had never known, dulled by time, but still suspended from their brightly coloured ribbons. Beside them were a pair of enamelled cufflinks, bought in Egypt in 1916 and made in the form of sarcophagi, a faded French flag souvenired from a Paris street on Armistice Day 1918, and my grandfather's stitched Medical Corps Red Cross sleeve patch.

After my grandmother's death, I became custodian of the suitcase's contents and, as time passed, slowly expanded the collection. At the time, it was not difficult to acquire military items brought back to New Zealand by veterans, but the oldest item in my collection came neither from France nor Egypt — it was a book on the Boer War published in 1900. Its spotted pages featured patriotic engravings, including depictions of British soldiers gallantly taking Boer positions at the point of a bayonet. As a child, I considered neither the accuracy of the images nor the book's repeated references to the Boer War, a title that both downplayed British involvement and implied that the responsibility for the death and suffering that occurred lay solely with South Africa's Boer population. The war now goes by a number of names, but in an attempt to correct this bias

I have chosen to refer to it simply as the South African War.

The Calvinist Protestant Boers were descendants of Dutch settlers who had emigrated from Europe to the Cape of Good Hope in the mid-seventeenth century and were later joined by French and German Huguenots. Although these settlers mainly spoke Dutch, over time the Boers developed their own language, Afrikaans, which combined Dutch with elements of other languages in the region. The early Boer settlers established Kaapkolonie (Cape Colony), which was administered for approximately 150 years by the Dutch East India Company. Concerned by the prospect of France securing a foothold in southern Africa, the British first took control of Cape Colony in 1795 following the Battle of Muizenberg, before returning it to the Boers in 1802 and then resuming control again in 1806 during the Napoleonic Wars.

Although some Boers had moved northwards in the eighteenth century, in 1834 the British increased resentment among the Boers by abolishing slavery in the colony. This, together with the imposition of the English language and British law, saw thousands of disaffected Boers embark on Die Groot Trek (the Great Trek), a migration to the north-east. This took them into regions that were largely uninhabited due to what indigenous Africans call Mfecane (the scattering) — the chaos and devastation caused by Zulu attacks on other African tribes living in the area. Nonetheless, the Boer migration increasingly brought them into contact, and sometimes conflict, with the indigenous African population. In 1843 the British annexed the Boer republic of Natalia, which became the British colony of Natal. In 1852 the Boers established the Suid-Afrikaanse Republiek (the South African Republic), followed by the adjoining Oranje Vrystaat (Orange Free State) in 1854.[2]

Although the British initially recognised the independence of the two Boer nations, concerns that the expansion of German interests in Africa could threaten British colonies in the region saw the British annex the South African Republic in 1877. Tension between the British and the Boers, who resented British attempts to again exert control over their affairs, led to the First Anglo-Boer War (1880–81), in which the British were defeated. The South African Republic, known to the British as Transvaal State, obtained full independence in 1884.

In late 1895, tensions again arose when Leander Starr Jameson, a colonial administrator and confidant of Cape Colony prime minister (and ardent imperialist) Cecil Rhodes, invaded Transvaal with a small force of predominantly British South Africa Company police. The invasion, known as the Jameson Raid, was an attempt to overthrow the Boer government of Paul Kruger, the president of the South African Republic. The raiders aimed to foment an uprising in Transvaal among the 'uitlanders' (the predominantly British immigrants living in and around Johannesburg), and wrest the region's extensive gold reserves from Boer control.

The raid ended in ignominious defeat when Jameson and his men were quickly overwhelmed and forced to surrender to the Boers. Though the British government denied any involvement in the ill-advised debacle, the Jameson Raid nonetheless proved acutely embarrassing. The raid, together with British demands that the republic's non-Boer population be granted the vote, and the Boers insisting on the withdrawal of British troops from the republic's borders, were catalysts for the South African War that broke out in 1899.

The conflict followed a period where the supremacy of the British Empire was assailed on several fronts. While the United Kingdom and its allies had finally negotiated an end to the costly Crimean War with Russia in 1856, later in the century the empire suffered humiliating defeats in Africa. First, British forces, trained and armed with modern weaponry, were comprehensively routed by Zulu warriors at the 1879 Battle of Isandlwana during the Anglo-Zulu War. Then, two years later, came the defeat by Boer forces during the First Anglo-Boer War. Equally chastening for the British public was the death in 1885 of Major-General Charles Gordon at Khartoum in Sudan, at the hands of the Muslim forces of the self-proclaimed Mahdi, Muhammad Ahmad bin Abd Allah.

Just as reports of the 'frightful disaster' at Isandlwana and the Boer victory at Majuba Hill in 1881 appeared in New Zealand newspapers, so too did stories of Gordon's demise in Sudan.[3] Closer to home, the competing designs of Germany and the United States in Sāmoa in the late 1880s caused consternation. With New Zealand heavily reliant on exports to and imports from the United Kingdom, any challenge to the British Royal Navy's ability to secure trade routes had serious implications for the empire's South Pacific colonies.

British military pride had been partly restored by the destruction of Zulu forces at Ulundi in June 1879, and was further reinforced by the comprehensive

Jameson Raid prisoners under Boer guard following their surrender at Doornkop in 1896. NATIONAL ARMY MUSEUM, LONDON, 1980-12-47-1

defeat of the Mahdist forces in Sudan at the Battle of Omdurman in 1898; but in the South African Boers, the British Empire faced a much sterner challenge. The men in the Boer kommandos were largely volunteers, but what they lacked in training they made up for in sheer determination. Armed with modern German rifles and supported by predominantly French and German manufactured artillery, the kommandos contained a leavening of battle-hardened veterans who had tasted victory in the First Anglo-Boer War and fought against the indigenous African population.

That New Zealand would support British actions in South Africa was never seriously in doubt. At the end of the nineteenth century, many Pākehā New Zealanders either were born in the United Kingdom or had relatives there. The cause was writ large during a parliamentary debate two months prior to the war when Seddon spoke of 'those of our flesh and blood in South Africa'.[4] Although there were New Zealanders in South Africa at the time, Seddon was most likely referring to the empire's wider European, English-speaking population.

Against this patriotic backdrop, *Our First Foreign War* considers the war's social, economic and political impact on New Zealand society. It does not pretend to be a comprehensive military history of battles and tactics, though where military actions influenced New Zealand public opinion they are discussed. In many ways, it is the story of individuals told through the accounts of New Zealand civilians and soldiers who were in South Africa, and their families and friends in New Zealand. To varying degrees, their actions influenced the nation, and despite hostilities ending, the war continued to exert its own influence on their lives and the lives of those around them.

The impact of the South African War on New Zealanders within their own country, including women, children, Māori, politicians, trade unions and the clergy, is of no less importance. When considered in the context of the larger conflicts that followed, New Zealand's contribution to the South African War was relatively small. At the time, however, the nation's role was most definitely not seen as insignificant.

From an historian's perspective, studying the impact of the war on New Zealand has distinct advantages. New Zealand's small population, coupled with the limited number of men and women who played an active role in the conflict, has allowed me to identify and contact the families of several of those who served in South Africa. Through their generosity, I gained access to information that until now has not formed part of New Zealand's historical record of the war. I have also relied on a number of other sources, including parliamentary reports and returns, archival records in New Zealand and overseas, letters and newspaper reports. The *Wanganui Collegian* proved an especially useful source, given that Wanganui Collegiate School Old Boys served in multiple contingents, as well as in irregular forces raised in South Africa.

Perhaps inevitably, New Zealand's role in the South African War and the impact of the conflict on New Zealand society were eclipsed by the much larger global conflicts that followed. For years, the sheer enormity of the two world wars has relegated the South African War to little more than a prelude to the main events. With notable exceptions, the primary focus of many existing texts that do consider New Zealand's involvement in the South African War is the actions of New Zealand men and women in South Africa, often with an emphasis on military operations. Until now, there has been no fine-grained analysis of the war's impact on New Zealand society as a whole. Given that nearly 120 years have passed since the conflict ended, an in-depth examination of its influence is long overdue.

Our First Foreign War seeks to address this imbalance by providing new insights into a number of areas, which include: the economic impact of the war; its influence on education in New Zealand schools; the behaviour of New Zealand troops (both within New Zealand and in South Africa); the role of those who opposed New Zealand involvement; and the role of the church. The war occurred at a time when New Zealanders were continuing to develop a sense of national identity while at the same time maintaining strong imperial links. In September 1899, Seddon informed Parliament that an 'emergency' had arisen in South Africa, adding that 'the occasion now exists for us to prove our devotion to the Empire'.[5] Two weeks later, Dunedin citizens perusing their *Otago Daily Times* learned that hostilities had commenced in an article titled 'War at Last'.[6]

CHAPTER ONE

'THE FLAG THAT FLOATS OVER US'

PATRIOTISM AND SOUTH AFRICA

Under normal circumstances, Dunedin residents being roused from their beds by the tolling of the town hall bell and the piercing shriek of steam whistles would be cause for general alarm. However, the circumstances surrounding this cacophony on 18 May 1900 were anything but normal. If further proof was required, the sight of Robert Chisholm, the mayor of the southern New Zealand city, repeatedly discharging his shotgun into the chilly morning air provided it. As lights appeared in windows across the city the noise increased with the addition of school bells, fire bells, explosions, and rockets arcing across the pre-dawn sky. In response, Dunedin's citizens spilled onto the streets, enthusiastically striking anything capable of producing a sound, from gongs to empty kerosene tins.[1]

The cause of these uncharacteristic displays was neither invasion nor emergency. It was confirmation that after a 217-day siege the British Army had finally liberated 'dusty, dirty, dilapidated Mafeking' — a remote and ordinarily insignificant way station of British imperialism in southern Africa.[2] A New Zealander who had been in the town during the siege described Mafeking as 'only a small place (about the size of Patea)', but for most its size was immaterial.[3] What Mafeking had come to represent was far more important, and the celebrations in Dunedin mirrored similar rapturous scenes across the British Empire. The relief of this nondescript town thousands of miles from New Zealand shores had been eagerly anticipated. When the news finally arrived, it unleashed a tumult of patriotism.

To an anxious New Zealand public, the prolonged siege of Mafeking by Boer military forces had seemed interminable. As weeks turned into months, newspapers closely followed the town's fortunes, with hopes of a British breakthrough dashed as rumours of the town's imminent relief came to nothing. Admittedly, the lifting of the sieges of the two other South African towns invested by the Boers — Kimberley and Ladysmith — had also resulted in feverish public outpourings. Nonetheless, by the time the first imperial troops trotted into Mafeking the town had become a symbol of British resolve in the face of adversity. That it was besieged in the first place was undoubtedly a British reverse, but the empire had been spared a morale-sapping capitulation. Even if holding out longer came largely at the expense of Mafeking's starving black African population, the refusal of the town's commanding officer, Colonel Robert Baden-Powell, to surrender was seen as an exemplar of British determination and pluck.

As the sieges showed, the war's initial progress had hardly been encouraging. While New Zealanders familiar with Boer tenacity had initially expressed reservations, once hostilities broke out in October 1899 the general expectation was for a swift British victory. With the overwhelming might of British arms brought to bear on the numerically smaller Boer forces, Britain would surely prevail. Three months prior to the declaration of war, the *Feilding Star* optimistically predicted that within a week of war's outbreak all of Transvaal would be part of the British Empire: 'England would crush the Transvaal as a giant would crush a worm.'[4]

But it soon became clear that Boer leaders had no intention of meekly accepting peace on imperial terms. Seizing the initiative, they took advantage of Britain's lack of preparedness. Before vessels carrying reinforcements could dock at Cape Town and Durban, disgorge their khaki cargoes and turn the tables in Britain's favour, the Boers hoped to use force to lever political advantage. For the British, the enemy's resolve proved as disturbing as it was unpalatable. There were no decisive victories cast in the mould of Lord Kitchener's 1898 rout of Mahdist forces at Omdurman in Sudan. Instead, in the initial stages of the war the British public was forced to subsist on a diet of humiliating defeats and inconclusive victories. At the battles of Magersfontein, Stormberg and Colenso in December 1899, British troops were repulsed with heavy losses. Rather than accept battle on British terms, the Boers engaged the enemy from concealed defensive positions. Their kommandos used their mobility and superior knowledge of the terrain to inflict British casualties and withdraw when their positions became untenable.

Shortly before the war, the New Zealand premier, Richard John Seddon, addressed the House of Representatives. Seddon claimed it was well known what New Zealand was prepared to do 'to maintain the good old British flag' should necessity arise.[5] He also spoke of wiping out the stains of the military defeats the British had sustained at the hands of the Boers at Majuba Hill and Bronkhorstspruit during the First Anglo-Boer War.[6] There was a widespread belief that the British had unfinished business in South Africa. In time for Christmas 1899, the British children's annual *Chatterbox* was sold in New Zealand bookstores.[7] It featured an account of the 'inglorious' fight at Majuba that also spoke of 'wiping out the stain of that defeat'.[8] A reporter who visited a Dunedin school classroom in December 1899 said that all the children in the class raised their hands when asked about Majuba.[9]

The New Zealand governor, Lord Ranfurly, echoed Seddon's views, telling Wanganui Collegiate School students that Majuba Hill and the death of General Charles Gordon at the Mahdist siege of Khartoum were stains on Great Britain's reputation.[10] However, by the early months of 1900 the overarching desire to avoid further costly defeats meant that if the British public could not have another Omdurman in southern Africa, they at least wished to be spared the ignominy of another humiliating Khartoum at Mafeking. Mayor Chisholm informed the Dunedin crowd that the relief of Mafeking was the best news they had received since the war began.[11]

In the preceding days the excitement had been palpable as British forces edged closer to Mafeking. The MHR for the City of Auckland electorate, George Fowlds, suggested that regardless of the hour when news of the relief was received guns in the city's forts should be fired.[12] Having first sought Lord Ranfurly's permission, Seddon instructed Lieutenant-Colonel Arthur Penton, the Commandant of Forces, to have 'royal salutes' at the ready.[13] The long-awaited news finally reached Wakapuaka Cable Station on the Nelson coast at two in the morning on 18 May and was transmitted to the *Otago Daily Times'* Dunedin offices. Despite the hour, the newspaper notified Chisholm who decided that although the news was not official the city should be informed at 6 a.m. by the tolling of the town hall bell. The newspaper portrayed the lifting of the siege as much more than simply a strategic reverse for a Boer enemy forced into retreat; 'Mafeking Relieved!' screamed the oversized headline.[14]

Across the empire, Baden-Powell became the 'Hero of Mafeking'. He may not have delivered a decisive victory, and a New Zealand churchman questioned the morality of his tactics, but at least he had prevented another demoralising defeat and restored British military pride following the earlier embarrassments.[15] In stark contrast to press portrayals of Baden-Powell, the *Otago Daily Times* characterised the besiegers as the 'refuse of the Boer army, together with the scum of Europe' and claimed General J. P. Snyman, the Boer commander, 'and his devilish crew' had 'put the very savages to shame by their campaign of systematised savagery'.[16]

After Chisholm's declaration of a half-holiday, normal business in Dunedin ground almost to a standstill. Following the delivery of patriotic speeches marking the occasion, both the stock exchanges suspended trading, with many of their members retiring to the Otago Club.[17] Work ceased at Port Chalmers, where vessels were decorated with flags and bunting, and Hillside Railway Workshops staff downed tools and readied their parade banners.

As the news spread, schools joined the festivities, with Arthur Street schoolboys ringing the school bell continuously for two hours. After erecting an image of Baden-Powell featuring the words 'British pluck for ever', the children joined the throng gathering in the city. At Union Street School, President Kruger was burned in effigy, after which the school's cadets fired a volley of blank cartridges. Noting Baden-Powell's loyalty, the headmaster of High Street School advised his children to follow the officer's example and remain loyal to their teachers, their school, their empire and their queen. By early afternoon, parade participants had gathered in marshalling areas in the crowded Octagon. The procession represented a cross-section of Dunedin's citizens, including bands, Volunteer and cadet corps, football clubs, city councillors, students, nurses, timber workers, jockeys, railway employees, butchers, paper mill workers, Fuller's Vaudeville Company and 'two niggers in a gig'.[18]

Press photographs captured the scale of the celebrations.[19] Amid the sea of humanity in the Octagon, boys climbed light standards to secure a better

A crowd throngs the Dunedin Octagon in May 1900 following receipt of news of the relief of Mafeking. WEEKLY PRESS, 30 MAY 1900, NEWSPAPER HISTORY OF THE BOER WAR IN SOUTH AFRICA, 1899–1902, COMPILED BY GEORGE FANNIN, ALEXANDER TURNBULL LIBRARY, P F968 FAN

vantage point. Others occupied the second storey of Gray's Oban Hotel or peered from the windows of Mills, Dick and Co. Printery above the Edinburgh Dining Rooms. Still more packed the windows of the Otago Cycling Club, while men viewing the spectacle from the roof of Jolly, Connor and Company Printing Works dangled their legs precariously over the edge.

Cheering themselves hoarse, flag-waving crowds lined the parade route. Students from the School of Mines carried a banner that read 'Bravo, brave Baden-Powell'. Outside the Grand Hotel, a Chinese entrepreneur with a Union Jack wrapped around his hat sold British flags, while one patriotic reveller painted his dog red, white and blue.[20] With a pipe band and scores of children in its wake, the fire brigade proceeded from the Octagon to a crowded patriotic meeting at the Agricultural Hall. The festivities continued well into the night, with Dunedin's main street brightly lit and bands of children carrying Chinese lanterns roaming the city centre striking drums, blowing whistles and singing 'Soldiers of the Queen'. Reporters and typesetters worked furiously to ensure a fitting account of the day's events appeared in the press, with the *Otago Daily Times* estimating the crowd at between thirty and forty thousand.[21]

Yet the public revelry and bonhomie had limits; there was also a darker intolerance on display. One banner depicted a member of the National Council of Women (NCW) embracing a Boer. The NCW had drawn public condemnation after members criticised the war during their national meeting in Dunedin the previous week.[22] A letter in the *Otago Daily Times* the day after Mafeking's relief dismissed the council's members as 'Boeresses in disguise'.[23] People also gathered menacingly outside the premises of a Dunedin tradesman suspected of harbouring 'pro-Boer' views.[24]

In towns big and small across the nation the news was received with a similar combination of relief and elation, although after months of disappointment some remained sceptical. When the news reached the small South Island mining town of Reefton its authenticity was initially questioned.[25] It was not until Saturday afternoon that the populace accepted that the siege had finally ended. In the central North Island town of Taihape the long-anticipated event reportedly caused 'intense excitement', but its citizenry also waited until Saturday to celebrate.[26] Napier crowds packed the wharf and breakwater, eager to pass on the glad tidings to passengers on the coastal steamer *Waihora* when she docked, only to discover those on board had already deduced from the abundance of flags visible on shore that the siege was over.[27] Wanganui Collegiate School's magazine accurately captured the prevailing mood: 'a universal burst of joy throughout the country hailed the news of the relief of Mafeking'.[28]

Long after weary revellers cleared the streets, Baden-Powell's star continued to shine. 'The Hero of Mafeking' was showered with gifts from a grateful empire and New Zealanders were determined not to be outdone. The town of Gore sent him an engraved, gold-mounted, greenstone paper knife; the mining region of Blue Spur debated over a gold trophy and a pair of gold spurs, before finally sending a gold medallion featuring a miniature gold spur and the inscription 'Our Trusty Knight of the Empire'.[29] While Gore's paper knife was engraved with New Zealand fern leaves, the medallion featured the rose of England, the Irish shamrock and the Scottish thistle. Even the children of Palmerston North's Campbell Street School sent a gold and greenstone pendant to Baden-Powell commemorating the end of the siege.[30] Other patriotic gestures were less tangible. In the wake of Mafeking's relief, there was a flurry of parents naming their children after the siege and the town's commander. Mafeking Baden Powell Gunn and Arthur William Baden Powell (author of *Native Animals of New Zealand*) were just two of several New Zealanders who carried a lifelong reminder of the siege.

Yet parades and patriotism told only part of the story; for some, events in South Africa had greater significance. Dannevirke sisters Hettie and Florence Tansley were singers with the Payne Family of Bellringers, a musical troupe that performed in New Zealand and Australian theatres.[31] By 1899 the troupe had added South Africa to its itinerary and in June the sisters found themselves in Transvaal with war rumours rife.[32] Despite the adventure of seeing the world, Florence Tansley made it abundantly clear that the appeal of her theatrical life had waned. In the South African Republic with war imminent, she claimed Dutch 'spies' were everywhere. Florence wrote, 'I shall be delighted to get back to New Zealand where there is civilization, and, when I do get back, I think I shall stay.'[33]

New Zealanders were also present in the towns besieged by the Boers. Wanganui Collegiate School Old Boys Sergeant-Major Edward Jollie and Sergeant Rupert Hosking were in Mafeking.[34] As garrison paymaster, Jollie

Right: New Zealander Sergeant-Major Edward 'Teddie' Jollie, who was besieged in Mafeking while serving in the British South Africa Police. In a letter home, Jollie described the misery of the town's black African population during the siege. THE COLLEGIATE SCHOOL, WANGANUI, IN SOUTH AFRICA, 1899–1900, WANGANUI: A.D. WILLIS, PRINTER, [1901], N.P., WANGANUI COLLEGIATE SCHOOL MUSEUM

Opposite: Edward Jollie, his second wife Sarah and their children c.1912. After the war, Edward spent time in England following the death of his first wife. He met Sarah, a widow from Lancashire, on the voyage back to New Zealand. Edward Jollie was killed in New Plymouth in 1925 after his bicycle collided with a motorcycle. LEONARD (TIM) JOLLIE

witnessed first-hand the impact of Baden-Powell's policy of reserving most of the town's limited food stocks for Mafeking's European inhabitants. Although stopping short of blaming Baden-Powell for the Africans' misery, Jollie described the malnutrition of Mafeking's black African population.[35] He claimed to have seen emaciated Africans drop to the ground as he paid them. Jollie also said he had witnessed Africans attempting to ward off starvation by cooking and eating the soles and heels of old boots as well as consuming dogs and horses that had died of sickness. According to Jollie, the indigenous Africans convinced themselves they were gaining weight when their bodies began to bloat with disease. Apparently unaware of the bitter irony of his remarks, Jollie noted that Baden-Powell had gone to Pretoria to collect tins of the queen's chocolate — a gift from Queen Victoria to all imperial soldiers fighting in South Africa.[36]

While Wanganui Collegiate School remained strongly supportive of the British war effort, Jollie's account belied a passage in W. Francis Aitken's 1900 book *Baden-Powell, the Hero of Mafeking*, which was sold in New Zealand during

Sergeant Rupert Vivian Hosking, who served in D Squadron of the Protectorate Regiment Frontier Force during the siege of Mafeking. Hosking sustained a serious gunshot wound to his leg during the Boers' final attempt to take Mafeking. THE COLLEGIATE SCHOOL, WANGANUI, IN SOUTH AFRICA, 1899–1900, WANGANUI: A.D. WILLIS, PRINTER, [1901], N.P., WANGANUI COLLEGIATE SCHOOL MUSEUM

the war. Writing without personal experience of conditions in Mafeking, Aitken claimed that tales of natives dying of starvation were gross exaggerations.[37] Reviewing the book, the *Evening Post* claimed that '[p]eople naturally wish to know all about the man who is the hero of the day'.[38] The plight of Mafeking's starving Africans was less newsworthy.

Some New Zealanders saw in Britain's African colonies and the Boer republics an opportunity to improve personal circumstances, while others took the chance to escape from their lives at home. To the south of Mafeking, New Zealanders played an active role in keeping the Kimberley diamond mines of Cecil Rhodes beyond Boer reach. Once war broke out, John Gillespie of Blenheim resigned his job at a Kimberley newspaper and served as a sergeant in the town's defence force.[39] Leaving an unhappy marriage in his wake, Willis Peat had travelled with his son from New Zealand to Cape Town in 1897 and reportedly served in the Kimberley Town Guard during the siege.[40] Following his father's death, contracting typhoid and seriously breaking his arm, 16-year-old 'Trooper' Louis Peat received £37 in financial assistance on Seddon's instructions for passage back to New Zealand.[41] Also in Kimberley, in the Kimberley Light Horse, was Patrick Madden from the West Coast settlement of Dillmanstown. Madden had been working at Rhodes' diamond mines for four years when hostilities broke out.[42]

Further east, in the Natal town of Ladysmith, the four New Zealand-born Melville brothers served in the Border Mounted Rifles during the siege.[43] Like Jollie and Hosking, the three eldest Melvilles were Wanganui Collegiate School Old Boys. Also in Ladysmith were former Tīmaru residents Harry and Rose Shappere.[44] New Zealand-born Harry served in the Royal Horse Artillery in India but was besieged in Ladysmith soon after his arrival in South Africa.[45] His sister Rose, formerly a nurse at Adelaide Hospital, made her own way to Johannesburg, where she served with the St John Ambulance Association. Rose had spent time in South Canterbury during her childhood and was reportedly one of the first nurses to travel to the front, initially treating both Boer and Briton.[46] On the outskirts of Ladysmith, she toiled in the disease-ridden and overcrowded Intombi Hospital, where she fell ill with jaundice.[47] The siblings were unusual in that both received the Queen's South Africa Medal, with Harry's award featuring the 'Defence of Ladysmith' clasp.[48]

A number of New Zealanders and men with New Zealand connections had taken part in the ill-fated 1895–96 Jameson Raid. The raiders' defeat and capture by the Boers undermined British prestige in the region and created a power vacuum in Matabeleland. Seizing their opportunity, disaffected Matabele, chafing under British South Africa Company control, attacked outlying farms and miners' camps in an attempt to drive the Europeans from their lands.[49] Ernest Monk, the son of New Zealand parliamentarian Robert Monk, and fellow New Zealander Charles Kirk accompanied Jameson and were captured by the Boers at Doornkop. Also among the prisoners were Frank Holloway and Robert Thompson, who gave New Zealand addresses when a nominal roll of the captured men was compiled.[50] Robert Jack, who joined the Second Contingent in 1900, spent three years in the Matabele Mounted Police and also appears to have taken part in the raid.[51]

These men were just some of the New Zealanders who had first-hand experience of South Africa. Hughes Lockett of 'Wangannie' (Whanganui) and Sergeant Allen Bell of the Ninth Contingent served in the 'Matabele War'.[52] Alexander Duirs from Hāwera settled in Johannesburg prior to the war and joined Gillespie in the Kimberley Light Horse.[53]

Trooper Monaghan, a former employee of Wairikeiki Station in Southland, was serving in the Matabele Mounted Police as war loomed.[54] With two months of his contract remaining, Monaghan intended returning to New Zealand but doubted he would be able to leave until the 'Transvaal trouble' was settled. In a letter home, he cautioned that fighting the Dutch wouldn't be the same as dealing with the Matabele.[55]

Also in Kimberley were Robert Grieve, his wife Elsie and their two sons.[56] Elsie was New Zealand born, while Robert's father lived in the tiny Southland community of Waianiwa.[57] When the siege was finally lifted, Elsie described her delight at having tea with a group of New Zealanders from the vanguard of Lieutenant-General John French's relieving column: 'I am almost off my head with excitement at seeing faces from home and hearing them speaking in the New Zealand tongue.'[58] The serendipitous nature of the gathering can hardly have been lost on the participants: Sergeant Hazlett of Dunedin, Corporal Grant of Ōamaru, Corporal McKegg of Henley, Trooper McConway of Marlborough, Trooper Johnston of Kaihiku and Trooper Mitchell of Balclutha sipped tea in

Elsie's Kimberley house thousands of miles from home.[59] After the monotonous siege diet, Elsie was clearly elated: 'I am so proud of my country, especially when our young New Zealanders were among the first to relieve Kimberley and bring ME a leg of mutton.'

Like these New Zealanders in Africa, the European populations of the British Empire were connected not only by their sense of imperial unity but also by the complex system of postal networks and undersea cables that disseminated news. Reinforcing this were the roads, railways and sea lanes that connected the empire, and the Royal Navy vessels protecting its trade routes. It was these all-important maritime links that allowed New Zealanders to escape their isolation, but the human traffic between New Zealand and Africa was not entirely one-way. From early 1902, all individuals wishing to visit Cape Colony and Natal were required first to obtain a permit;[60] the reasons given by applicants provide an insight into the links between New Zealand and the South African colonies. Nurse Louisa Hallam had accompanied an invalid to New Zealand from South Africa and wished to return home.[61] Dunedin resident Maria Colvin, formerly of Ladybrand, applied for a permit so she could rejoin her family in South Africa.[62] Elizabeth Donald wished to marry her fiancé, a King William's Town coachbuilder, while Fred Arnott wanted to rejoin his father, a contractor in Cape Colony. Annie Wattam planned to live with her married sister in Natal; Jane Nielsen wanted to join her husband serving in the Johannesburg Mounted Police; and Fanny Marsh sought permission to travel to South Africa so she could act as housekeeper for her son.[63]

The patriotism displayed in 1899 did not develop in isolation; many New Zealanders had links to Africa that pre-dated the conflict, and, like the Melvilles, moved freely about the British Empire in pursuit of opportunities and adventure or to maintain family ties. Individuals from Great Britain and its colonies spent time in New Zealand for similar reasons. The importance of maintaining these bonds and honouring a perceived obligation to 'the Motherland' were recurring themes during the First Contingent debate in 1899.[64] In Parliament, Seddon said some would ask why New Zealand should involve itself in a distant war, and he offered a justification:

> *The answer is simple. We belong to and are an integral part of a great Empire. The flag that floats over us and protects us was expected to*

> *protect our kindred and countrymen who are in the Transvaal. There are in the Transvaal New Zealanders, Australians, English, Irish and Scotch; and others from British dependencies; they are of our own race and our kindred.*[65]

New Zealanders were, Seddon claimed, a portion of the 'dominant family of the world . . . the English speaking race'.[66]

CHAPTER TWO

'RALLY TO THE CALL OF HOME AND COUNTRY'

DOMESTIC REACTION TO THE WAR

Warning that the empire faced an emergency, in September 1899 Premier Richard Seddon sought parliamentary support for a resolution offering a New Zealand contingent for service in Transvaal. At the time, New Zealand's defence force largely consisted of locally raised Volunteer corps and a small Permanent Force. In 1898, the Permanent Force consisted of 286 men, many of whom were responsible for manning the artillery defences of New Zealand's principal harbours.[1] The month before the possibility of sending a contingent was discussed in Parliament, Seddon had tabled an Imperial South African Association pamphlet in the House detailing the grievances of the Transvaal uitlanders.[2] The Association, dedicated to upholding 'British supremacy', included numerous peers, politicians, and the fervent imperialist Rudyard Kipling. In particular, it objected to President Kruger's restrictions on uitlanders obtaining the vote.[3] It was common knowledge, Seddon claimed, where New Zealanders' sympathies lay as South Africa was 'an integral part of the Empire'.

Seddon's resolution was seconded by opposition leader Sir William Russell.[4] Claiming it was a privilege to support the resolution, Russell said that it was not for him to enquire deeply into the origins of the conflict.[5] Party rivalries were temporarily set aside as Russell echoed the views of many parliamentarians on both sides of the House.[6] With few exceptions, they too showed little interest in examining the moral justification for the impending war; their patriotic support was no more than what was expected of loyal members of the British Empire.[7]

Riccarton MHR William Rolleston referred to bonds of kinship and affection that linked New Zealand and the United Kingdom.[8] This was reinforced during the debate in which the United Kingdom was described as the 'Home country', the 'Old Country', the 'Old Land', the 'Mother Country' and the 'Motherland'.[9] Though concerned that a desire to emulate other colonies was pulling New Zealand into a war, parliamentarian Richard Monk expressed his loyalty to the 'parent state'.[10] When John Graham, MHR for the City of Nelson, addressed Parliament he stressed that he was more than just a New Zealander — he was a patriotic British subject.[11]

Legislative Council member Thomas Kelly expanded on the parent–child analogy, claiming it was the duty of children to assist their parents.[12] After the British Secretary of State for the Colonies, Joseph Chamberlain, learned that the resolution had been adopted with 'the greatest enthusiasm' by 54 votes to five, he commended New Zealand's patriotic spirit.[13] The *Wanganui Chronicle* later claimed that New Zealanders were not particularly concerned about the origin of the war, or the military authorities' mistakes: 'What the people felt was that the supremacy of the Empire was at stake.'[14] Once New Zealand's offer of troops was formally accepted, the government was inundated with applications from men eager to serve in South Africa. Before hostilities had even broken out, Lieutenant-Colonel Stuart Newell called for tenders to provision the contingent's Karori camp, where successful applicants from across New Zealand had begun to gather.[15]

The power of the press

Though Seddon noted before the war that there was 'limited' South African information in the New Zealand press, throughout the conflict newspapers played a central role in both reporting and interpreting events in South Africa, and shaping New Zealanders' attitudes.[16] The remoteness of the conflict, coupled with censorship and the expense of transmitting news, largely dictated the flow of information. With the larger newspapers drawing on a variety of sources, including the United Press Association (UPA) and Reuters, the press remained the primary conduit of war-related news.

The first news cable link between London and Wellington had been established in 1876. Messages cabled from London passed through Portugal and

Gibraltar, then via Malta to Egypt and on to Aden and India, followed by the British-controlled Straits Settlements of Penang and Singapore. From there they travelled via the Dutch East Indies to Australia, and finally on to Wakapuaka Cable Station at Cable Bay, north-east of Nelson.[17] A Napier newspaper explained how four press agencies were involved in the transmission of South African news. Two were reportedly based in London, one in Australia and the other in New Zealand. An Australian Press Association (APA) office in London accessed South African news, while the New Zealand Press Association (NZPA) kept an office in Sydney.[18] This allowed the NZPA to access APA cables.

At first, press coverage was largely partisan, with the demand for news forcing newspapers to adapt. Those with sufficient resources sent correspondents to South Africa while also getting news from international agencies, official dispatches, soldiers' letters and parliamentary reports. The *Evening Post* dispatched two correspondents, James Shand and John Moultray, with Moultray contributing drawings and articles.[19] Dressing like a trooper and spending time in military camps, Shand also acted as a correspondent for Otago newspapers and was a forerunner of the 'embedded' reporter.[20] The UPA initially arranged for First Contingent captain William Madocks to send accounts of his contingent's activities, though the practice was discontinued.[21] On at least one occasion, Madocks, Captain Davies, Colour-Sergeant Cardale and the First Contingent's commanding officer, Major Alfred Robin, also produced phonographic recordings that were shipped to 'Mr Turnbull' in Wellington.[22]

W. D. Campbell, the *Auckland Star*'s war correspondent in South Africa, rejected the criticism of Second Contingent captain Norman Smith, who professed an 'undisguised loathing' for the articles of New Zealand reporters in South Africa.[23] Smith claimed the correspondents were often miles from the events they described and exaggerated New Zealand soldiers' performance while downplaying that of their British counterparts. Campbell alleged that colonial correspondents were pressured into writing articles that enhanced the reputations of certain officers, while suppressing other information.[24] Though Campbell often relied on officers' accounts, he accused some New Zealand officers of breaching regulations by submitting their own articles to the press. It was 'duty', he claimed, that compelled him to emphasise New Zealanders' actions, but he contrasted the helpfulness and professionalism of the British officers he encountered with the 'ignorant manners' of New Zealand's 'amateur officers'.[25]

As well as reporting wartime events, some newspapers also made tangible contributions to the war effort. Trumpeting its devotion to 'the Motherland', the *Press* established a 'More Men Fund' and shipped chocolates, socks and tobacco to First Contingent troopers.[26] Though Seddon assured Parliament that newspapers supported the war effort, the Catholic *New Zealand Tablet* remained ambivalent.[27] It implied that commercial interests were behind recent British wars, while claiming the press was responsible for 'raising the popular temperature to the fever heat'.[28] In the initial stages of the war, this was a responsibility few New Zealand newspapers shirked.

The Christchurch *Weekly Press* saw its circulation rise to 40,000 by the end of 1899, making it New Zealand's largest-selling newspaper.[29] When Ladysmith was relieved, major newspapers featured extensive war coverage, while resources, locations and printing schedules meant smaller newspapers struggled to compete and often featured less information.[30] Nonetheless, New Zealanders flocked to press offices seeking confirmation that the siege had been lifted, as they did at the rumoured relief of Mafeking.[31] While the number of newspapers fluctuated, in January 1900 there were 11 more registered newspapers in the colony than in the previous year.[32] Yet increased circulation did not always offset increased costs. War, a newspaper claimed, was a curse for metropolitan newspapers because it meant vast expense, disruption of routines and increased strain on workers.[33] The Ashburton Publishing Company emphasised that war-related expenses made it '*absolutely necessary*' that debtors settle their accounts with the *Ashburton Guardian*.[34]

To reduce expense while meeting public demand, editors turned to letters from New Zealanders in South Africa; these offered a cheap alternative to costly cables. Though most recorded daily events, some men described their military service as furthering the empire's interests. In a letter from Mafeking, Rupert Hosking claimed: 'The loyalty here is tremendous. Everyone is willing to do another five months such as those just gone, if by doing it the Empire will derive benefit therefrom.'[35] Nonetheless, not all soldiers wanted their letters published. Seventh Contingent trooper Hobart Tennent told his mother that 'on no account' should his letters be published, while Lieutenant Michael Canavan declined to give details of his experiences due to unspecified 'trouble' over soldiers' interviews and letters.[36] Family members could also be selective about which letters were submitted to newspapers.[37]

Postal volumes, shipping schedules and the constant movement of soldiers made maintaining a reliable mail service between South Africa and New Zealand challenging. Shipping delays meant Shand's reports were normally published approximately one month after he had written them, and in April 1902 a trooper in the Seventh Contingent received mail sent the previous Christmas.[38] Though it was categorically denied by the Post Office, Moultray claimed that mail sent to him in South Africa had been rifled, while war correspondent Alf Morton reportedly found Fourth Contingent mail among rubbish in a South African post office.[39]

Troopers Claude Jewell and William Saunders and Corporal George Wilson all worked for newspapers as civilians.[40] A newspaper described Jewell's letters as 'amongst the most readable epistles received from South Africa'.[41] He criticised upper-class officers and accused his British commander of incompetence.[42] Saunders alleged that Cape Colony newspapers were minimising British casualties, and wrote a *Cape Argus* article criticising the management of military hospitals.[43] George Wilson, of the Tenth Contingent, worked for the *Waimate Advertiser* prior to the war. Trooper James Christie of Milton served in the Bushveldt Carbineers, a corps that gained notoriety when lieutenants Harry 'Breaker' Morant and Peter Handcock were court-martialled and shot after being found guilty of summarily executing Boer prisoners.[44] Christie, who after the war became the editor of the *Bruce Herald*, wrote newspaper accounts of his version of the incidents.[45]

Reports in 1902 that the German *Tägliche Rundschau* newspaper had accused British soldiers of raping Boer women and girls and forcing them into prostitution triggered widespread antipathy towards Germany within New Zealand and revived public support (which by then had begun to wane) for further contingents.[46] Seddon ordered government buildings closed so civil servants could attend a patriotic meeting protesting the claims.[47] Though Seddon opposed a boycott of German goods as 'un-British', he considered imposing tariffs, and some New Zealanders indicated that they would avoid purchasing German-made products.[48] Other foreign publications were also viewed with suspicion; City of Wellington politician George Fisher warned of the 'insidious power' of an American pamphlet criticising British actions in South Africa and recommended the government publish two pro-British pamphlets as 'an antidote'.[49]

With notable exceptions, newspapers depicted New Zealanders' involvement in the war in heroic terms that largely avoided the unpleasant realities of warfare.[50] Following his escape from Boer captivity, multiple newspapers described Sergeant Andrew Peterson as a 'Plucky New Zealander', while other press accounts described the First Contingent's 'gallant combat' and the Seventh Contingent engaging the Boers 'with the utmost dash'.[51] This contrasted with many press portrayals of the Boers. Shortly before the war, the *Evening Star* referred to 'the malice of those who spoke of the stupid Boers', yet once hostilities commenced the same newspaper used 'Stupid Boers' as titles for two articles.[52] The Boers were frequently portrayed as a cowardly, treacherous and brutal enemy who routinely abused the white flag and used expanding 'dumdum' bullets, with the *Press* calling the enemy 'murderous scoundrels'.[53] Newspapers reported that Dunedin schoolchildren believed the British use of high explosives was no worse than Boers abusing the white flag or firing on ambulances, and an Invercargill kinematograph exhibition included a representation of 'white flag treachery' that showed 'the cowardly tricks of the Boers to perfection'.[54] While the Presbyterian magazine *The Outlook* criticised newspaper bias, the *Free Lance* claimed the press had exhibited tolerance and 'rebuked Briton and Boer' alike when their conduct warranted it.[55]

In an attempt to provide a more balanced account, the Wellington Peace and Humanity Society reportedly considered producing its own newspaper to publish war-related material rejected by the mainstream press.[56] Occasionally newspapers and school journals contained graphic depictions and accounts of warfare where little was left to the imagination.[57] The *Otago Witness* publishing a photo of men killed at the Battle of Spion Kop would have done little to assuage the concerns of soldiers' families.[58] A trooper's letter to the *Christ's College Register* contained a disturbingly candid description of a New Zealander's death.[59] The *Wanganui Collegian* contained one of the most harrowing accounts when it published a letter from one of the school's Old Boys, Patrick Fitzherbert, in which he described the Boer dead following the Battle of Paardeberg:

> *I shall never forget the sight of the killed and wounded Boers after the surrender of [General Piet] Cronje. Hundreds lay about the bed of the river, mingled with dead horses, cattle and mules. Most of the killed were killed by our shells, which simply blew them to pieces. All more or less mutilated, with eyes swollen to abnormal size (by the heat) bulging from their sockets — heads missing, limbs missing! Some blown clean in two. Then the burying! Men and horses thrown into their rifle pits and covered over. My God, it's awful and I never wish to see the like scene again. It really does seem like murder.[60]*

In stark contrast to the engraved images in children's books with their idealised, heroic and bloodless depictions of warfare, such letters made little attempt to shield children from the brutality of battle. Cadet corps numbers at Christ's College and Wanganui Collegiate School suggest, however, that these accounts did little to dent cadet enrolment during the war years.

British actions during the war were far from beyond reproach, and the plight of Boer women and children held in British concentration camps became a divisive issue. The *New Zealand Tablet* described their alleged starvation as systematic, deliberate and unnecessary, while *The Outlook*'s editor said the camps were a 'most painful subject'.[61] A Southland newspaper dismissed the reported ill-treatment of Boer women and children as lies, and the editor of the *Grey River Argus* claimed that Boers in the camps would concede 'they were never better off in their lives'.[62] In response to the editorial, a contributor to the *Otago Witness* claimed the camp death toll would haunt the national conscience.[63]

Many of those in the camps had been displaced through the British policy of destroying the homes of Boer civilians suspected of assisting the enemy, and either slaughtering or removing their livestock. New Zealanders played an active role in these operations; Hobart Tennent described burning Boer homes and taking animals, but claimed that the Boer women welcomed incarceration.[64] This alleged acquiescence did not silence critics such as Dunedin MHR Alfred Barclay. When Barclay described the Boers' treatment in the camps as 'infamous', many of his constituents demanded his resignation, claiming his actions were an affront to New Zealanders fighting in South Africa.[65] Members of the Dunedin Presbytery vehemently denied the accusations of *The Outlook*'s temporary editor, William Hutchison, who claimed that Boer women and children had been 'ill-used'. Instead, a Presbytery member asserted that the camps were the one thing 'that shone out brightly' in the conduct of the war.[66]

Some New Zealand soldiers were uneasy about the displacement of Boer families. Former Wanganui Collegiate School schoolmaster Joseph Orford complained that he and his fellow troopers were 'dead sick of making war on women and chickens and cattle'.[67] Orford's actions hardly fulfilled his former school's hope, expressed in the *Wanganui Collegian*, that he would return 'covered with glory'.[68] Fellow Old Boy Lieutenant John Montgomerie described his men driving the Boers before them as they advanced on the Transvaal town of Rustenburg, burning and looting Boer homes flying the white flag.[69] Other soldiers, too, seemed indifferent to the Boer civilians' fate.[70] Trooper Charles Tasker said that he was 'living high' on their livestock after collecting 'refugees' and burning their homes.[71]

Several New Zealand soldiers' accounts belied the image of the enemy as 'murderous scoundrels'. Jewell was treated courteously when he was captured, adding that 'it is only the strict discipline of the army that keeps [British troops] from worse atrocities than the much maligned Dutch farmer'.[72] A wounded trooper from the Fourth Contingent also praised the treatment he had received when captured, but accused Boer general Christiaan de Wet of ill-treating prisoners.[73] Even this was disputed; when Blenheim trooper Francis Morrison was captured by De Wet's kommando he said he was treated with kindness, though a newspaper claimed that Morrison had 'ingratiated himself' with the general.[74]

A common complaint among soldiers in South Africa was their limited access to current news.[75] The challenge of supplying highly mobile forces was reflected in Trooper Luke Perham's claim that apart from occasional copies of the *Press* and the *Canterbury Times* his contingent had not seen a newspaper in months.[76] Similarly, in a letter published in the *Otago Daily Times* Lieutenant John MacDonald said the New Zealand public knew far more about the war's progress than his men and claimed the Fourth Contingent had not received mail in 10 months.[77] Montgomerie also suggested that people would be astonished by how little he knew of the war: 'It is you in New Zealand that get all the news.'[78] When they did arrive, newspapers and mail came as a welcome respite from the rigours of campaigning. In some cases newspapers forwarded bundles of their

Lieutenant (later captain) John Montgomerie, a clerk at Levin and Company in Wellington who served in the Second and Eighth contingents. In August 1900 Montgomerie described his men driving the enemy before them as his contingent advanced on the Transvaal town of Rustenburg. After removing any goods from Boer homes suspected of sheltering the enemy, the troops burnt the houses and left the Boer women and children 'weeping over the wrecks'. THE COLLEGIATE SCHOOL, WANGANUI, IN SOUTH AFRICA, 1899–1900, WANGANUI: A.D. WILLIS, PRINTER, [1901], N.P., WANGANUI COLLEGIATE SCHOOL MUSEUM

publications to contingents; Saunders said he had seen the *Otago Daily Times* in Bloemfontein.[79] The *Press* claimed that cigarettes and cigarette papers were the only items that were more popular than news from home.[80] Some families in New Zealand also had access to South African papers, with Trooper William Farquharson's father regularly receiving dated copies of the *Cape Town Times*.[81]

As the war progressed, press support became less unequivocal as many grew weary of the conflict. In early 1901 the *Taranaki Herald* noted the 'almost ridiculous lengths to which khaki fever is carrying the colony' and cautioned that support for the war was in danger of degenerating into a 'spurious kind of patriotism', while a poem titled 'The Cult of Khaki' lamented the nation's obsession with the 'dreary, dispiriting' hue.[82] Some newspaper editors began questioning the need for further New Zealand contributions.[83] The *Thames Star* claimed that most returning soldiers would not settle down to work on the land and would instead seek special treatment in the form of pensions or less arduous positions in government departments.[84] This was at least partially true. Although some men were government employees before the war, when a list of the occupations of former contingent members appeared in the First New Zealand Mounted Rifles Association's journal, it included men employed by the Defence Department, the Post Office and the Railways Department.[85]

Despite the volume of war news in New Zealand newspapers, complaints continued to surface about the quality and quantity of coverage. The blame, however, did not lie entirely with the press. For newspapers, military censorship was yet another hurdle to overcome.

'The merciless hand of the military censor'

In September 1900 William Herries, Bay of Plenty MHR, asked Seddon whether information contained in contingent dispatches that was not confidential could be tabled in the House. Seddon replied that he had instructed Lieutenant-Colonel Penton to examine the South African dispatches of majors Robin, Jowsey and Cradock and, after removing sensitive information, to distribute the remainder to newspapers. Nevertheless, the premier described the issue as 'delicate'. It was not, Seddon claimed, in accordance with military etiquette to publish dispatches as he feared it might result in New Zealanders being court-martialled.[86]

In 1899 the press blamed censorship for the reduced content of war-related cables.[87] In 'The Press Censorship: Why the Cables are Unsatisfactory' the *Nelson Evening Mail* sought to explain why cable news from South Africa was not always as consistent, accurate and complete as desired, and claimed the situation was not entirely the responsibility of newspapers or their correspondents.[88] Soon after war was declared, newspapers in New Zealand reported that strict censorship was in force in Natal, while the Cape of Good Hope administration had suspended all cable messages sent in cipher.[89] *The Outlook* also claimed that letters from Johannesburg were subject to strict military control, and a letter sent to Dunedin by Private Robert Lawrence bore a 'Passed by Censor Johannesburg' stamp on the envelope.[90] Another letter sent by Lieutenant George Leece also featured the censor's stamp.[91] Even within the Presbyterian Church there were attempts to influence the content of *The Outlook*. After *The Outlook* published articles criticising the war, the Otago Synod reportedly mooted the idea of forming a committee to 'actively interfere' with war-related content in the paper.[92] Discussing the conflict's social impact, the *New Zealand Tablet* noted that wartime communications had long been subjected to 'the sharp eye and merciless hand of the military censor'.[93]

Early in 1900 James Shand complained about postal and telegraph departments being under military control, while another correspondent alleged the censors' actions were proof of a deliberate attempt to 'hoodwink' the British public.[94] A 1902 Wellington newspaper article provided an insight into the type of material the censors wished to suppress. It claimed censors had redacted all mention from a *Daily Mail* correspondent's telegram of British commanding officer Lord Kitchener ordering the summary execution of Boers captured wearing khaki.[95] The correspondent circumvented the censors by including the report in a letter that evaded their attention. A Sixth Contingent soldier also claimed that the capture of over one hundred members of his contingent had been kept secret for a period.[96] Even seemingly innocuous objects could attract official scrutiny; when Lieutenant Henry Heywood sent his father toys made by a Boer prisoner, all reportedly bore the censor's stamp.[97]

While New Zealand soldiers' letters occasionally contained sensitive information, unless they were intercepted en route to African ports their value to the enemy was limited. By the time they reached New Zealand the information was dated and the men who had written them had usually moved on. A letter Trooper Bert Stevens posted to his father in Hāwera took 41 days to arrive, while a 'delayed' letter from Trooper Alex Wilkie reportedly took 69 days to reach Ashburton.[98] The *Ashburton Guardian* noted that Wilkie's letter had not been opened as it claimed others had been. Even though the Boers could not intercept seaborne communications, during an attack on Roodewal, General de Wet seized 2000 British mail bags.[99] New Zealand letters were among the haul, with Luke Perham complaining to his mother that De Wet had 'burnt all our mails'.[100] Although the flow of letters continued unabated, Sixth Contingent orders for May 1901 reminded soldiers of king's regulations relating to the publication of military intelligence: 'It must be clearly understood that officers and soldiers are held responsible for any communications to their friends which may subsequently be published in the press.'[101]

Newspapers themselves appear to have been willing to at least temporarily suppress sensitive information. Following Trooper Charles Tasker's court martial and imprisonment in England for sleeping on duty in South Africa, the War Office reluctantly accepted the clemency pleas of Seddon and Tasker's parents.[102] The permanent under-secretary for the colonies stressed that Tasker's release was 'very exceptional'.[103] A grateful William Pember Reeves, the New Zealand Agent General in London, assured the under-secretary that he had asked New Zealand newspaper correspondents in London to say as little as possible about Tasker's release.[104] Many newspapers confined their initial coverage to a single sentence, and it was not until the month after Tasker's release that the *Evening Post* gave more detailed accounts.[105]

After Ninth Contingent trooper Leolin Arden was found dead near his South African camp, a court of inquiry found he had either been hit by a stray bullet or intentionally killed.[106] Lord Ranfurly informed the acting premier that he had received a telegram reporting Arden's death from an 'accidental' gunshot wound, but made no mention of the possibility that his death may have been murder.[107] The misadventure version of Arden's death appeared in the press and the commandant of the New Zealand forces informed the trooper's stepmother that he had lost his life 'for the good of the Empire'.[108] It was not until the month after Arden's death was first reported in the press that his *Wanganui Collegian* obituary noted foul play may have been involved.[109] Citing the school magazine

Hāwera farmer Trooper Bert Stevens of the Eighth Contingent with his mother, Delia Stevens, and his younger brother, Howard Waldo Stevens. In October 1914 Howard enlisted in the New Zealand Medical Corps. NATIONAL ARMY MUSEUM TE MATA TOA, 1999.3239

as their source, several newspapers then reported that Arden's death might not have been accidental.[110] That a school magazine could provide a more detailed and accurate account of the incident than official cables says much about the impact of censorship during the war.

'An undesirable spirit of militarism'

Despite the initial popularity of New Zealand's involvement in the conflict, there was concern about where it might lead. While public suspicion of militarism pre-dated the war, some in New Zealand believed national security demanded a strong armed force. A 1900 report by the secret Joint Defence Committee proposed a raft of changes estimated to cost over £370,000 including increasing Volunteer membership to 18,000 and purchasing 30,000 rifles.[111] To offset the £105,000 needed to buy the weapons and equipment, the committee chairman estimated £42,000 could be recouped by selling 12,000 rifles to members of defence rifle clubs.[112] The committee also discussed Seddon's proposal for a 2000-strong Imperial Reserve — a trained military force that could serve overseas if and when required.[113] The *Evening Post* dismissed the reserve as a threat to democracy and a 'recruiting ground for an Imperial or local standing army' at the disposal of the executive.[114]

While the idea of a reserve generated debate about the country's defence needs, for some the number of contingents being sent to South Africa was of more immediate concern. In 1900 rifle club member Penelope Farquharson expressed concern that if the government raised a sixth contingent few young men would be left in New Zealand.[115] And although the *Auckland Star* praised New Zealand's patriotism in raising four contingents, it also questioned the need for more.[116] A newspaper decried the prevailing 'effervescence of khakiism' and what it termed 'khaki fever', though the *Otago Witness* argued that the danger was not that New Zealanders' military spirit would grow, but rather that it might prove unsustainable.[117] Although its members were dismissed as 'Pro-Boer Cranks', combating militarism was a key objective of the Auckland Peace Association, and outspoken Legislative Council member Henry Scotand also criticised the wave of militarism he saw washing over New Zealand.[118]

As the war entered its second year, Chief Justice Sir Robert Stout added his voice to those concerned about the dangers of 'overindulgence in military enthusiasm'.[119] A *Nelson Evening Mail* editorial titled 'Imperialism Run to Militarism' claimed New Zealand needed to discriminate between the two, and warned of the economic impact of sending out young men who, having acquired 'the military habit', would only reluctantly return to civilian life.[120] Similar views were expressed by William Napier, Auckland City MHR, who observed that it was work, rather than militarism, that sustained nations.[121]

These concerns had little impact on the enthusiasm of men eager to serve. Early in 1900 Ranfurly informed Joseph Chamberlain that the Fourth Contingent could be almost any size as men were volunteering across the country.[122] As the war progressed, the size of the forces New Zealand dispatched to South Africa grew incrementally from the approximately 200 men of the First Contingent. Nonetheless, by 1902 contingents were still attracting far more applicants than required, with Seddon maintaining that 5400 men had applied for the 1000 positions in the Ninth Contingent.[123] Despite the Joint Defence Committee noting equipment and training facility shortages, incomplete harbour fortifications and a reliance on imperial funding, Seddon remained characteristically bombastic.[124] He assured the Heretaunga Mounted Rifles that there was little chance of an enemy landing on New Zealand's shores and claimed any invading force would be wiped out within 24 hours.[125] *New Zealand Times* journalist Pierce Freeth expressed a similar level of confidence in 1901 when he appeared before the Royal Commission on Federation with Australia.[126] An opponent of militarism, Freeth believed adequately armed New Zealand Volunteers employing Boer tactics could repel an invasion.[127]

While the nation's wartime preoccupation with the 'Cult of Khaki' was largely driven by a desire to display solidarity with the British Empire, it also reflected deeper concerns about the nation's vulnerability to German, Russian, French and Japanese expansionism. The new century saw the United States and European nations established in the South Pacific. To counter their influence, Seddon advocated extending New Zealand's territorial boundaries to include the Cook Islands, Fiji, Tonga and the Society Islands.[128] A European resident in Fiji who supported the proposal claimed that Fijian colonists were 'like the defenders of Mafeking' — they were 'widening the outposts of the Queen'.[129] Seddon raised the British 'surrender' of Sāmoa with the Secretary of State for the Colonies. He claimed that despite assurances that neither Germany nor the United States desired territory in the Pacific, Britain had allowed them to divide Sāmoa.[130]

Seddon also wanted to ensure a military role for New Zealand in any future conflict involving the empire. When asked in Parliament whether he would seek colonial representation at a possible inquiry into the South African War, Seddon rejected the idea, believing that making such demands might adversely affect New Zealand's chances of being asked to contribute militarily if the empire's interests were threatened.[131]

The conflict provided a fillip to the school cadet system, which had been in existence for more than a decade prior to the outbreak of the South African War.[132] In 1901 Seddon supported legislation making drill compulsory for schoolchildren. He envisaged boys learning drill at school then continuing in the cadets before enrolling in Volunteer corps as adults and finally joining rifle clubs. When Wellington City MHR John Hutcheson, who voted against sending the First Contingent, opposed the bill and warned of the 'demon of militarism' in Germany, Seddon suggested that if militarism, conscription and military training were giving Germany an advantage, New Zealand should follow suit if it were to meet rivals on equal terms.[133] Wellington Education Board inspector Robert Lee claimed that one of the most important duties of a citizen was to learn how to shoot, and he asserted the argument that the cadet system engendered 'an undesirable spirit of militarism' had been discredited by educational authorities.[134]

Cadets at more affluent schools practised with real weapons, and the Parnell Lady Cadets, composed almost entirely of girls and young women, also drilled with rifles.[135] In a 1900 letter to Seddon requesting more suitable weapons, Wanganui Collegiate School principal Walter Empson noted his cadets were intimidated by the formidable recoil of the obsolete Snider rifles they used (a government report also noted that the Snider's 'antiquity and kicking powers' rendered it useless as a cadet weapon).[136] Nonetheless, Collegiate Old Boy Sergeant Duncan Blair of the Second Contingent acknowledged the benefits of his cadet training and credited the cadet corps with instilling 'military aspirations' in the students.[137]

In 1902 Invercargill MHR Josiah Hanan asked the government to provide Invercargill cadets with captured Boer rifles.[138] Arguing that Invercargill was

Members of the Parnell Lady Cadets in 1901. *New Zealand Graphic and Ladies' Journal*, 23 November 1901, p. 985, Auckland Libraries Heritage Collections, NZG_19011123_985_2

likely to become 'one of the leading cities in the colony', Hanan claimed it was entitled to its fair share of the weapons.

In 1900 the *Christ's College Register* reported a substantial increase in cadet recruits, adding that displays of patriotism at the Christchurch school were at 'fever heat'.[139] When Christ's College Old Boy Lieutenant Arthur Neave was killed at the Battle of Paardeberg in the same year, he was lauded in the school magazine for maintaining the school's honour, and for 'meeting the proudest end a man could desire'.[140] Such deaths were presented to children as both honourable and manly, and even Queen Victoria requested a photo of Neave. Whether such portrayals influenced cadet membership is unknown, but by early 1902 there were 5712 cadets in 92 corps nationwide.[141]

In addition to the cadets, the situation in South Africa provided the main impetus for wartime increases in New Zealand Volunteer Corps membership — a 'remarkable military revival' that was noted during the 1902 conference of colonial prime ministers.[142] Volunteer mounted corps rose from 13 in 1898 (with a total membership of 808) to 71 corps in 1902 (with a total membership of 5467).[143] Expenditure on the Volunteer force rose from £26,171 in the 1897–98 period to £95,069 for 1901–02, while membership increased from 6820 in 1898–99 to 12,504 in 1901–02.[144] On paper these figures looked impressive, but sporadic attendance often reduced corps effectiveness, with some lacking sufficient officers to function efficiently and only coming together as battalions during annual camps.[145]

Defence rifle clubs also grew during the war, with clubs increasing from 72 in 1901 to 114 in 1902.[146] In the Wellington district, the number of clubs almost doubled, from 39 in 1901 to 62 in 1902.[147] Before the war, Arthur Penton, the commandant of forces, had been critical of rifle clubs.[148] He believed that as club members only fired at static targets they lacked field shooting experience. By contrast, he cited the Boers' former dependence on fresh meat, and related the story of a Boer who gave his son three cartridges and promised to thrash him if he failed to return with three deer.[149] An *Evening Post* editorial defended the New Zealand Rifle Association from Penton's criticism, accusing him of wanting to force the European military system upon New Zealand, and of failing to understand that rifle clubs were 'essentially democratic, and in every respect alien to militarism'.[150]

'This demon of jingoism'

The impact of New Zealanders who opposed the war was undermined by their failure to coalesce, with most groups remaining fragmented, regionally based and, in the main, ineffectual.[151] Instead of providing the public with a convincing alternative viewpoint to the dominant narrative, dissenting voices were largely overwhelmed by patriotic and jingoistic outpourings (although, as noted earlier, the intensity of public feeling would diminish as the war dragged on).

The 1899 declaration of war was greeted with wild enthusiasm by the First Contingent men living under canvas at Karori military camp.[152] The soldiers cheered themselves hoarse and sang patriotic songs until the bugler sounded 'lights out'. Even prior to hostilities commencing most newspapers cast the Boers as the antagonists. In Dunedin, the *Evening Star* alleged that defenceless women and children had been killed by Boer forces in an attack on an armoured train.[153] Another newspaper accused the Boers of spitting in the faces of British women preparing to leave the Boer republics and alleged that 70 women and children had been kept in cattle trucks for 30 hours without food or water, resulting in two children dying of exposure.[154]

As tensions rose in 1899, newspapers initially referred to the 'Transvaal troubles', the 'Transvaal difficulties' and, as the situation worsened, the 'Transvaal crisis' — all phrases that had been used since the 1880s.[155] But from the first day of fighting until long after the guns fell silent, across the empire the 'Boer War' was the favoured name for the conflict.[156] Of New Zealand newspapers, the *Marlborough Express* alone recognised Britain's role in the conflict by frequently referring to the 'Anglo-Boer War'.[157]

Like the politicians who voted to offer a contingent, for the majority of New Zealanders the question of responsibility was immaterial; the empire was at war and their duty was to rally round the flag. Residents of towns across the country formed committees and established funds to raise money to support the troops. In Dunedin, coins rattled in 'a silver shower' on the stage of the Princess Theatre during a recital of Kipling's poem 'The Absent-Minded Beggar'.[158] It was a rare patriotic event where this jingoistic work was not featured together with the equally popular 'Soldiers of the Queen'. An estimated 13,000 people attended a Dunedin fundraising floral fête, with Fitzgerald Brothers' Circus rescheduling its matinee show to avoid affecting the fête's attendance, and in Auckland £500 was raised during a military display.[159] Even some who had initially opposed the war were carried along by the prevailing patriotic mood. Politicians Robert McNab and James Thomson reportedly delivered 'stirring speeches' at a concert to raise money for the Fourth Contingent in Clinton.[160] As well as fundraisers, long lists of patriotic contributions from businesses, clubs and individuals filled newspaper columns.

Two soldiers returning from the war to their Waikouaiti homes were reportedly met by the largest crowd ever gathered in the small Otago town.[161] A year later, Waikouaiti trooper John Townsend also returned from South Africa.[162] In anticipation of his arrival, triumphal arches were erected and the streets were decorated with flags and lanterns. When Townsend's train finally arrived he was carried aloft by well-wishers to a buggy drawn by enthusiastic supporters.[163] The newspapers documented similar scenes that played out across the country.[164] The Palmerston correspondent of the *Otago Witness* reported that residents in the region were determined not to be outdone by other districts in their Fourth Contingent contributions.

This fascination with the war was driven in part by a sense of collective ownership. Speeches, newspaper articles, advertising and correspondence were peppered with references to 'our boys' and 'our sons'.[165] Only days after war was declared, Auckland pharmacist Wilfred Manning suggested that each of 'our boys' should take a bottle of his toothache remedy with them to South Africa, and Griffin and Sons Ltd produced 'Our Boys' chocolate.[166] Even as the war entered its final months, Wellington jewellers Stewart Dawson and Company advertised flasks, field glasses and wrist watches for 'our boys going to the front'.[167]

Appealing to patriotism for commercial gain may have been disingenuous, but the public's engagement with the conflict was sincere. After Jessie Whitehead's brother died in a train accident at Potchefstroom that killed several New Zealand soldiers, she compiled a scrapbook of photos relating not only to her deceased brother but also to the other victims.[168] In early 1900 the general manager of New Zealand Crown Mines at Karangahake wrote to Lieutenant-Colonel Banks. The manager claimed that everyone at the company was delighted with the praise the New Zealand contingents had received from General French. The manager was certain 'our boys' would do their duty in the battles ahead and claimed the world must view 'our army in Africa as one composed of heroes'.[169]

'She pro-Boers' and 'women patriots'

Comparatively few New Zealand women travelled to South Africa during the war, but within New Zealand they were an integral part of the nation's response to the conflict.[170] In the tiny Canterbury settlement of Springburn women raised £24 for New Zealand troops in South Africa through a patriotic concert and dance at Staveley Hall that was so crowded many had to stand outside.[171] Even lowly paid Dunedin domestic servants, many of whom were women, contributed sufficient funds to purchase a horse for a contingent member.[172] North Island women were equally involved, with the Wanganui Ladies' Club sending clothing to the troops.[173] Waipawa resident Alice St Clair Inglis also sent clothing with instructions that Hawke's Bay men were to be given preference when it was distributed, and in Auckland Jessie Bodle sent medical supplies, hoping few men would need them.[174] Many family members who shipped gifts to South Africa were motivated as much by concern for their relatives' comfort as they were by patriotism. The First and Second contingents were sent liquorice, jam, eau de cologne, cakes, meat, newspapers, preserves, apples, figs, stationery, tobacco and, in one case, a salmon.[175] Women like Penelope and Leonie Farquharson of North East Valley in Dunedin who had relatives serving in South Africa followed the soldiers' fortunes with interest and regularly corresponded with them.[176]

New Zealand women's experience of the conflict was not confined to fundraising and supplying comforts for the troops. For some, the inevitable wartime separations meant straitened circumstances. Private Clement Totman sailed for South Africa leaving his wife without financial support.[177] When the military authorities became aware of her predicament, Totman was discharged and sent home, though Rebecca Totman stressed that she had no desire to see her husband discharged and simply wanted financial security.[178] An Eighth Contingent recruit appeared in the Paeroa Court charged with attempting to leave the country without providing for his unborn, illegitimate child.[179] After the baby's 17-year-old mother took legal action, the recruit was ordered to support the infant and did not sail with his contingent.[180] Other attempts to apply civil law were less successful. When a detective attempted to arrest a trooper aboard the troopship *Gymeric*, Lieutenant-Colonel Sommerville informed the officer that he had no jurisdiction aboard the vessel. The trooper was accused of trying to leave New Zealand without providing maintenance for his wife, but he was nonetheless permitted to sail for South Africa.[181]

Politically active women came to the forefront of the New Zealand public debate surrounding the conflict in May 1900.[182] Though the National Council of Women (NCW) was not united in condemning British actions in South Africa, its 1900 Dunedin conference provided a forum for the expression of anti-war sentiments. Delivering a paper on peace and arbitration, Wilhelmina Bain was openly critical of Britain's role in the conflict.[183] In response to this, prominent suffragette and NCW president Kate Sheppard expressed disappointment; stressing that her caution was not motivated by a fear of public opinion, Sheppard said she had advised delegates to avoid specific reference to the war.[184] She also asserted that the *Evening Star* had misinterpreted Bain's comments in a scathing editorial critical of 'pro-Boer speakers' at the conference.[185]

Bain remained unapologetic, claiming she had only been told shortly before delivering her speech to confine it to peace and arbitration. She said she stood for the interests of humanity and was as much pro-British as pro-Boer.[186] Nonetheless, her address angered the editor of the *Otago Daily Times*, members of the Otago community, and Mayor Chisholm, who in response declined to chair an NCW meeting.[187] Although not all NCW members agreed with Bain's stance, the council believed that women had both a right and a duty to promote peaceful conflict resolution and shape the political decision-making process.[188]

Bain claimed the rise in militarism was intended to transform men into 'automatic killing machines' and accused New Zealand of sending young men to Africa to kill boys as young as 16 and men as old as 70.[189] Though she could not have known it at the time, the age range Bain attributed to the Boer forces was also that of the New Zealand contingents. Bain's motion supporting arbitration and rejecting militarism was seconded by Marianne Tasker, whose son Charles served in the Sixth Contingent.[190] In an apparent contradiction, Marianne later served on the committee organising the Eighth Contingent send-off in Wellington.[191] According to the *Evening Star*, she had also spoken in support of New Zealand sending troops to South Africa.[192]

During the NCW conference, Margaret Bullock, the younger sister of former

Marianne Allen Tasker, National Council of Women member and founder, and from 1895 leader of the Women's Democratic Union. Tasker's son served in the Sixth Contingent and was imprisoned in England for falling asleep at his post while on duty. ALEXANDER TURNBULL LIBRARY, PAColl-6882-1

Wanganui MHR Gilbert Carson, also championed the British cause despite her brother's earlier opposition to sending the First Contingent.[193] One newspaper claimed Bullock was the only NCW member who understood the situation in South Africa. She believed that England's prestige would have been irrevocably damaged had the British not confronted the Boer republics.[194] Despite Bullock's views, Bain received support from like-minded New Zealand women such as Margaret Sievwright of the Gisborne Women's Political Association (GWPA), who saw the war as the result of capitalist machinations.[195]

Sievwright contrasted the British response in South Africa with the inertia of Christian European nations when Armenians were 'butchered' in 1896 during the rule of the Ottoman sultan Abdul Hamid II. The difference, she claimed, was that the South African dispute involved territory, gold and diamonds.[196] Shortly before the war Sievwright tabled a resolution at an NCW meeting deploring growth in the armaments industry.[197] It was seconded by temperance worker and suffragist Annie Schnackenberg, who opposed the involvement of Australasian countries in armed conflicts and accused the British authorities of viewing the colonies as a recruiting ground for 'Imperial militarism'.

For many New Zealanders, Sievwright's views were no more palatable than Bain's, and even within the GWPA she attracted criticism. Agnes Scott, who had originally been elected to represent the association at the Dunedin conference, applauded Chisholm's refusal to chair the NCW meeting and attempted to distance the GWPA from Sievwright's comments.[198] She claimed Sievwright did not speak for the association and stressed it remained loyal to queen, flag and country.[199] 'An English Married Woman' who had planned to attend changed her mind as her 'true British patriotism could never tolerate the atmosphere of so many (apparently) she pro-Boers'.[200] Florence Brewer was equally disdainful of the NCW and declined to sing at its conference due to what she described as the 'unpatriotic and disparaging remarks' of its members. She did, however, sing the following week before an estimated audience of 3000 during celebrations marking the relief of Mafeking.[201]

A correspondent to the *Evening Star* praised its condemnation of Bain's address and claimed the vast majority of New Zealand women supported the newspaper's stance.[202] The NCW, Ina Clifton believed, had cast a slur on the character of all New Zealand women. The editor of the *Otago Witness* was more ambivalent. He downplayed the public outcry over Bain's comments, claiming it was only marginally more ridiculous than her opinions. The editor did not believe Bain was a Boer sympathiser. Instead, he claimed, she had succumbed to the attraction of deviating from mainstream thought and was more concerned about her public image than whether her views withstood public scrutiny.[203]

Like those of Marianne Tasker, Jessie Williamson's actions during the war appeared contradictory. The Irish-born Williamson served as the NCW's vice-president, supported arbitration, considered warfare outdated (except for controlling 'savages'), and had been the only woman on the first Transvaal Refugees' Fund Committee.[204] She also took part in contingent preparations, served on the Wanganui General Committee responsible for raising troops, and said she was proud to be a citizen of a country that had sent its sons to fight in South Africa.[205] Williamson and Bullock's loyalty was strongly defended by Katie Newcombe and Susie Wright of the Wanganui Women's Political League, who rejected accusations that the two NCW members were Boer sympathisers. They noted that the pair had played a pivotal role in sending a trained nurse to South Africa and had provided refreshments and entertainment for a contingent send-off.[206]

The level of patriotic support among women leaves little doubt that individuals such as Sievwright and Bain were in the minority. In Kūmara a woman returned her purchases in protest after a local shopkeeper voiced support for the Boers, while women from the same South Island town sent gold and greenstone chains to British generals Buller and French.[207] A mother reportedly wrote to the *Evening Star* asking why women were barred from serving as her daughter wished to enlist.[208] In 1902 women also rejected the German allegations of British wartime excesses. Edith Statham, the secretary of both the Society for the Protection of Women and Children and the Young Women's Christian Association, refuted the German 'insults' and encouraged New Zealand women to support the empire by forming a national league.[209]

At a patriotic meeting Deputy Premier Joseph Ward read a resolution on behalf of 'the women of New Zealand' refuting the 'slanderous' German allegations that reflected on 'the honour of our sons and husbands and the sons and husbands of the Mother Land' fighting in South Africa.[210] It was seconded by Mary Seddon, who claimed that New Zealand women were willing to sacrifice their brothers and sweethearts if it benefited the empire.[211] Seddon, the premier's daughter, served as a lieutenant in the Young Ladies' Contingent (YLC).[212] The group, which dressed in khaki and performed military drills at patriotic events, had been formed by Lady Mary Douglas (the wife of the under-secretary of defence, Sir Arthur Douglas), who arranged a Government House fête for the Transvaal Contingent Benefit Fund.[213] The YLC's fundraising activities extended beyond the war effort: in 1900 the 'girls in khaki' staged a Wellington concert in aid of the Indian Famine Relief Fund.[214]

The popularity of the YLC led to the appearance of similar corps in other regions. A newspaper claimed the premier had issued instructions for the weapons and accoutrements used by the 'Wellington Amazon Contingent' to be sent to Dannevirke for the town's Huia Ladies Khaki Contingent, and an Auckland Ladies Benevolent Society fête included 'a charming regiment of Amazons in khaki'.[215] The highlight of the 1900 Hauraki Band Bazaar was 14 women dressed in the khaki uniform of the Hauraki Rifle Volunteers and armed with real rifles and bayonets conducting drill exercises.[216] Whanganui had the 'Wanganui Amazon Carbineers', and drapers in Mataura sold fabric in 'the Fashionable Khaki Shades'.[217] Similar khaki-clad women's patriotic groups appeared in Maungakaramea, Thames, Gisborne, Picton and Temuka.[218]

Premier Seddon applauded the patriotism of these groups and was willing to provide limited assistance, but he was not prepared to be their financial benefactor. In the West Coast town of Greymouth many young women joined the town's patriotic Khaki Corps, and in 1900 the secretary of the Greymouth Oriental Bazaar Committee requested khaki fabric, 43 khaki hats and the loan of rifles and 'some cartridges' for corps members, presumably for display purposes.[219] Noting that the women would pay for the hats and cloth, the secretary asked that the Defence Department keep the price as low as possible.[220] The bazaar for a hospital was a success but failed to meet financial expectations, and requests from the Defence Department seeking the return of the rifles and payment for the hats and khaki cloth initially went unanswered.[221] The rifles were eventually returned, but Seddon refused to cancel the debt and instructed the Defence Department to begin legal proceedings to recover the money.[222]

Days before the war began, the Defence Department declined offers from several nurses, as well as the Red Cross Brigade, to serve in South Africa, as the

Members of the Greymouth Khaki Corps. *Otago Witness*, 30 January 1901, p. 26, Alexander Turnbull Library, N-P-2173-26

British Army was responsible for soldiers' medical care.[223] In a letter to the *Press*, 'A Nurse' claimed numerous trained nurses in New Zealand were 'fired with patriotic zeal' and several volunteered, including Nelson Hospital matron Melita Jones.[224] In early 1900 Seddon relented and agreed to four nurses going to Africa, informing a nurses' deputation that the imperial authorities were prepared to accept colonial nurses on the proviso that the British government was not responsible for their expenses.[225] Once again, women were at the forefront of fundraising and represented the vast majority of contributors to the Nurses' Fund.[226] Provinces vied for the limited number of positions available, with Wellington and Christchurch eager to contribute and the Otago and Southland Nurses' Fund raising more than £1200.[227] Among the contributors was an Ōamaru woman who gave the proceeds of the sale of her bicycle.[228] Shortly before the war ended, support remained for inexperienced New Zealand nurses being given an opportunity to hone their skills and 'do their duty' in South Africa.[229]

The first four nurses selected — Jane Peter, Grace Webster, Gertrude Littlecott and Annie Hiatt — sailed from Lyttelton aboard the *Lincolnshire*.[230] They were accompanied by Charles Pierson of Eichbaum's Chemist in Temuka, who served as an ambulance 'dresser', and Mabel Brooke-Smith and Constance Geraldine Jeffreys, who paid their own passages and later served aboard the hospital ship *Orcana*.[231] Nora Stevens, a Nelson nurse whose father was the Manawatū MHR and whose brother John served in the Second Contingent, received patriotic contributions from Nelson and Ōtaki and also made her way to South Africa.[232] Involvement in the war clearly appealed: in a letter to family in Tīmaru, Laura Woollcombe told her mother, 'This is a fine life!'[233]

Nurses represented the largest group of New Zealand women who served during the war, yet despite the widespread support they received at least one doctor remained sceptical. Robert Bakewell, who had served during the Crimean War (1853–56) with Florence Nightingale at Scutari, questioned sending nurses to South Africa.[234] While acknowledging that nurses at Scutari had performed well in difficult conditions, Bakewell dismissed them as either 'gushing enthusiasts who don't know what they are about' or strong-willed women who went to war in pursuit of money, marriage and improved social status.[235] The *Press* accused Bakewell of misogyny and stated that his claims were untrue of 'the majority of volunteer nurses'.[236] By 1905 Bakewell's views had moderated;

describing Florence Nightingale as a tactless and 'thoroughly disagreeable woman', he claimed that '[a] dozen Miss Nightingales were not worth one sister like ours'.[237]

In 1902, meanwhile, the Education Department sought applications for 20 female teachers' positions in South Africa on one-year contracts.[238] Successful applicants received an annual £100 salary to work in the concentration camps that largely housed Boer women and children, with Lord Ranfurly noting there had been a total of 222 applicants for the 20 positions.[239] Some questioned the teachers' actions, and in an article titled 'An Undesirable Sacrifice' the *Observer* dismissed as idiotic recklessness the decision of the 400 teachers the newspapers claimed had applied for the South African positions.[240] It predicted that, instead of the 'alluring excitement of life in a military camp', the teachers would encounter the squalor and misery of the concentration camps where the British were despised.

Another newspaper disagreed with those who believed young New Zealanders should be confined to their own country and considered they could hold their own in any part of the world.[241] Neither the modest pay nor the difficulties they would face dissuaded the independent-minded New Zealand women, dubbed 'the Learned Eleventh', who successfully applied.[242] When their contracts ended some took the opportunity to travel further afield, while several who remained in South Africa maintained connections to home through letters and social interaction with fellow New Zealanders.[243]

'Youthful patriots'

In 1900 George Squirrell, the chairman of the Auckland City Schools Committee, addressed Wellesley Street School students. Squirrell reminded them that where the Union Jack flew, 'there reigned freedom and liberty, and woe be to the nation or individual who dared to insult it'.[244] He added that in the future New Zealand would look to the colony's public schools for soldiers to fight the country's battles. For his students, the importance of New Zealand's links to the British Empire hardly needed reinforcing; while scenes of jubilation took place when holidays marking significant wartime events interrupted normal school routines, many children also took an active interest in the contingents and participated in war-related activities.[245]

With adults so heavily invested in the war and school education already having a strong imperial focus, it was inevitable that the conflict's influence would spill into New Zealand classrooms. Children's ages largely dictated the extent to which they were exposed to the war, but schools, teachers and curricula also played key roles. At Wanganui Collegiate School students conducted regular Parliamentary Union debates on a range of war-related topics. One debater chose to represent the 'electorate' of Pitsani — the starting point for the Jameson Raid.[246] In another debate the motion that 'the conduct of the British Army, as a whole, in the present War is more deserving of censure than praise' was overwhelmingly opposed.[247]

An *Evening Star* article titled 'The Empire's Nursery Ground' illustrated the war's impact on students. The reporter described a visit to a High Street School classroom in Dunedin decorated with a map of Transvaal, cartoons of Kruger, and portraits of Boer general Piet Joubert and First Contingent commanding officer Major Robin. When questioned, students located Ladysmith on the map, identified the Boer president, explained the grievances of the British uitlanders, and named Baden-Powell as the defender of Mafeking. The children also knew that Boer weapons were manufactured in Germany and described Boer general Piet Cronjé as 'cunning, crafty and cruel'. One student claimed the war was being fought because the Boers 'beat us last time'.[248]

In Wellington, Robert Lee noted that nearly 80 per cent of student compositions in the 1899 English examination dealt with the Boers and claimed the essays were characterised by ardent patriotism.[249] The war also influenced a number of other examination questions: students were asked to spell 'besiegers' and in the geography section had to locate Mafeking and other significant war-related towns.[250] Teachers were also expected to keep abreast of events in South Africa. The 1899 Pupil Teachers' Examination featured essay topics on 'British rule in South Africa', 'The Boer war now going on' and 'The New Zealand Contingent for the present war in the Transvaal'.[251] Even children's humour was influenced, with the *Wanganui Collegian* publishing a number of students' war-related jokes, including, 'Why does Kruger walk on grass? Because he can't walk on Rhodes.'[252]

More than just enthusiastic spectators, New Zealand children were actively involved in fundraising.[253] A group of children in Wellington calling themselves the Ellice Street Rifles raised money for the war effort while dressed as contingent members, and in Canterbury an entrepreneurial boy made an effigy of Kruger and charged passers-by to pelt it with wooden balls.[254] The patriotism of one Christchurch 15-year-old landed him in court. The boy was sentenced to six months' probation after being found guilty of forging a cheque in a failed attempt to raise sufficient funds to go to Transvaal.[255]

Though the war dominated newspaper coverage of international events, the famine in India and the Boxer Uprising in China also attracted children's attention.[256] Using the pseudonym 'Lord Roberts' (the commander-in-chief in South Africa), a child described the war with the Boers and the fighting in China as terrible.[257] Of 30 youthful correspondents to Dot's Little Folk column, 15 referred to the relief of Mafeking.[258] 'Poppy' complained that there was no 'fresh' news about the Boers: 'I think all the people are more concerned about affairs in China.' Like 'Lord Roberts', many young contributors used war-related pseudonyms such as 'Lady Smith' and 'Kimber Lee' in letters to the press.[259]

'God's cause'

With the outbreak of hostilities, New Zealand clergy faced the challenge of reconciling their spiritual beliefs with warfare. For some, this reconciliation seems to have been comparatively easy; at a religious service Frederic Wallis, the Anglican Bishop of Wellington, was unequivocal.[260] From a makeshift pulpit draped with the Union Jack, Wallis addressed the crowds that had made the journey to the Karori paddock where the First Contingent was encamped. He wished the men success and said they would fight for justice and truth because the conflict was 'God's cause'.[261]

In Lyttelton, Reverend Coates decried the horrors of war, but believed the situation in South Africa represented an occasion where war was justified, and Reverend McNicol prayed that if war could not be avoided, it might in some way benefit mankind and further God's cause.[262] Reverend Chambers, the senior chaplain for the Permanent Force, acknowledged the claims of other nations regarding Britain's 'territory-grasping proclivities', but refuted the charges and assured his congregation that Britain was defending her empire while protecting the weak.[263]

A Timaru Anglican archdeacon saw the war as a mysterious extension of

God's work, believing it was 'directed by Divine Providence for good purposes, which would yet appear'.²⁶⁴

A vocal supporter of both the empire and British actions in South Africa, Anglican Reverend Curzon-Siggers cautioned his Dunedin congregation that if the war were lost it would not only spell the end of the empire but also see New Zealand paying tribute to a foreign power.²⁶⁵ Curzon-Siggers was, however, not entirely unsympathetic of the Boers. In an 1899 article titled 'The South African Problem', he claimed that 'with some justice' the Boers had complained that as pioneers they had to keep black Africans in 'a state of technical slavery, but not of human servitude'.²⁶⁶

The Anglican Church also actively promoted militarism among New Zealand youth. Following the example of the church in England, Church Lads' Brigades in Hastings, Masterton, Carterton, Feilding, Hāwera and Wellington were run on military lines, the boys drilling with rifles.²⁶⁷ In a 1901 letter to the Defence Department, Lieutenant Donald of the Masterton Brigade noted that two of the corps members were officers serving in Transvaal, and he claimed the corps was increasingly recognised as 'the future military training school for the British Nation & Army'.²⁶⁸

Comparatively few clergy voiced opposition to British actions in South Africa, and the objections of those that did were based more on the conflict's immorality than a denominational stance.²⁶⁹ In October 1899 Reverend Rutherford Waddell, the editor of *The Outlook*, questioned what was drawing New Zealand into the conflict, and what was to be gained from it.²⁷⁰ Waddell accused Rhodes' British South Africa Company of being driven by insatiable greed and a thirst for power, and claimed the Secretary of State had 'jockeyed the nation into an unnecessary war'.²⁷¹ Believing the colonies had an obligation to consider the morality of the war, Waddell noted that other clergymen unreservedly supported the move to armed aggression.²⁷² In 1900 he declared that 'the flame of patriotism has burst forth and is burning up everything else'.²⁷³ Waddell was soon to experience its heat. Warning that those who spoke out against the conflict could expect public vilification, he advised his readers that 'the crowd is rarely mannerly, and when it is on the war path [sic] it is sometimes brutal'.²⁷⁴ In a letter published in *The Outlook*, 'Veritas' claimed others would share the writer's sense of indignation at Waddell's comments, while a *Tuapeka*

Reverend Rutherford Waddell, the outspoken editor of the Presbyterian *Outlook*, who openly questioned both the reasons for the South African War and the manner in which it was being conducted. PRESBYTERIAN RESEARCH CENTRE (ARCHIVE), P-S15-22

Times correspondent accused Waddell of slander for suggesting Cecil Rhodes and Chamberlain had orchestrated the conflict, and the *Southland Times* hoped the Presbyterian organ would 'step into line'.[275] Refusing to be cowed, Waddell described the tactics of war as cruel and immoral and pointed to a *London Spectator* article that called Rhodes the most dangerous enemy of the empire.[276] Waddell referred to reports that Baden-Powell had feigned retreat to lure Boer forces over a powerful mine, and said that, if true, the officer's actions would be 'atrocious'.[277]

Despite his opposition to the war, Waddell published letters criticising his views. A correspondent to *The Outlook* called Waddell's comments an injustice to the generals defending the empire and no better than jingoistic calls for the war's extension, while 'Briton' said it was outrageous for Waddell to suggest Britain was pursuing the war for financial gain, adding that unless he changed his position the veiled threats directed at him would become real.[278] Though Waddell considered the South African War unjust, he applauded the patriotic fervour it generated.[279] Commenting on the public's fixation with the conflict, he assured his readership that he was not complaining: 'It is a right and proper thing.'[280] He also claimed the war roused 'the manhood of the nation', increased patriotism and showcased imperial unity.[281] Waddell enjoyed a degree of support within the Presbyterian synod, but this centred more on freedom of speech than on sympathy for his views. Seconding a motion expressing regret at Waddell's comments, Reverend Davidson claimed many were outraged by the editor's stance, though Reverend Hewitson proposed an amendment commending Waddell on his independence of opinion and willingness to publish views that ran counter to his own.[282]

Waddell was not alone in questioning the reasons for the war. Reverend Bates, an Anglican, said war would not have broken out had it not been for the Transvaal goldmines, while Reverend James Gibb, the Presbyterian minister of Dunedin's First Church, said that his support for the war was constrained by his uncertainty regarding its causes.[283] Gibb identified protecting national rights and alleviating the suffering of the downtrodden as legitimate incentives for war but doubted whether either existed in the South African situation. Though initially critical of the martial fervour engulfing New Zealand, Gibb had an apparent change of heart, later claiming Britain had no choice in 1899 other than 'to draw the sword, and fling the scabbard afar'.[284] Even Waddell acknowledged the need for victory once fighting began.[285]

In January 1902 *The Outlook*'s temporary editor William Hutchison expressed regret that the wasteful 'war of extermination' was still continuing.[286] Hutchison was especially critical of the deaths of Boer children in the concentration camps, which he described as a dark page in the history of the conflict. Like Waddell, Hutchison attracted vociferous criticism, with the *Star* calling *The Outlook* a pro-Boer newspaper.[287] It claimed that following Hutchison's editorial the Mataura Presbytery had passed a resolution drawing the attention of *The Outlook*'s Publishing Committee to Hutchison's disloyal and anti-British sentiments.[288] In 'A Pro-Boer Editor', the *Star* announced Hutchison's resignation, claiming the Publishing Committee believed his opinions were antagonistic to Presbyterians and other churches.[289] 'Staunch Presbyterian' George Denniston, who had succeeded Robert Chisholm as mayor of Dunedin, was so infuriated by Hutchison's views that he cancelled his subscription to *The Outlook* in protest.[290]

While the war exposed divisions within the Presbyterian Church, reports of Irish volunteers fighting alongside the Boers increased tensions between Protestants and Catholics. The *New Zealand Tablet* noted that war and religion never made 'good bedfellows' and claimed capitalists and 'Conservative organs' had rejected diplomacy by calling for an 'avoidable and unnecessary' war that was 'aggravated by the presence of overwhelming numbers of dark-skinned pagans who are already preparing to introduce the methods of barbarian or savage warfare'.[291] It added that the financial cost of the war would ultimately be borne by the working classes.[292] In Parliament, Patrick O'Regan, the MHR for Buller, vehemently denied the 'malicious slander' that Irish were assisting the enemy.[293] Although the Irish fought with distinction in South Africa as part of the British Army, an Irish volunteer corps did fight alongside the Boers, and a New Zealander lamented narrowly missing an opportunity to face them, claiming that had they not withdrawn the Irish volunteers would have rued the day 'they turned renegades'.[294]

A number of Catholic clergy also staunchly defended Irish loyalty. At a Dunedin patriotic meeting Father O'Neill noted that whenever British supremacy was challenged Irish Catholics stood ready to shed their blood in its defence.[295]

Drawing attention to the number of Irish names in the Fourth Contingent, Father Ryan told his congregation that as long as liberty and civil and religious equality were respected, there would be no more devoted and faithful subjects than 'the sons and daughters of Ireland'.[296] A significant number of Catholics did serve in New Zealand contingents, and Lord Ranfurly made a point of speaking to men who 'hailed from the Emerald Isle' during a Fourth Contingent inspection.[297] *The Outlook* acknowledged the Irish contribution but remained critical of Irish clergy, maintaining that with some exceptions they were 'anti-national and anti-patriotic'.[298] It also attacked its Catholic counterpart: 'The natural trend of the Romish Church,' *The Outlook* claimed, 'may be judged from the disloyal attitude of its official organ.'

Entrenched religious prejudices also affected New Zealand's Jewish community during the war, despite Jews' commitment to the empire's cause. Early in 1900 Dunedin rabbi Adolph Chodowski asked his congregation to pray after Kimberley's relief.[299] Although Jews such as Percy Cohen, Albert Samuel and South African Light Horse trooper Isodore Cohen also served in New Zealand contingents and irregular forces during the war, their participation did not stem the flow of anti-Semitic content in newspapers and contingent members' letters.[300] A Presbyterian trooper claimed Mafeking was 'not much of a place — full of Jews, like all African towns', and a Christchurch trooper described Transvaal as a gathering place for Jews and gamblers.[301] A New Zealander visiting Ladysmith also noted complaints about Jewish merchants being permitted to 'fleece' British soldiers.[302]

A correspondent to the *Auckland Star* accused Jews of selling Transvaal bullion to the Boers, and profiting from the war 'without risking a drop of their precious Hebrew blood'.[303] In response, Jewish merchant and first mayor of Auckland Philip Philips produced an official return of Jews serving in the British Army and claimed that Jews in England had contributed £37,000 towards the war in a single week.[304] Despite overt anti-Semitism, Jewish religious leaders such as Chodowski and Auckland's Rabbi Goldstein remained supportive of the British cause, and Ethel Benjamin, an Orthodox Jew who in 1897 became the first woman in New Zealand to graduate with a law degree, offered to draft wills for Fourth Contingent members.[305]

Soldiers' access to spiritual guidance remained a priority for religious groups, and the Dunedin presbytery distributed New Testaments to Otago and Southland members of the Fourth Contingent.[306] Soldiers in Durban sang from the hymn and prayer books they received before their departure, and Catholic members of the Tenth Contingent received prayer books and rosaries from Dean Foley.[307] The extent to which the war influenced soldiers' piety is difficult to quantify, but it did affect some worship practices. Fifth Contingent trooper John Burnett enjoyed a Presbyterian service conducted by two British officers, and in his diary wrote that 'it seems as if the Almighty is with our cause', but when Frank Perham, an Anglican, attempted to attend a service at the Worcester 'Dutch and German' church he claimed to have been unwelcome.[308] He instead opted for an English church, which was 'crammed with khaki'.[309]

Yet to other New Zealand soldiers faith was markedly less important. En route to South Africa, Daniel Dutton, a Presbyterian and one of the Ninth Contingent chaplains, combined Presbyterian and Anglican rituals by getting the vessel's captain to read part of the Anglican service.[310] Dutton observed that there were 'many good lads' in the Ninth Contingent 'with a sprinkling among them of men of a different stamp'.[311] At an Auckland ceremony prior to the Sixth Contingent's departure, Reverend Ready noted with disapproval that during prayers soldiers were passing around beer.[312] Once the contingent reached Pretoria, Farrier James Egan, a Catholic, was sentenced to 14 days' imprisonment for 'using grossly insulting language to a female' and breaking away from a group of Roman Catholic soldiers on their way to a church service.[313]

Perhaps in the hope of avoiding such actions, New Zealand churches remained committed to providing spiritual and moral guidance to the troops. When the Otago Presbytery offered to fund a scripture reader for the Fourth Contingent, the Presbytery's preferred choice, Robert Tennent, volunteered his services.[314] Tennent, whose brother was already serving in the Australian Bushmen's Contingent, claimed that New Zealand soldiers being sent to South Africa would face not only the Boers but also 'temptations to sin'.[315] Seddon remained unconvinced of the need, but the Presbytery again attempted to send a chaplain with the Eighth Contingent — though Tennent drowned in the Waikato River shortly after the offer was made.[316] His grieving father admonished the government, claiming that of the colonies, New Zealand alone lacked its own chaplains, while also hinting at ramifications at the ballot box.[317]

Seddon finally relented in 1902 when Opunake Mounted Rifles chaplain Arthur Compton was appointed captain-chaplain of the Eighth Contingent.[318] Compton was followed by Daniel Bates, an Anglican from Invercargill who later became the government meteorologist, and Daniel Dutton (both appointed chaplains to the Ninth Contingent); and Sydney Hawthorne, a Devonport Anglican, and Lyttelton Wesleyan John Luxford, who both served in the Tenth Contingent.[319]

'Entirely in the interests of capitalists'

Presaging First World War suspicions about unions, the loyalty of some unionists was questioned during the conflict.[320] In 1900 Wellington resident John Williamson wrote to Lieutenant-Colonel Penton claiming that Danish contingent member Julius Petersen had gone to South Africa 'for the purpose of joining the Boers, or acting as a spy'.[321] In a confidential response to a query from Penton, Petersen's employer, Harrison Brothers of Kaitoke, accused him of being 'a very deep union fellow' and an 'out & out rank socialist'.[322] The company warned that if Petersen 'really sympathized with the Boers he would undoubtedly require watching very closely'.

Labour groups had the capacity to disrupt troop movements and trade with South Africa, but within the labour movement the war proved divisive. While some workers questioned the conflict's legitimacy, there was no concerted attempt to disrupt maritime traffic to and from South Africa. Port workers relied on uninterrupted trade with Britain, and to a lesser extent the other territories of the British Empire, for a portion of their wages. In a single month in 1899 three vessels that docked at Port Chalmers in Otago had loaded in London, while one had loaded in Glasgow and another in Liverpool.[323] Once hostilities commenced and trade volumes to Natal and Cape Colony soared, there was little evidence of unions favouring disruptive industrial action. During the first years of the conflict, widespread union support for British actions in South Africa ensured the New Zealand labour movement played a leading role in the country's response. Even when individual unions became critical of the war, this criticism was undermined by equally strident expressions of union support.

In October 1899 trade unionists took part in patriotic displays at Wellington's Labour Day parade, where Seddon apologised to workers carrying out alterations to the First Contingent's vessel.[324] He noted that, like contingent members, the tradesmen were doing their duty.[325] In the south, management and labour divisions were temporarily set aside when the president of the Trades and Labour Council mingled with politicians and the president of the Chamber of Commerce at a Dunedin patriotic fundraiser.[326] Support for the contingents also came from at least one women's labour organisation, with the Tailoresses' Union making equipment for the Third Contingent, which together with the Fourth Contingent were known as the 'Rough Riders'.[327] The Dunedin branch of the union also contributed £25 to purchase a contingent horse.[328]

Men employed at Wellington docks were an integral part of celebrations marking the relief of Ladysmith, and when the siege of Mafeking was finally lifted railway employees in Christchurch and Dunedin also joined parades where Wharf Lumpers Union members carried a banner reading 'When Duty Calls Zealandia's Sons Obey'.[329] After news of Mafeking's relief reached Port Chalmers, the captain of the *Aberfoyle*, which had recently arrived from Glasgow, assembled his crew and the watersiders unloading his vessel and proceeded to the Terminus Hotel where they drank toasts to Queen Victoria and her generals.[330] The previous month a watersider at Bluff was reportedly instructed to leave the docks by his co-workers after voicing support for the Boers.[331] Setting a precedent for First World War concerns about saboteurs infiltrating New Zealand ports, an Auckland resident wrote to the acting prime minister in 1901 about guarding contingent vessels.[332] W. Morton claimed that 'well paid spies, agents and traitors' were operating in Australasia and advised Joseph Ward to prevent 'Pro-Boers' getting aboard the *Cornwall* and 'firing, scuttling, disabling her engines, or sinking' the troopship.[333]

By late 1901 some unionists had become disgruntled by war-related government spending at a time when government employees reportedly faced redundancies.[334] The disaffected unions moved a strongly worded resolution opposing the dispatch of the Eighth Contingent 'to assist in waging against the Boers a hideous and unholy war of extermination, which we believe was begun, and now carried on, entirely in the interests of capitalists'.[335] The Operative Sausage Case and Skin-makers Union expressed its 'pleasure' at the resolution, while Wellington unionist T. Lynch claimed that the conflict was not in the interest of workers.[336] But when the resolution was passed, Trades Council president William Naughton resigned in disgust, claiming it did not reflect Labour Party views or those of most unionists, with the plumbers, bookbinders, butchers and saddlers unions backing his views.[337]

The Bakers' Union and the Timber Yards Workers' Union stated that their continued Trades Council affiliation depended on it rescinding the resolution, while the United Furniture Trades Union described it as 'ill-timed and viciously worded'.[338] A delegate accused the Carpenters' Union of being behind the resolution, claiming it had been passed during a poorly attended meeting. Bowing to these pressures, the Trades Council rescinded the resolution shortly after its passage. Instead, the Wellington Typographical Union carried its own resolution: 'That this union has no sympathy with any disloyal sentiments in connection with the Boer War.'[339] Similar tensions surfaced within individual unions; delegates John Cole and William Noot supported the anti-war resolution only to have their fellow union members vote to rescind it. Irked by this reversal, Cole and Noot resigned as Trades Council delegates and members of their union's management committee.[340]

After Alfred Barclay condemned the use of concentration camps, the Workers' Political Committee rejected press accusations that he was a Boer sympathiser.[341] The committee considered the attacks on Barclay unjustified and passed a resolution expressing its confidence in him.[342] Yet members of the public and unionists remained divided, with voters in Ravensbourne electorate calling for Barclay's resignation, and employees at Dunedin's Hillside Railway Workshops unanimously deciding not to invite him to their annual picnic, considering his presence 'undesirable' because of his alleged pro-Boer views.[343] Barclay was later confronted by Union Jack-waving railway employees who passed a resolution denouncing his 'traitorous utterances'.[344] Their attitude was hardly surprising given that workshop staff served in a number of different contingents.[345] The *Nelson Evening Mail* claimed the Dunedin politician's labour support base had evaporated due to his alleged Boer sympathies, with Barclay denying the accusations, but conceding he was 'anti-jingo'.[346]

Dissent and disinterest

As noted earlier, in an address to the Legislative Council Henry Scotland also expressed distaste for jingoism. 'It is painful to see,' Scotland said, 'how people are persecuted, and their loyalty questioned, because they express condemnation of this war.'³⁴⁷ Gilbert Carson spoke in opposition to Seddon's resolution during the First Contingent debate but chose to leave the House rather than vote against it.³⁴⁸ During a speech shortly before the 1899 general election Carson attempted to explain his actions. He said that he had initially considered it unnecessary to send a contingent, but once the decision was made he shared his fellow parliamentarians' desire to dispatch the men.³⁴⁹ Many Whanganui voters remained unconvinced and Carson's stance contributed to his single-term parliamentary career ending in 1899.³⁵⁰

The *Evening Post* claimed that Carson and fellow politicians James Kelly and Thomas Taylor owed their election defeats to their opposition to Seddon's resolution.³⁵¹ But opposing Seddon did not necessarily sound the death knell of political careers, with Taylor being re-elected in 1902.³⁵² Robert McNab also opposed sending soldiers to South Africa, but abstained from voting and successfully contested the Mataura electorate in 1899 and 1902.³⁵³ John Hutcheson won the Wellington City seat in the same year as he opposed the dispatch of the contingent.³⁵⁴ The failure of Michael Gilfedder and James Thomson to support Seddon's resolution did not prevent them retaining their parliamentary seats in 1899, and Barclay also won a Dunedin seat in the December 1899 election before losing it in 1902.³⁵⁵

The enthusiasm of some politicians who had initially supported the war waned as it entered its final year. When Seddon canvassed parliamentarians on sending additional men to South Africa, Richard Monk was among those who opposed the dispatch of the Ninth and Tenth contingents.³⁵⁶ Though Monk had supported New Zealand's original military commitment, he believed New Zealand had done its share and felt other colonies needed to contribute more. Franklin MHR William Massey said that unless New Zealand received a specific request, he could not agree to the dispatch of further contingents as he, too, felt New Zealand had done enough.³⁵⁷

The politicians' concerns reflected doubts in the wider community as New Zealanders tired of the war.³⁵⁸ The *New Zealand Tablet*, which considered Seddon high-handed in offering a Tenth Contingent prior to receiving a formal request, questioned both the war's conduct and the dispatch of additional troops.³⁵⁹ The editor of the *Nelson Evening Mail* claimed that not only had the war become an intolerable drain on New Zealand but the country had also done enough in the cause of imperialism and patriotism 'without further wrenching the hearts of mothers'.[360] The *Observer* noted the degree of indifference that greeted the 1902 return of Sergeant-Major Walter Callaway. According to the newspaper, despite Callaway serving in more than one contingent, being wounded in action and receiving recognition for meritorious service, there was 'not even a cheer to welcome the hero home'.[361] A newspaper correspondent reporting on a medal presentation for two returned troopers said he thought such ceremonies were 'a downright piece of nonsense'.[362] Although less common, similar displays of indifference also occurred earlier in the war, with the audience at a 1900 Orchestral Society concert attended by members of the Fourth Contingent reportedly showing little interest in the soldiers.[363]

Nonetheless, even in the conflict's final months significant support remained. A South Island newspaper suggested military service would satisfy soldiers' craving for change and adventure, and maintained that new contingents could be raised without testing New Zealanders' patience or patriotism.[364] It claimed the Ninth Contingent send-off showed that patriotism had not abated, and while it noted that the festive atmosphere that characterised earlier departures had been replaced by a grim determination, the paper claimed this new mood did not bode well for those expressing pro-Boer sentiments: 'It would not have been well with the Wellington "Peace and Humanity" Society, or any other narrow-minded organisation, to have been on the scene.'[365]

'Traitorous pro-Boer utterances'

While dominant voices in New Zealand overwhelmingly supported Britain's actions, as we have seen, not all New Zealanders gave ringing endorsements of their country's involvement. Pacifism, opposition to capitalism, Gladstonian liberalism, Irish nationalism and, as the war progressed, war-weariness may all have been factors influencing opposition to the conflict.[366] Of the five members of Parliament who voted against sending a contingent to South Africa, only two spoke during the debate. Thomas Taylor and John Hutcheson said that, while they did not oppose war, like Carson they felt the empire's existence was not in sufficient danger to warrant New Zealand's participation.[367] Taylor went further, claiming it was a territorial war motivated by capitalism.[368] Like Reverend Bates and Margaret Sievwright, Taylor believed that had it not been for the discovery of gold in Transvaal the independence of the Boer republics would never have been threatened.

Though supportive of Seddon's resolution, Legislative Council member John Rigg also harboured reservations. Rigg expressed sympathy for the Boers and claimed the imperial authorities had been goaded by capitalists.[369] Henry Scotland was the sole Legislative Councillor who spoke in opposition to the resolution. When it was passed and members rose to sing the National Anthem, Scotland incensed council members by remaining seated, and he continued to express opposition once the war started: 'Imperialism is in the air. Sir, I hate the very word "Imperialism" and the very name of "Empire".'[370]

During a parliamentary debate on the proposed Libel Act, Alfred Barclay accused the *Evening Star* of becoming carried away by 'the exuberance of its patriotism'.[371] Barclay, who in 1899 had published a pamphlet on the theories of Karl Marx, had been repeatedly criticised in the newspaper and was advised to resign his parliamentary seat by 'A Worker' after allegedly making 'pro-Boer' remarks in a speech.[372] Barclay told Parliament that no other newspaper possessed the *Evening Star*'s 'keenness of scent' for exposing pro-Boers.[373] Noting that the paper's editor was Jewish, Barclay told the House that given the Jews' experience of bigotry and fanaticism Cohen should have been more tolerant of Boer sympathisers, and claimed the Spanish Inquisition was moderate compared to the *Evening Star* attacks.[374] McNab and Thomson supported the passage of the Libel Act, with McNab claiming he too had been falsely labelled a pro-Boer in the press.[375]

In Waihī, a religious group came under attack during the Mafeking celebrations when Captain Jansen of the Salvation Army (reportedly acting under orders) refused to hoist the Union Jack at the local barracks.[376] When Jansen attempted to address the angry crowd he was pelted with eggs and drowned out by derisive hoots.[377] In a letter to the *Auckland Star*, 'Miner' claimed that respectable citizens had reacted to an insult to the flag and it was not Jansen's refusal to hoist it that had enraged the Waihī crowd but the Salvation Army's subsequent lowering of the Union Jack raised by those who had forced their way into its barracks.[378] He also referred to traitors in South Africa and claimed there were equally treacherous individuals in New Zealand. Though stopping short of accusing the Salvation Army of treason, the inference was clear.

In New Zealand's bigger cities overt opposition to the war was confined to comparatively small groups, who often became objects of ridicule. The Auckland Peace Association (APA), which was formed in May 1899 and by 1900 had approximately 50 members, commended politicians who opposed Seddon's 1899 resolution.[379] Its actions were widely reported in the New Zealand press and in at least one Australian newspaper.[380] Association member Joseph Peckover was better qualified than most to comment on the South African situation. He had lived in South Africa during the 1880s when the British were defeated at Majuba and he blamed the current conflict on the Jameson Raid and the imperialist ambitions of Cecil Rhodes.[381] The APA's membership also included members of the clergy, and Reverend A. H. Collins blamed the Waihī 'riot' on the press for encouraging militarism among the populace.[382] At an association meeting, Reverend W. S. Potter complained of 'shallow, thoughtless people' rejoicing in the horrors of war.[383] A *New Zealand Herald* editorial obliquely accused the group of 'short-sighted treason' and appeared to suggest that Collins was an enemy emissary.[384] 'If this be "peace",' the editor claimed, 'we want none of it.'[385]

Like its Auckland counterpart, the Wellington Peace and Humanity Society remained small and ineffectual but was also vilified in the press. In February 1902, the society passed a resolution supporting the leader of the British Liberal Party who was a vocal critic of British actions in South Africa.[386] Criticising the resolution, the *Evening Star* claimed that it would do more 'to harm the cause of peace than a million jingoes' and would be seen internationally as reflecting the views of the New Zealand populace.[387] The society had reportedly sent Seddon a copy of another resolution opposing the war. The *Free Lance* dismissed the 'Wellington peace-and-humanity-stop-the-war sort of patriot', claiming they were characterised by the rhyme:

> We don't want to fight but by jingo if we do,
> We won't go to the front ourselves but we'll send the mild Hindoo.[388]

Those who voiced opposition to the war ran the risk of physical assault, and in Whanganui an individual who publicly agreed with German criticism of the conflict was labelled a pro-Boer and reportedly received a 'hammering'.[389]

Another press report claimed a passenger aboard a vessel in Lyttelton who made disparaging comments about the British following the relief of Ladysmith was assaulted and unceremoniously hoisted onto the wharf using the ship's crane.[390] There was also the case of Upper Hutt storekeeper and Justice of the Peace Vilhelm Jensen.[391] In the months prior to the war Jensen wrote a number of lengthy letters to the editor of the *Evening Post* that raised the ire of his fellow Wellingtonians.[392] In one, Jensen praised the impartiality of an August 1899 *Evening Post* editorial questioning British intentions in massing forces in the Rhodesian town of Bulawayo.[393] With the Jameson Raid still fresh in their minds, what, the editor asked, could the Boers be expected to think of such 'insulting' actions and a 'hectoring attitude that is not creditable to the English name'.[394]

Shortly after war was declared, Jensen addressed an Upper Hutt public meeting. He said that while he had never been to South Africa and did not personally know any Boers, he was the only person in Wellington prepared to speak in their support. The meeting quickly descended into farce after Jensen claimed that Cecil Rhodes would resort to any means to achieve his goals. When Jensen blamed the war on Chamberlain, the crowd broke into 'For He's a Jolly Good Fellow', and when he suggested that Britain intended seizing Transvaal once it had resolved the uprising in China the crowd vociferously sang 'Rule Britannia'. Jensen was finally drowned out when the crowd erupted into 'Soldiers of the Queen' after he claimed the war was against fair play, humanity and Christianity.[395]

Jensen was again in the press in 1901 after members of the Seventh Contingent visiting Trentham mistook the Danish ensign Jensen flew over his Upper Hutt premises for the Boer Vierkleur, tore it down and trampled it underfoot.[396] The troopers had reportedly been goaded by local residents who informed them that Jensen was a pro-Boer with the effrontery to publicly display the Boer flag.[397] In a letter to Francis Bell, the Danish consul, the aggrieved Jensen described himself as 'a well-known Boer sympathiser' and said he had almost been mobbed by around 40 contingent members when he retrieved the flag from the ground.[398] The incident was finally resolved when Jensen received an apology from the contingent's commanding officer and the government formally apologised to Bell.[399] Seddon dryly observed that similar unpleasant occurrences could be avoided if the men received a lesson on national flags.[400]

While the combined press coverage provided Jensen with a platform for his views, this was not always the case for Boer sympathisers in other regions, with newspapers reluctant to publish material seen as supporting the enemy.[401] Waverley resident D. Fleming asserted that the *Wanganui Chronicle* had refused to publish letters he had written during the war.[402] A self-confessed 'pro-Boer' and outspoken critic of British actions in South Africa, Fleming claimed to have spent five years living among the Boers before the war.[403] He had found them to be reasonable, more trustworthy than Europeans and dedicated to retaining their independence.[404] Within weeks of the war's outbreak, the Whanganui newspaper dedicated an entire editorial to discrediting Fleming's views.[405]

In Napier, the principle of free speech was sidelined altogether by a 1900 Borough Council resolution intended to gag council employees who criticised British actions in South Africa. Passed unanimously, the resolution appeared in the local newspaper under the ominous title 'A Warning'. It instructed overseers to inform council workers that making remarks 'derogatory to the British Government in connection with the war in South Africa' would result in instant dismissal.[406] Similar threats did not pass unchallenged; a Christchurch editorial titled 'Should Englishmen be Muzzled' supported the right of those with pro-Boer sentiments to express their views regardless of whether they ran counter to public opinion.[407] While stressing that the *Press* was not 'pro-Boer', the editor claimed that freedom of speech was under attack. Following claims that several Westport Harbour Board employees had expressed 'disloyal sentiments', the board passed its own resolution instructing its engineer to instantly dismiss Boer sympathisers.[408] The *Manawatu Evening Standard*, a frequent Seddon critic, claimed that the resolution was the worst kind of autocracy and criticised the three 'Seddonite' politicians it claimed were on the board and behind the resolution.[409]

Though teachers were largely supportive of the war, they too could find their jobs in jeopardy if suspected of disloyalty. In 1901 the New Zealand Ensign Act formalised the use of the Royal Navy Reserve blue ensign with the addition of the Southern Cross as the nation's official flag.[410] Under the act, anyone found to have defaced the flag was liable to a penalty not exceeding £5. However, as Kirikiri School headmaster James Murray discovered, failure to comply with Education Board directives concerning the flag could prove far more costly.

Members of the Kirikiri School Committee accused Murray of refusing to declare a holiday following the occupation of Pretoria by British forces.[411] While conceding this was true, Murray believed he lacked the authority to send his students home. He stated he had been willing to give them a holiday the following day on receipt of a written instruction from the School Committee but claimed this had not been forthcoming. While Murray's initial refusal was anathema to the area's Loyal and Patriotic Committee, it was his refusal to salute the flag that set the headmaster on a collision course with the Education Board.[412] Matters came to a head when the board delivered an ultimatum: Murray was to salute the flag and instruct his students to do the same or tender his resignation.[413] After declining both options, Murray was summarily fired from his £175-per-annum position.[414]

A petition signed by 64 residents of the Kirikiri district (including two school committee members) urged the board to reconsider. They believed Murray's unwillingness to salute the flag was driven by conscience rather than disloyalty and called for his reinstatement. One petitioner noted that obeisance to the flag was neither an obligation under the Education Act nor part of the school curriculum.[415] He believed Murray was fired not for refusing to salute the flag but because of suspicions he was either a Fenian or a pro-Boer.[416] Murray's case was also taken up by the Women's Christian Temperance Union and the Women's Political League, who considered his punishment unduly harsh.[417]

The headmaster's refusal seems to have stemmed from his personal (and possibly religious) aversion to saluting an object, which he suggested could be considered akin to idolatry. In a letter to the Education Board, Murray rhetorically asked,

> *Is it not degrading to the free-born children of this colony to try to compel them to an act of the most servile homage which is never required of the English, Scotch, or Irish children, and such as would scarcely be expected from the pagan, barbarian slaves of a Central African chief? Could anything be devised more utterly foolish than to salute an inanimate, useless piece of drapery in the same manner as an intelligent human being . . .?*[418]

Had a parliamentary bill proposed in 1900 been enacted, Murray's students would have been doing more than just saluting the flag. The Ashley MHR Richard

Meredith joined fellow parliamentarian John Millar in opposing the passage of the State-School Children Compulsory Drill Bill 1900. Meredith claimed the object of the proposed legislation was to introduce a spirit of militarism into schools — a claim that was difficult to refute.[419] The bill obliged education boards to ensure boys and girls over the age of eight attending public schools undertook both military and physical drill.[420] Though it was subsequently struck out, a clause in the proposed legislation allowed the government to withhold funding from any school that failed to comply with its provisions.[421]

Meredith's opposition to the bill came as no surprise given that he opposed the establishment of standing armies and was equally wary of fostering militarism and imperial jingoism.[422] Like Meredith, Millar was also critical of the spirit of militarism he saw sweeping New Zealand and considered the bill an attempt to legalise a modified form of conscription.[423] Following discussion in the House, the proposed legislation underwent various amendments, including the removal of the word 'compulsory' from the title. More importantly, Millar successfully moved an amendment that effectively expunged all reference to military drill.[424] In its place, the House passed the Physical Drill in Public and Native Schools Act 1901, which came into force early in 1902. Providing they received ministerial approval, the new act allowed school boards to dictate the mode in which physical drill was conducted. This meant that boards could still ensure children received instruction in regimented exercises such as marching.

The type of threats that were aimed at Murray may have silenced some dissenters, but others remained willing to publicly express sympathy for the Boer position, albeit under the cloak of anonymity. In a letter to the editor, 'Colonial Irishman' criticised a 'vicious and intolerant' *Otago Daily Times* columnist for his attacks on pro-Boers.[425] 'Colonial Englishman' expressed some support for this view and accused the columnist of being bloodthirsty.[426] Other correspondents dismissed the views of Boer sympathisers, with one advocating the expulsion of pro-Boer agents he claimed were operating within New Zealand.[427] Discussing pro-Boers in New Zealand society, 'Gold Miner' from Lake Wakatipu reflected the prejudices of the time. He claimed 'these vipers would become pro-Chinaman as readily as pro-Boers if there was money in it'.[428] Though the war polarised British supporters and Boer sympathisers, there were also those who were more ambivalent yet still found the strident patriotism of the 'Khakisian' repellent. 'Colonial Native', who did not consider that harbouring Boer sympathies and being an ardent patriot were mutually exclusive, claimed a significant section of the New Zealand public had lost their senses.[429] 'Cannot,' he asked, 'we hammer our enemies and yet feel for them?'[430]

Like Jensen and Fleming, Takapuna resident Charlotte Bewicke refused to be cowed by public opinion and used her real name when expressing support for the Boers in the press. Under the title 'In Defence of the Boers', the *Auckland Star* published a letter from Bewicke in which she expressed pleasure that other likeminded individuals took 'a right view of these much-abused Dutch farmers'.[431] Bewicke offered to accept contributions on behalf of sick and wounded Boers and contributed £5 to start the fund. She repeatedly advertised it, providing detailed lists of contributors to the press (one of which incongruously appeared in the same column as an advertisement claiming that Britain would 'wipe out' the Boers like Flora Soap wiped out dirt).[432]

By April 1900 Bewicke had collected a total of £21.[433] Whether she sought the consent of the 41 individuals who gave money before she had their names published in the press is unclear. While there were English names among the contributors, in a possible reflection of Irish and German sympathies for the Boer cause, there were also names such as Boyle, Geraghty, Turley, McGuire, McGowan, Winklemann and Mosier.[434] Though the *Observer* rejected Bewicke's claim that Britain's involvement was driven by a desire to seize Transvaal, it commended her for having 'the courage to think for herself'.[435] Bewicke's activities also attracted the interest of Australian newspapers, the *West Australian Sunday Times* facetiously observing that it was only reasonable to collect money for the Boers as they would be the ones most in need of assistance.[436]

Like Bewicke, Auckland resident Flora Nicholls also came to the defence of the Boers, though in Nicholls' case she was speaking from personal experience. She considered the Boers a pious and honest people, and dismissed claims of Boer hatred for the British. Nicholls had lived in South Africa for two years and claimed to owe her life to Boer women who had taken turns to ride 50 miles to provide 'loving kindness and tender nursing' when she was suffering from fever.[437] Though Bewicke and Nicholls seem to have avoided public censure, many New Zealanders were less forgiving of individuals such as Alfred Barclay and Thomas Taylor. Barclay claimed in Parliament that his four-year-old son had been ostracised by classmates who told the boy his father was a pro-Boer who ought to be shot.[438]

Even dedicated imperialists like New Zealand's Agent General in the United Kingdom, William Pember Reeves, were not immune to suspicion and public rebuke.[439] After dispatches Reeves sent from London appeared in several papers, he too was accused of being a pro-Boer and attacked in the press.[440] The contentious dispatches, which estimated both the size of Boer forces and the extent of losses on both sides, enraged some New Zealanders who believed they were intentionally misleading. The mayor of Gisborne cabled Seddon expressing his disapproval of Reeves, while a correspondent to the *Otago Daily Times* wrote that Reeves' sympathies were 'transparently with the enemies of his Queen and country'.[441] The following month, Reeves was accused of sympathising with Boer advocates in England including radicals, socialists and Irish nationalists.[442]

In Parliament Seddon tabled Reeves' detailed rebuttal of the accusations. The Agent General claimed that news received by the New Zealand press went through a process of expansion and embellishment that resulted in the public being misled by inflated and inaccurate figures.[443] He added that due to the expense of sending cables he was obliged to convey information to the government as succinctly as possible.[444] He was not, he added, likely to engage in 'flowers of sentiment' when it cost almost four shillings per word to send cables to New Zealand. Despite his condemnation of newspaper practices, Reeves had his supporters in the press. A *Tuapeka Times* editorial claimed Reeves was the victim of a 'patriotic heresy hunt', adding that the Agent General had been denounced and insulted by jingoists.[445] Nonetheless, like President Kruger after the relief of Mafeking, Reeves suffered the ignominy of being burnt in effigy and accused of being an 'anti-patriot'.[446]

Few individuals fuelled more controversy during the war, however, than James Grattan Grey. The Hansard chief reporter, Grey also acted as a foreign correspondent for the *New York Times* and had reportedly received government permission to engage in independent journalism during the parliamentary recess.[447] Shortly after war broke out, Grey wrote a *New York Times* article criticising New Zealand's response to the conflict in which he claimed the country and other democratic British colonies had become 'infected with imperialism of the most pronounced type'.⁴⁴⁸ In the article titled 'Jingoism in Australasia: A Wave of Imperialism Sweeps the Colonies', Grey wrote that other countries would think it odd that self-governing colonies thousands of miles from South Africa would send men to fight people with whom they had no quarrel and assist in their subjugation, despite the Boers also claiming the right of self-government. But, Grey claimed, 'the jingoistic spirit at the Antipodes is too inflamed just now to care anything about the rights or wrongs of the question'.⁴⁴⁹

After extracts from the article appeared in the *Evening Star*, Seddon wrote to Grey asking if he had written it. Refusing to be intimidated, Grey readily admitted to being the author. He indicated he had no intention of recanting and regretted the 'wave of jingoistic hysteria' in New Zealand.⁴⁵⁰ Irked by Grey's intransigence, Seddon advised the House to refer the matter to the parliamentary Reporting Debates and Printing Committee.⁴⁵¹ After Parliament followed Seddon's advice, Grey appeared before seven of the committee's 10 members. Four had supported Seddon's 1899 contingent resolution, one had opposed it, one had abstained, and the last was not an MHR at the time. When Grey informed the committee that he stood by his views on the war, the committee recommended his dismissal, claiming he had flouted an earlier committee recommendation that Hansard staff abstain from involvement in politics by writing articles for publication.⁴⁵²

Following one of the longest parliamentary debates of the war, Grey's fate was sealed.⁴⁵³ Some parliamentarians who opposed sending the First Contingent supported his dismissal. Others who supported New Zealand's participation in the war either felt Grey had not breached his employment contract or saw his case as a challenge to freedom of thought.⁴⁵⁴ Though James Thomson and Michael Gilfedder voted against sending the First Contingent, they supported the termination of Grey's government employment.⁴⁵⁵ William Collins, the Christchurch City MHR, claimed he was ashamed at the content of Grey's article but opposed his dismissal, considering it an attack on liberalism.⁴⁵⁶ Grey's termination was ultimately supported by a majority of 36.⁴⁵⁷ The *New York Times* presented Grey as a champion of free thought and speech and claimed he had been threatened with physical violence.⁴⁵⁸ The newspaper claimed New Zealand was the most jingoistic of the Antipodean colonies, possessing a level of intolerance that was inconceivable in a democratic country.[459]

Grey received support from William T. Stead, the editor of the British journal *The Review of Reviews* (who would later be among those lost in the 1912 sinking of the *Titanic*).[460] Stead was vilified in the New Zealand press for his opposition to the war and was described as an 'arch-traitor' by the *Evening Post*.[461] He commended Grey for his stance 'in the midst of the semi-delirious sentiment which has submerged common sense, reason, justice and humanity'.[462] In 1901 the imperial authorities approached the government with a proposal to locate Boer prisoners on Rakiura Stewart Island.[463] A newspaper suggested that if Boer prison camps were to be established on outlying New Zealand islands, 'traitors' such as Stead, Lloyd George (the British Liberal politician and outspoken critic of British policies in South Africa), Alfred Barclay and William Hutchison should be incarcerated there as well.[464]

Grey and his wife left New Zealand in 1900 for the United States. Prior to their departure, Charlotte Bewicke and other supporters attended a function in their honour where Seddon and the New Zealand Parliament were accused of driving the Irish-born Grey from his adopted homeland.[465] Even on their voyage to San Francisco controversy followed the couple, with Grey's wife reportedly being ostracised by her fellow passengers after refusing to sing 'God Save the Queen'.[466]

CHAPTER THREE

'AN ESPECIALLY FINE LOT OF FIGHTING-MEN'

THE PERFORMANCE OF NEW ZEALAND SOLDIERS DURING THE SOUTH AFRICAN WAR

Although interest levels varied, with war-weariness inevitably becoming a factor, for much of the South African War press coverage reflected the New Zealand public's fascination with the contingents. Public expectations were in part met by praise for the New Zealand soldiers from high-ranking British officers. Though these accolades may have been motivated in part by a desire to ensure the continued flow of colonial manpower and resources, senior officers including Major-General John French, Brigadier-General Bryan Mahon, Major-General Edwin Hutton and Lord Kitchener nonetheless had a largely positive impression of New Zealand troops.[1] In a letter to Cabinet minister William Hall-Jones, General Ian Hamilton wrote: 'I have soldiered a long time now, but I have never in my life met men I would sooner soldier with than the New Zealanders.'[2]

Yet despite the number of men who had some experience with both firearms and horses, it would be a mistake to view the average contingent member prior to leaving for South Africa as an experienced shot who was completely at home in the saddle. In truth, the contingents comprised men from diverse walks of life, many of whom learned the essentials of their military field craft during operations on the African veldt. Furthermore, as in any comparatively large group of soldiers, individual skill levels ranged from proficient to bordering on inept.

Of conspicuous service to the empire

For New Zealand newspaper editors eager for stories favourably comparing the First Contingent's performance with that of the imperial troops, the defence of a strategic kopje (a small hill) near Slingersfontein exceeded expectations. When a party of Boers attempted to rush a stone breastwork cresting one of the hilltop's steep faces, New Zealanders came to the support of Yorkshire Regiment men who had reportedly become 'disheartened' after their sergeant-major was killed and their officer seriously injured.[3] Not only did newspapers describe the New Zealanders' courage in the ensuing action, they also claimed that without the initiative shown by their officer, Captain William Madocks, in leading the New Zealanders over the wall and driving the Boers back at bayonet point, his men and the British regulars would have faced probable annihilation.[4] In London the *Times* further burnished the New Zealanders' reputation, with Lord Roberts reportedly commenting on 'the excellent conduct and bearing' of the New Zealand Mounted Rifles.[5]

These accolades, however, came at a price, as Sergeant Samuel Gourley (the son of Legislative Council member Hugh Gourley) and Private John Aitken-Connell were both killed in the action.[6] Noting that the kopje was known as 'Zealanders' Hill', the *Times* exhibited a poor knowledge of New Zealand's political landscape by referring to Gourley as the son of the 'Premier of New Zealand'.[7] The following month the newspaper claimed that the distinction won by Australians and New Zealanders in the operations around Rensburg was further evidence of the value of colonial troops.[8] Few newspapers mentioned that Madocks was in fact a British Royal Artillery officer on a five-year secondment.[9]

Estimates of Boer casualties sustained at Slingersfontein varied: Private Walter Callaway claimed over 36 were killed and '50 or 60 wounded'; while Private Joseph Culling, who was also present, put the figure at 100 Boer dead.[10] Culling claimed to have remained calm despite Gourley and Aitken-Connell's deaths and implausibly told his mother he had single-handedly killed 'nine or ten'.[11] Another account by Trooper William Macpherson made no mention of Culling's role, but though Culling's version was almost certainly embellished, few in New Zealand seemed to question its veracity.[12]

When Lord Ranfurly read aloud a report of the skirmish during a review of the Second Contingent at Island Bay in Wellington, cheers erupted among the assembled soldiers and civilians.[13] The duty of informing Hugh Gourley of his son's death fell to Lieutenant-Colonel Penton, who said Samuel had died gallantly fighting for the empire and that the New Zealanders' actions at Slingersfontein reflected glory on the colony.[14] Joseph Chamberlain joined the British commanders in praising the New Zealanders' conduct, while James Shand reported the troopers' nonchalance and bravery under fire.[15] The Slingersfontein incident was also raised in the New Zealand Legislative Council, where George Whitmore commended Madocks but grossly exaggerated the Yorkshire Regiment's losses by claiming nearly all its men were killed.[16] A London cable put the Yorkshire casualties at six dead and five wounded, and the *Times* gave the same total.[17] While checking Aitken-Connell's body for personal effects, Trooper Frederick Shaw reportedly came across a packet of letters and a photograph of a young woman. Aitken-Connell was prepared for the worst; the photo had the words 'Please bury with me' written on it, and a letter to his sweetheart had the written request 'Would some kind friend please post this'.[18]

Illustrated papers portrayed wartime actions exemplifying British determination, and Slingersfontein became a colonial extension of this, with photos of New Zealand Hill and the graves of Gourley and Aitken-Connell appearing in the press.[19] A decade after the event, an Auckland newspaper included a suitably heroic depiction of the First Contingent men repelling the Boers.[20] Though Slingersfontein was touted as a victory won through the heroism of the nation's soldiers, newspapers were soon reporting a skirmish with a very different outcome.

British general Robert Broadwood's withdrawing column was ambushed by General Christiaan de Wet and his men at Koorn Spruit, a tributary of the Modder River near Sanna's Post. Twenty-two New Zealanders were reportedly captured during the 'disaster', including Quarter-Master Sergeant Prosper Berland, though five subsequently escaped.[21] Berland reported that gunfire erupted after a British soldier did not immediately obey a Boer instruction to dismount and was promptly shot. A ferocious exchange followed as the British force attempted to prevent the loss of its artillery. Caught in the crossfire, Berland and his fellow prisoners (including George Miller, the son of the Speaker of the Legislative Council Henry Miller) were ordered by their captors to take cover under the wagons.[22] When the shooting finally subsided, the captives began the long march to a Boer prisoner-of-war camp. Berland said the prisoners initially raised few complaints, but as their captivity dragged on he claimed their behaviour changed. In a report to Major Robin, Berland noted a complete lack of discipline among the New Zealand prisoners where 'everyone considers himself as good as the other'.[23]

In August 1900 New Zealand newspapers reported another disaster — this time a skirmish at Ottoshoop, near the western border of the South African Republic, that left Captain John Harvey of Balclutha and Trooper Septimus McDougall of Pirinoa dead and 12 other New Zealanders wounded.[24] The Fourth Contingent men had set out from Mafeking as part of a 2500-strong column that included an artillery battery manned by New Zealanders. When Boers blocked the road to Zeerust the force began dislodging the enemy from a succession of strongly defended positions.[25] Dismounting, the New Zealanders advanced up a steep kopje under heavy fire, forcing the Boers to withdraw, but not before sustaining casualties including Trooper Oscar Bottom, who later succumbed to his injuries at Kimberley.[26]

Also among the casualties were Michael Canavan, Trooper William Vinsen and Lieutenant Robert Collins, who later named his Wellington home after the battle that left his wrist shattered.[27] In Parliament, Seddon told the House he had received word from South Africa that the wounded were recovering well, while the *Otago Witness* published photos of some of the Ottoshoop casualties.[28] The intensity of the Boer fire was described by Corporal Dudley Hewitt. The corporal sheltered for the night with fellow Wanganui Collegiate School Old Boy Sergeant-Major Conrad Saxby and Sergeant-Major William Jickell (who in 1898 had captained the Nelson College team against Hewitt's old school).[29]

In November 1900 it was the Second and Third contingents' turn to make newspaper headlines. Leaving camp before dawn to attack positions at Rhenosterkop, the New Zealanders intended giving the Boers 'a drubbing' but were instead confronted by a determined enemy.[30] Upon encountering resistance, they dismounted and advanced a short distance before being ordered to lie down. The men lay in the open for the entire day, exposed to both the fierce sun and enemy rifle and artillery fire. The ferocity of the engagement was recounted in newspapers and school magazines; Trooper Fred Delany saw his friend Corporal Rodney Devereux shot down, and after Private George Hyde was also fatally hit, Sergeant Piers Tudor called for Waipukurau surgeon

Sidney Godfray to tend to the gravely wounded Sergeant Frederick Russell.[31] Although Russell was lying nearby, the intensity of the Boer fire meant Tudor was powerless to help the sergeant, who died before the doctor's arrival.

The men held the position until eleven o'clock that night, when they were relieved by Australians. When Tudor finally reached camp at eleven-thirty he had not eaten since three in the morning and claimed to have had only a pint of water all day.[32] Five New Zealanders were killed during the battle with numerous others wounded, including five officers.[33] Trooper Albert Beath received a bullet wound to his leg, a bullet graze to his elbow and a gunshot wound to his shoulder, while Godfray was himself shot while attending to Lieutenant Frederick Tucker.[34] Despite the carnage, in a letter written shortly after Rhenosterkop, Lance-Corporal Francis Ryan wrote, 'I felt proud to be a New Zealander that day.'[35]

Though they suffered casualties at engagements such as Slingersfontein, Ottoshoop and Rhenosterkop, the New Zealand contingents had been largely spared the heavy death and injury toll that characterised the major battles of the war. That changed at Langverwacht Hill. Lord Kitchener's much vaunted 'drives', intended to trap Boer kommandos between advancing British forces and lines of wire fences and fortified blockhouses, had met with limited success. Nonetheless, in early 1902 the British commander drew the net ever tighter in the hope of landing two big fish — General de Wet and Marthinus Steyn, the Orange Free State president.

De Wet had long proved a thorn in the British side, repeatedly attacking British forces while avoiding capture. He had little intention now of surrendering and following General Piet Cronjé into exile on the island of Saint Helena. Instead, he chose to make a desperate attempt at Langverwacht Hill to break through the encircling forces. A section of the British line was held by men of the New Zealand Seventh Contingent occupying a series of defensive positions that stretched up the flank of the hill to its crest, where they met Lieutenant-Colonel Charles Cox's men of the Third New South Wales Mounted Rifles. Acting on his scouts' reports, De Wet identified these largely unsupported outposts as offering his kommando its best chance of escape.[36]

The week prior to the Langverwacht engagement, Seventh Contingent member James Kirkwood was killed and Corporal John Ashton was wounded while on outpost duty, though neither man was felled by Boer bullets. Both were shot by Private Peter Carmichael, a fellow Seventh Contingent member, who was occupying an adjacent outpost. Giving evidence at the inquiry into Kirkwood's death, Sergeant William Dobson stated that he was in his own outpost soon after midnight when a shot rang out, followed soon after by another.[37] The bullets passed close to Dobson, who heard Ashton in the next outpost immediately cry out that he had been hit. Dobson claimed that despite Ashton's warning a further two shots were fired.

The men were understandably jittery as 10 minutes earlier all the outposts had opened fire on two mounted men spotted in the darkness moving in front of the New Zealand pickets. Corporal Robert Carr, who was in charge of Carmichael's position, woke to the sound of gunfire. Realising that the direction of Carmichael's shots was at right angles to the front, Carr reportedly told Carmichael that he was firing on his own men.[38] When Dobson reached Ashton and Kirkwood, he found the corporal shot through the hand and Kirkwood lying dead. The *Manawatu Evening Standard* listed Kirkwood as 'killed in action', while other newspapers reported that he had been killed and Ashton 'slightly wounded' at Groot Rietspruit Farm.[39] The newspapers made no mention of the friendly fire. Following the war, a government report listed Kirkwood as 'accidentally killed'.[40]

Barely a week after Kirkwood's death, members of the Seventh Contingent in outposts on Langverwacht Hill were alerted by the sound of approaching cattle.[41] It was not the first time the Boers had used this ploy in an attempt to conceal their movements, and the New Zealanders opened fire on the herd and the enemy they correctly assumed was advancing in its wake.[42] With any chance of surprise now lost, pandemonium erupted. Trooper George Klee, whose brothers, Trooper Victor Klee and Lance-Corporal Louis Klee, also served in the Seventh Contingent, described the ensuing fighting as 'hell on earth'.[43] After overwhelming the first outpost, the Boers turned left and advanced uphill toward a pom-pom (a model of quick-firing cannon) on the summit, storming one New Zealand position after another. Those who survived the initial onslaught and were able to move rallied around the pom-pom, but the gap in the line they left behind them was exploited by members of the Boer kommando, who managed to escape.[44] Staff Sergeant Frederick Crespin claimed that once the breach was made, several Boer women rode through screaming, 'Kill the English bastards! Kill the English dogs!'[45]

The engagement received widespread newspaper coverage not only in New Zealand but also in the United Kingdom and Australia.[46] Lieutenant Charles Phair sustained five bullet wounds, all of which were classified as severe, and was one of 16 Seventh Contingent men badly wounded at Langverwacht who received silver-mounted pipes from King Edward VII's wife, Queen Alexandra.[47] In his account of the engagement, Corporal Edmund Foster said that after he and Walter Stevenson were surrounded they both surrendered, but he alleged that Stevenson, who was killed in the fight, 'was shot with his hands up'.[48]

As was so often the case, press accounts of the action varied, with the *Times* estimating Boer casualties ranged from 15 to 35 killed.[49] In the *Auckland Star* a Seventh Contingent trooper claimed the New Zealanders found 32 dead Boers after the fight, while the *Otago Witness* put the figure at 35.[50] Readers of the *Southland Times* could have been excused for thinking Langverwacht had been a New Zealand victory. Quoting Sergeant Charles Minifie's account, the newspaper put the enemy's total casualties at 60 killed and 90 wounded.[51] Estimates of the size of the Boer force also varied wildly, with the *Times* putting the Boer strength at between 600 and 800, and Sergeant Kenneth Malcolm later claiming that 2000 Boers were involved.[52] Crespin left little doubt that the New Zealanders had put up a stiff resistance during what a newspaper termed their 'glorious stand'.

Trooper Roland Westropp claimed that when the gunners operating the pom-pom were shot, Minifie (who was also wounded) and two other soldiers manhandled the weapon into a gully to prevent it falling into Boer hands.[53] Crespin reported that after Minifie, Lance-Corporal Duncan Anderson and Sergeant Walter Miller were wounded, Trooper William Warner and two or three others pushed the pom-pom over the edge of the hill. Trooper Harry Tatton claimed that he saved the pom-pom by running it down the hill into the New South Wales camp.[54] Tatton added that after the battle he and Minifie had been commended by Cox for saving the gun and hoped to receive a Distinguished Conduct Medal (DCM) for his actions.

With daylight finally illuminating the carnage, the Red Cross began removing the injured from the field. Crespin, describing the scene, claimed dead and wounded lay in every trench, with piles of empty cartridges a grim testament to the fight's ferocity.[55] A parliamentary report listed 23 New Zealanders killed in action during the fight, though Malcolm put the number at 24 in a book he wrote after the war.[56] The bodies were buried in a single communal grave on the veldt, with rocks outlining the freshly turned sod. For many of the wounded, however, the ordeal was far from over. Westropp, who had been shot through the instep of his foot by a bullet that then lodged in his thigh, described being painfully jolted during the three-day journey to Harrismith in an ambulance wagon.[57] When Akaroa accountant Lytton Ditely enlisted soon after the outbreak of war in 1914, he still bore the scars of the bullet wound he had received 12 years earlier at Langverwacht.[58] Like Westropp, Ditely described 'three killing days' spent en route to Harrismith where a procession of officers including Lord Kitchener visited the wounded. While recuperating, Ditely wrote, 'I can tell you I feel proud to belong to the Seventh — they fought like tigers.'[59] Malcolm, who was among the wounded, would later proudly state that he had 'fought & bled, for the Empire's cause'.[60]

Among the dead were two childhood friends: lance-corporals Percy Nation and Duncan Anderson. They had grown up together, enlisted on the same day and had reportedly promised their parents that they would look out for each other.[61] The Seventh Contingent's junior officers were also well represented on the casualty list. Lieutenant William Forsythe was killed during the engagement, and lieutenants Charles Phair, James Colledge, Stapylton Caulton, and the Irish-born Daniel Hickey were all wounded.[62] Born in Manchester in 1840, Caulton was 61 years old with a wife and at least six children when he enlisted in 1901.[63] He had previously served as a private in the Napier Colonial Defence Force during the New Zealand Wars, and in 1886 was a lieutenant in the Cook County Rifle Volunteers.[64]

A posed photo of Langverwacht survivors captures their mood. Brandishing their rifles menacingly and festooned with bandoliers, the men stare grimly at the camera. Apart from their weapons and the cartridge-laden belts that criss-cross their chests and encircle their waists, there is little marking them as soldiers. Scarcely an item of military uniform is discernible and, as Major Robin had earlier observed, British troops encountering them in the field could easily have mistaken the New Zealanders for their Boer enemy.

The Seventh Contingent may have sustained heavy casualties and failed to prevent De Wet's escape, but in Parliament as well as in New Zealand schools and newspapers the men were fêted as heroes. Their actions at Langverwacht were also mentioned in English and Australian newspapers, with the *Sydney Morning Herald* quoting Lord Kitchener's dispatch praising the New Zealanders' gallantry and resolve.[65] Chamberlain also commended the men's 'heroic defence' in a dispatch to Ranfurly accepting the services of another thousand New Zealanders in a tenth contingent.[66] In their annual report on South Canterbury schools, the region's two school inspectors noted children's awareness of what had taken place at Langverwacht Hill:

> *Remote as they are from the great heart of the Empire, the children have been deeply interested in the notable events of the year. They followed with close attention the exciting incidents in the final operations against the Boers, and recognised with pride the gallant stand made by South Canterbury lads during the wild night rush of De Wet's riders at Bothasberg.*[67]

In a sanguinary depiction of Langverwacht, the *Southland Times* claimed that those who lost their lives 'died fighting against heavy odds, with their rifles hard gripped, smoking, hot, and their bayonets running red'.[68]

While significant numbers of injuries to contingent members occurred in battle, some of the most serious resulted from a 1902 train collision. The wreck killed 13 men of the Eighth Contingent's South Island Regiment and grievously injured 13 more, three of whom later died.[69] The incident occurred when a goods train ploughed into another train carrying the contingent near Machavie, a station between Klerksdorp and Potchefstroom.[70] Trooper Herbert Glenie riding in the second wagon escaped injury, but many of his fellow contingent members were less fortunate. In a letter that was possibly an attempt to comprehend the enormity of what had taken place, Glenie gave a disturbingly graphic account of the accident's aftermath and told his parents, 'It was a terrible sight to see the poor fellows.'[71]

Lieutenant Charles Brebner claimed one trooper was so severely injured that he begged his comrades to end his suffering.[72] Brebner, whose own father was a stationmaster in New Zealand, maintained the accident was caused by a young and inexperienced stationmaster directing both trains onto the same line. On realising his mistake, the stationmaster reportedly attempted to shoot himself, but he was restrained and placed under arrest. Having survived the wreck, 'Corporal' Hunter from Invercargill (probably Provost Sergeant Edwin Hunter) reportedly had a second narrow escape. While galloping back to Potchefstroom to seek assistance, Hunter was fired on by British troops in a blockhouse who mistook him for a Boer.[73]

Though in many instances numerically weaker, the Boer forces' superior knowledge of much of the terrain over which military operations were conducted often gave them a tactical advantage. To counter this, British commanders needed to collect accurate information about the enemy's location and movements. Scouts operating ahead of columns were tasked with finding the enemy and testing its defences. It was in this role that the New Zealanders excelled. While addressing the Third Contingent prior to its departure, Lieutenant-Colonel Penton reportedly claimed that New Zealanders had the reputation of being the best scouts in the colonial forces.[74]

Whether the claim was accurate or not, Sergeant Duncan Blair, who would later become a decorated lieutenant-colonel in the First World War, described the dangers of operating ahead of supporting forces.[75] Blair recounted an incident in South Africa where he was acting as an advanced scout for his brigade together with Corporal Robert Aldworth and Trooper John Stevens.[76] Blair claimed the trio approached a house flying a white flag and were fired on by the occupants and other Boers on a nearby kopje. They were fortunate to avoid injury, although Aldworth's luck ran out when all three men were involved in another clash at Roode Kopjes, with far more dire results.

In many ways, the later action was typical of scouts' encounters with the enemy. Acting in concert with the Third Contingent, the Second Contingent headed south from Rustenburg towards Krugersdorp aiming to intercept the elusive De Wet. Describing the subsequent engagement, Blair wrote that after a 37-mile night 'march' the contingent (once again in its familiar advanced scouting role) crossed the Crocodile River in the early morning light and soon encountered the enemy. With orders to reconnoitre the low kopjes that the Boer forces were thought to have fallen back on, Blair and his comrades crossed an irrigation ditch blocking their path, unaware of the danger that lay ahead.[77]

Duncan Blair, a Whanganui farmer who served as a sergeant in the Second Contingent before receiving a lieutenant's commission in the Eighth Contingent. Blair was a lieutenant-colonel during the First World War. THE COLLEGIATE SCHOOL, WANGANUI, IN SOUTH AFRICA, 1899–1900, WANGANUI: A.D. WILLIS, PRINTER, [1901], N.P., WANGANUI COLLEGIATE SCHOOL MUSEUM

An explosion of gunfire at close quarters announced the Boers' presence among rocks and bushes on the Third Contingent's right flank, and four men were immediately shot from their saddles. Amid the confusion, and facing continuous volleys of rifle fire, the New Zealanders retreated across a field. Dismounting and taking cover in a mud hut, Blair and several other men attempted to provide covering fire. Though he reportedly galloped on for some distance before falling, Lieutenant Henry Bradburne was mortally wounded, while Trooper Luke Perham was killed instantly.[78]

Private John Heasley was seriously wounded, and Aldworth was shot as he passed his horse's reins to Stevens (who was later mentioned in Kitchener's dispatches).[79] Blair claimed Aldworth continued to fire several more shots before being ordered out of the firing line by Major Cradock. Bradburne died during the night and Aldworth was subsequently invalided home due to the severity of his wounds; Bradburne and Perham were buried at Bokfontein Farm the following morning.[80] Unlike many New Zealanders buried in the field, their graves were later marked with an iron cross and an engraved brass name plate erected by the soldiers of the Third Contingent.[81]

Although public confidence in the way the war was being conducted diminished as the conflict entered its final year, New Zealanders' pride in their soldiers' achievements remained largely intact. In 1902 William Hall-Jones cabled Seddon expressing support for the dispatch of a tenth contingent. Hall-Jones's attitude reflected both a growing self-belief regarding the contingents' martial proficiency and an unrealistic view of their capabilities. Reports of the capture of British Lieutenant-General Lord Methuen at Tweebosch had shaken Hall-Jones, who was shocked by what he considered the humiliating position of the British in South Africa.[82] For the parliamentarian, the answer was clear — combine the New Zealand contingents, place them under New Zealand leadership and then use what he saw as their scouting and fighting prowess to bring the Boers to heel. He advocated combining the Eighth, Ninth and Tenth contingents under New Zealand command, properly equipping them and giving them 'a free hand with either De Wet or De la Rey, leaving the Imperials to look after garrison work & blockhouses'.[83]

'A damned lot of cocktails and bullet-shy wasters'

Within days of news of the Machavie rail accident reaching New Zealand, the *Otago Witness* published photos of men from the region who had been killed or injured.[84] Newspapers were also quick to emphasise the New Zealanders' bravery in engagements such as Slingersfontein, Koorn Spruit, Ottoshoop and Langverwacht. Accusing British officers of 'gross carelessness and criminal neglect' at Koorn Spruit, the editor of the *Otago Daily Times* contrasted this with the 'coolness and bravery' of the New Zealand men.[85] Once again, British newspapers acknowledged the courage of colonial troops, with the *Times* referring to the gallantry displayed by the New Zealanders at Langverwacht.[86]

One of the more unusual responses to Langverwacht came from the premier himself. In his capacity as defence minister, Seddon advocated sending a number of sheep and cattle dogs with the Ninth Contingent. He suggested that if New Zealanders accustomed to stock-driving and mustering had the services of suitable dogs they might use the animals to head off the mobs of cattle that the Boers drove before them to break through British cordons.[87] The war's end denied the contingents the opportunity of attempting this manoeuvre under a withering enemy fire.

Newspapers later reported the dispatch of six sheep and cattle dogs to the Ninth Contingent to assist in moving Boer livestock.[88] The dogs were carefully chosen by Captain Arthur Trask of the Wakatu Mounted Rifle Volunteers, with four going to the Ninth Contingent, and the premier's son, then serving as a captain in the Eighth Contingent, receiving the remainder. As some newspapers facetiously observed, if the dogs had any military value it would be in mustering the Boer flocks rustled by the New Zealanders.[89] Seddon also accepted an offer of 10 homing pigeons from the Great Barrier Pigeongram Agency to carry dispatches in South Africa, with the birds shipped aboard the *Waiwera* in a box marked 'Homers for the Transvaal'.[90]

Although the New Zealand contingents were widely complimented on their performance in the field, relations within the various corps were not always harmonious. Major Robin ended a 1900 report to Parliament on a cautionary note; while assuring politicians that the contingents would continue to perform well, Robin claimed many had become weary of military service.[91] Some men appear to have been less tired of the war itself than of fighting under the barrel-chested Robin's command for little financial reward. When calls were made for men to join the South African Police, many seized the opportunity, with the *Press* reporting the 'defection' of more than 150 men.[92]

Although they had to keep their own horses, enlisted men in the South African Police received 10s per day compared to the 4s per day paid to troopers (though imperial pay was supplemented by the New Zealand government, which increased soldier's wages to the higher colonial level).[93] The pay disparity, coupled with the widespread belief that the war was drawing to a close, made police service an attractive option. Among the New Zealanders who joined were Private Edward Moore, Lieutenant (later Captain) Piers Tudor, and Private Bertie Willis, the son of the Wanganui MHR.[94] Archibald Willis noted in Parliament that his son had enlisted in the police despite his own objections.[95] Ronald Saxby, who served in the Third and the Ninth contingents, reported that 137 New Zealanders jumped at the chance of joining the police and noted that others were unsuccessful.[96] Though political rivals, William Massey and Seddon both objected to contingent members becoming police recruits. Massey criticised the enrolments, saying that men would remain in South Africa instead of returning.[97] In the House, Seddon strongly opposed giving discharges in Africa before the war had ended, claiming soldiers who enlisted in the police violated their terms of service: 'they were there for the war, and until the war was over we were not agreeable to give them their discharges'.[98]

In a letter sent from Johannesburg, Trooper Frederick Harcourt informed Seddon that reports of the premier's decision angered soldiers wishing to remain in the field.[99] According to Harcourt, many soldiers joined the contingents on the understanding that they could obtain their discharges in South Africa. These individuals had, he claimed, given up good positions in New Zealand to go to Africa, intending to remain there and take advantage of the opportunities they believed would appear following the end of hostilities. One trooper from the Fourth Contingent reportedly left New Zealand with £2000 given to him by his father to use as capital should he decide to settle in South Africa.[100] Harcourt added, 'Many of us can now see our way to good positions in the Mines, Railways, Police or trade but our hopes have been destroyed by your order.'

Luke Perham, who admitted he had been tempted to join the police, claimed Major Robin had accused police recruits and other soldiers who had joined the railways of being 'bullet shy'.[101] A New Zealand trooper who enlisted in the police also claimed Robin was irked by police recruits' 'desertion', adding that he placed every possible obstacle in front of men wishing to enrol.[102] Trooper Frederick Wood defended his decision to join the police and said he and his fellow police recruits were proud to be entrusted with keeping the peace. Wood said he had heard of Robin's accusations but claimed that when the men had left the contingent Robin told them joining the police was a 'splendid opportunity' and jokingly said he hoped they would return to New Zealand to pay off the national debt.[103] Hokitika trooper William Leslie further fuelled the controversy. He repeated Perham's claim, alleging that Robin had told the Commissioner of Police that New Zealanders who joined the constabulary from the contingents were a 'lot of cocktails' and cowardly 'wasters' who were scared of returning to the front.[104]

As contentious as these claims were, it was what Leslie went on to say that proved particularly incendiary. He attributed the contingents' good reputation to Major Davies and Captain Madocks. He claimed that while his contingent was under fire he had personally overheard Major Cradock telling Robin not to be afraid.[105] Incensed by Leslie's claim, Robin wrote to Lieutenant-Colonel Penton denigrating Leslie's own performance, dismissing the trooper's accusations as absurd fabrications and claiming Cradock would verify that no such conversation had taken place.[106] Robin's supporters sprang to his defence in the press; following publication of Leslie's letter, W. R. Douglas, who knew Robin personally, dismissed the trooper's claims.[107] A letter from Brigadier-General Mahon that was reported in several New Zealand newspapers also expressed astonishment at Leslie's comments.[108]

There were also concerns about recruitment activities within New Zealand aimed at attracting men to corps in South Africa. In March 1901 the *Ashburton Guardian* advertised a presentation on the war by Lieutenant Frederick Hughes. Hughes, described in the newspaper as 'the Hero of 118 Battles', had worked as a photographer in Rangiora and Christchurch before joining Brabant's Horse in South Africa.[109] He sailed for South Africa aboard the *Knight Templar* in 1900 as war correspondent for the *Lyttelton Times* and other New Zealand newspapers, leaving his wife and three young sons in Christchurch.[110] After serving in Brabant's Horse, Hughes became concerned about his family and returned to New Zealand on leave.[111] A month after his Ashburton presentation Hughes auctioned his collection of South African curios including Boer weapons, coins, 'a flogging whip, known as a "sjambok"', and British and Boer badges reportedly taken from battlefield dead.[112] The auction house described the collection as 'interesting to all British-born subjects'. Although Hughes' speaking activities and battlefield collection attracted a degree of local attention, it was his subsequent actions that piqued public curiosity.

In April 1901 Hughes, now affecting the rank of captain, placed advertisements seeking 'intelligent men of good character' as recruits for Brabant's Horse.[113] The advertisement reportedly attracted an immediate response with 200 men interested in joining the corps.[114] Hughes cabled Colonel Owen Thomas in South Africa informing him that he had located 100 men prepared to pay their own passages. Although the advertisement sought recruits for Brabant's Horse, Thomas had reportedly offered Hughes a captaincy conditional on the men enlisting in the Prince of Wales' Light Horse.[115] To prove his credentials, Hughes posted the cable from Thomas in the window of a Christchurch business.[116]

Nonetheless, the press remained sceptical, claiming Hughes had no love for the Defence Department after being denied a position in earlier New Zealand contingents.[117] An *Evening Post* article ambiguously titled 'Irregular Recruiting' noted that Hughes' captaincy appeared unofficial and premature.[118] Newspapers cautioned potential recruits that neither the government nor the Defence Department had any prior knowledge of Hughes' activities, adding that Lord Kitchener had stressed that no private individual had the authority to recruit men in New Zealand for South African service.[119] Those considering accompanying Hughes were warned that as he was acting in an unofficial capacity there was no guarantee of enlistment on arrival and they could find themselves stranded in South Africa without official status. Lord Ranfurly reportedly informed Hughes that his actions were illegal and would not be tolerated by the government.[120] Seddon also withheld support for his recruitment activities, while Lieutenant-Colonel Penton informed Hughes that the government would try to prevent men accompanying him. It could not, however, block them travelling to South Africa as private individuals.[121]

Hughes remained intransigent, claiming he was entitled to recruit men in New Zealand for the Prince of Wales' Light Horse. He eventually sailed for South Africa, and in a letter from Kroonstad reported that 25 New Zealand recruits had accompanied him from Wellington. This number, he claimed, was augmented by eight Australians from Sydney and Melbourne, with nine others working their passages.[122] Apparently confirmed in his captaincy, Hughes stated that all the recruits were immediately accepted, with one receiving a commission: 'All are well and happy, and glad of the chance of doing something for our King and country.'[123] In 1902 Hughes' wife, Blanche, applied for a permit to join him, though the couple did not remain in South Africa.[124] By 1914 they were again living in Christchurch when Hughes enlisted as a captain in the Canterbury Mounted Rifles.[125]

'The smartest among horses'

In a 1900 letter, wealthy New Zealand businessman George Stead encouraged other provinces to follow Canterbury's lead in supplying and equipping troops for South Africa. Stead claimed that the value of a fighting force in South Africa depended on its mobility and described what he believed were ideal contingent members well suited to the hilly terrain: 'Hardy young riders, selected from farmers, stock drivers, boundary riders, rabbiters, musterers, etc.'.[126] In truth, however, contingent members came from diverse walks of life, and many did not have a rural background. The Ninth Contingent was indicative of the cross section of New Zealand society represented in South Africa: members' occupations included accountant, draper, bacon-curer, chemist, dentist, bank clerk, civil engineer, tram conductor, boilermaker, grocer, fireman, postmaster, share-broker, soap-maker, hairdresser, letter-carrier, cycle mechanic, jeweller, railway porter and pastry cook.[127] Occupations did not necessarily reflect levels of horsemanship, and even in the hands of experienced riders, horses could be unpredictable. Several men were injured in riding accidents before their contingents left for South Africa. Troopers Charles Chapman and Edward Signal were crushed when their mounts reared and fell backwards.[128] Chapman was a Napier clerk while Signal, a Marton farmer, was described as an excellent horseman.

Before the New Zealand contingents could make an active contribution to the war effort they first had to get their mounts to South Africa. Given the distances involved and the cramped conditions aboard the transport vessels, this was no small undertaking. Veterinary officers did their best to prevent horses dying on the long voyages, and Seddon commended Veterinary Surgeon-Captain Towers for his attention to the horses aboard the *Drayton Grange*.[129] Of the 560 horses aboard the vessel, only eight reportedly died at sea. Major Thomas Jowsey of the Third Contingent also praised the work of Veterinary-Surgeon Henry Wilkie aboard the *Knight Templar*.[130] In his diary, Jowsey lamented the plight of the horses: 'It is heart-breaking to see these fine animals suffering, and I can do nothing for them.'[131] This was not entirely true, as Jowsey did his best to improve conditions for the animals. After discussions with the vessel's captain, he arranged for gratings that restricted air flow to be cut away and had wooden flues constructed to improve the air quality for horses below deck.[132] Both of these measures increased ventilation in the stifling lower hold where many of the animals were stabled. Jowsey commended the actions of the men of No. 1 Company who voluntarily moved 200 tons of coal in the ship's bunkers to increase air flow in the hold and alleviate the misery of sick horses by making room for them to safely lie down.[133]

Jowsey was less impressed by the 'combination of ignorance and inability to do anything connected with stable-work' on the part of the officers and men of No. 2 Company.[134] Accusing them of having no love for horses, he claimed they were unaccustomed to the animals and felt their duty ended once the horses were fed and watered.[135] As the vessel crossed the Indian Ocean, conditions below deck on the *Knight Templar* were hellish, with stifling temperatures in the dark, airless hold combining with the stench of ammonia from the horses' urine.[136] Nonetheless, of the 278 horses shipped aboard the *Knight Templar* only 10 died during the voyage, with another 10 listed as gravely sick on arrival at East London.[137] The remaining 258 survived their ordeal to be landed, weak and unsteady, on the Durban docks.

Some troopers realised that their comfort and safety depended to a large extent on the horse they rode and did their best to select a mount with the desired qualities. While crossing the ocean, former schoolmaster Private Joseph Orford descended into the hold of the *Gymeric* in an attempt to select a suitable mount. By the light of a match he found some horses had travelled well and put on condition, while others were 'mere clothes-racks'.[138] Orford also criticised the treatment some animals received from troopers: 'it's horrible the way some fellows neglect their horses'.[139]

One of the more shocking features of the war was the attrition rate of horses in the field. To maintain pressure on the Boers, a highly mobile style of warfare evolved, but mobility came at a price. A New Zealander in Roberts' Horse claimed pursuing the enemy was useless, as the imperial troopers' horses were fatigued and the Boers, on sturdy ponies unencumbered by weighty equipment, moved significantly faster.[140] Horses were often inadequately tended and fed while being relentlessly worked in climatic extremes of scorching sun and freezing rain. Describing conditions on a 250-mile trek, Sergeant Edward Fitzgerald said there was insufficient food for both the men and their horses, with the latter subsisting on a small quantity of crushed wheat and inferior hay.

On this meagre diet the horses were expected to travel 30 miles a day carrying heavy loads. Fitzgerald was appalled by their mortality rate, which he attributed to a combination of poor nutrition, sickness and sore backs.[141] Sergeant Blair also commented on the inadequacy of rations for both men and horses operating at a distance from their transports.[142] A government report noted that horses transported aboard the *Tropea* and the *Ormazan* fared badly following their arrival in Africa. The horses, which were off their feed and had not had their shoes removed before the voyage, were immediately put to work on arrival. Of the 257 horses that were unloaded at Port Elizabeth, only 222 made it to Pretoria, with the remainder too weak to complete the journey and being left along the way.[143]

Despite Orford's comments, the contingents did include skilled riders experienced in handling horses, with troopers such as Frank Perham and Conrad Saxby expressing pride in the New Zealanders' horsemanship and recording New Zealanders winning mounted military sports competitions in South Africa.[144] Of the British commanders, few were more qualified to comment on mounted troops than cavalry officer Major-General French. Trooper William Bunten claimed that French expressed his satisfaction with the First Contingent and used the New Zealanders extensively in operations.[145] A newspaper claimed that French believed the New Zealand troopers were 'the smartest amongst horses he had ever seen' following an incident involving horses being unloaded from a train.[146] French reportedly shook hands with Hugh Smith, George Mitchell, Daniel Johnston and George Arnold after they got four horses back on their feet in a railway truck. Mitchell was a painter who had served with Arnold, a Balclutha farmer, in the Clutha Mounted Rifles, while Johnston was a dentist's assistant who had served in the North Otago Hussars. Smith was a 'horse breaker'.[147]

Charles Baré of the Seventh Contingent was a horse trainer, and Hugh 'Smoke' O'Hagan had been a jockey in New Zealand.[148] Being disqualified for life by the Manawatu Racing Club for 'foul riding' was no impediment to O'Hagan enlisting in the Imperial Light Horse and the South African Light Horse. From his Monte Cristo camp in Natal, O'Hagan wrote to the club in an unsuccessful bid to have the ban lifted.[149] O'Hagan told his parents that he had been present at the Battle of Colenso and claimed to have been under fire 35 times without sustaining injury.[150] His fortunes changed, however, when he was shot twice during a 1901 skirmish.[151] Though injured, O'Hagan reportedly shot a Boer he saw removing possessions from New Zealand dead and wounded.[152] The Boer's leg had to be amputated, but the jockey, who newspapers claimed was a 'splendid horseman', befriended his former adversary while recuperating in the same Johannesburg hospital.[153] O'Hagan died aboard the *Britannic* en route to New Zealand in July 1901 and was buried at sea.[154]

Nineteen Sixth Contingent men were farriers in civilian life, while William

Thomson, a Māori member of the Ninth Contingent, had previously served as a farrier in the Fourth Contingent.[155] Also in the Ninth Contingent were men such as Farrier-Sergeant Patrick Fraser, Farrier-Sergeant John Voight, and Farrier Frank Goldsmith (who, like Thomson, were blacksmiths by trade), and James Vintiner, George Campbell, Lance-Corporal Walter Allen and Farrier-Sergeant Archibald Hutton, all of whom worked as grooms in New Zealand.[156] There were also farmers, station hands, shepherds, cattle drivers, carters, teamsters and ploughmen, to whom horses were second nature, and men like Lieutenant Robert Richards who gave his occupation as 'horse-dealer'.[157] Leo Matthews of Lower Hutt was another accomplished jockey.[158] After arriving in South Africa with the Fourth Contingent, Matthews won the Beira Cup in Portuguese East Africa, but like O'Hagan he had ridden his last race in New Zealand; he died of enteric fever in Cape Town in June 1901.[159]

A nation of marksmen?

While praising his men's riding ability during the military sports day at Klerksdorp, Conrad Saxby noted the New Zealanders exhibited less skill in shooting events.[160] In some cases where New Zealanders succeeded in hitting the enemy, it was attributed more to luck than aptitude. Trooper Wilfred Wilson claimed that a trooper in his contingent had hit a Boer at 1500 yards but added that he believed it was accidental, as the man was a poor shot.[161] Wilson noted, however, that men from his contingent had killed and wounded several Boers. A *Wanganui Herald* article titled 'Unable to Shoot' was scathing in its criticism of the marksmanship of Wellington Third Contingent applicants and claimed several had completely missed the target.[162] The men reportedly blamed their weapons and ammunition, but the newspaper claimed many recruits were unfamiliar with rifles.

While the percentage of rifle club members who served in South Africa is unknown, an increased awareness of defence issues during the war contributed to the number of rifle clubs in the country rising from 72 in 1901 to 114 in 1902.[163] Nonetheless, an *Evening Post* editorial was especially disparaging. It claimed there were insufficient rifles in the colony for the nation's defence and a similar shortfall in experienced shots. New Zealand could not, the newspaper alleged, provide 1000 marksmen, and until the situation was addressed those responsible for the nation's defence would be no more effective than 'a mob of sheep'.[164]

In 1899 Seddon had indicated that the Defence Department would give preference to skilled marksmen seeking to enrol in the First Contingent.[165] However, accidental injuries, deaths and dangerous practices involving firearms indicated the inadequate training many troopers received prior to departure for South Africa. In 1901 a correspondent to a Wellington newspaper maintained that men were gaining places in contingents more through personal connections than ability, and that there were men in camp who were poor riders and 'who could not hit a haystack at 100 yards range'.[166]

Warning in 1899 that the New Zealand public would not forgive a government that sent its men to fight with inferior weapons, parliamentarian George Hutchison noted that the modern rifles used by the Boers had a superior range to the Martini-Enfield carbines issued to the men of the First Contingent.[167] The premier, sensitive to anything that reflected on his defence portfolio, responded that the imperial authorities would probably distribute magazine rifles on the contingent's arrival in South Africa.[168] The following year, Auckland City MHR William Napier drew Parliament's attention to the variety of weapons used by the colony's Volunteers, including the Martini-Henry, the Martini-Enfield and the obsolete Snider.[169] Napier claimed that in South Africa the contingents' .303 Martini-Enfield rifles were being replaced by the British Army with modern Lee-Enfield or Lee-Metford rifles and recommended that New Zealand arm its defence force with these weapons.[170] Seddon conceded that if contingent members returned from South Africa with modern weapons it would benefit the colony, adding that there had been a plan to send the Martini rifles to England for conversion, but this had not occurred.[171] While admitting the government still held stocks of Snider rifles, the premier indicated it had also obtained new rifles. He added that a sizeable order had been placed for magazine rifles, though the order had not been filled due to increased wartime demand. Seddon assured Parliament that the government was aware of the need for a 'large number of modern rifles'.[172]

In fact, the First Contingent did not immediately receive newer weapons on reaching South Africa. In early December 1899 Major Robin reported that despite Webley revolvers being issued to officers, the imperial authorities had no .303 magazine rifles available, though the contingent's carbines were considered adequate.[173] At the end of the month, he implied his men had finally received magazine rifles, which 'any one [sic] can shoot'.[174] Robin noted the Boers often remained concealed among rocky kopjes until they were either dislodged by artillery fire or had their flanks threatened by mounted troops.[175] Though claiming good shooting was still necessary, Robin said that Boer tactics meant New Zealand soldiers had little opportunity to get clear shots.[176]

In an example of unsafe firearms practices that were relatively commonplace among the contingents in South Africa, troopers Alexander O'Keefe, James Jack, George Nichol and Frank Withers were each sentenced to 96 hours' confinement for discharging their weapons while on a train in South Africa.[177] Although no injuries resulted, other incidents had far more serious consequences. Captain Fred Abbott recorded a fatal accident that occurred in camp at Goedgevonden Farm in October 1901.[178] While Trooper Alfred Whitney was handling a captured Mauser rifle, he accidentally discharged it, killing his friend, Corporal William Byrne of the Seventh Contingent.[179] At a regimental inquiry, Whitney stated that he had asked to examine the rifle, then worked the weapon's bolt a couple of times and when no cartridges were ejected pulled the trigger, fatally shooting Byrne in the chest.[180]

Trooper Ellis Wrigley, who was also present when the incident took place, described Whitney taking the rifle from a mounted Canterbury contingent member, turning until the rifle pointed directly at the group of seated men and firing the shot that killed Byrne.[181] Wrigley made no mention of Whitney working the bolt or checking the weapon's breech and magazine. Whitney's three-year service as a sergeant in the Waimate Mounted Rifles had failed to instil in him even the most rudimentary of safe firearms practices.[182] Following the shooting, Whitney, who would later be killed in action at Langverwacht, was 'in a state of prostration through grief' and had to be sedated.[183]

Dangerous handling of weapons was not the sole preserve of New Zealand contingent members. William Smith, a New Zealander from Havelock who studied engineering in Scotland and then joined the Lanarkshire Yeomanry, was shot in the thigh when a corporal in an adjoining tent (described in the press as 'the most useless man in the regiment') accidentally discharged his weapon.[184] Smith survived long enough to write to his father without mentioning his injury, but finally succumbed in Ladybrand Hospital. He was praised in the New Zealand press for his 'manly unselfishness' in not informing his family of the severity of his wound.[185]

Together with Byrne, Kirkwood and Smith, Private William Raynes of the Sixth Contingent was fatally injured at Mokari in Orange River Colony in an incident highlighting the dangers of jittery soldiers firing blindly at unidentified targets.[186] Newspaper accounts claimed that the 'regrettable incident' had occurred while Raynes was on outpost duty and shouted a warning to his comrades that Boers were approaching.[187] Alerted by his cries, the Otago squadron turned out to reinforce the picket, only to find Raynes had been accidentally shot. Following Byrne's death, Lieutenant-Colonel Porter of the Seventh Contingent ordered squadron leaders to emphasise the need for care when handling weapons. Porter had received complaints about 'the common practice' of discharging weapons in camp and stressed the importance of keeping rifle muzzles elevated when weapons were being examined or cleaned.[188] The following day, Trooper Thomas Crawford of the Canterbury squadron was accidentally shot in the arm at close range by a fellow contingent member.[189]

Four months earlier, heavy firing had erupted among outposts of the Seventh Contingent at Tabakplaats in the pre-dawn hours.[190] Henry Houchen was shot in the thigh, while Arthur MacFarlane, Alfred Thom and John Sellers were also wounded after scouts mistook them for Boers and opened fire.[191] Porter blamed the incident on a brigadier who he alleged had sent out advanced posts without notifying the officers, and the brigade-major, who had allegedly ordered the outposts relieved before sunrise. Another contingent member and seven Africans were reportedly injured after cartridges were thrown in a campfire, and Ninth Contingent Private Horace Baker was shot in the foot while making his bed.[192] There was a degree of inevitability that injuries would occur, and unsafe practices in New Zealand Volunteer corps highlighted the dangerous combination of firearms and inexperience. In January 1902 Manchester Rifles Volunteer Percy Crawford was fatally shot in his tent at Feilding Racecourse after John Thompson accidentally loaded his weapon with live ammunition instead of blanks during a practice night attack.[193] The exercise reportedly involved men with limited firearms experience who were not asked if they were carrying live ammunition and whose weapons were not checked.[194]

When New Zealand soldiers reached South Africa they found that warfare on the veldt presented a number of challenges. Alexander 'Sandy' Peddie of Napier, who served in Brabant's Horse, highlighted the difficulties of engaging the enemy: 'The kopjes are awful places to get at, and you can never see the Boers to have a shot at them.'[195] Joseph Orford claimed he had not had a shot under 1200 yards, adding that he did not think he had 'even scared a Dutchie yet'.[196] Nonetheless, Peter McDonald, a noted Wairarapa deerstalker, still had an opportunity to show his skill.[197] Orford said an officer gave McDonald (who had not been a Volunteer) permission to shoot at a Boer outpost estimated to be two miles away. McDonald's first shot reportedly made a mounted Boer move position, the second caused him to duck below his horse's neck and the third forced him to gallop away. At least two Māori members of the contingents were also competent shots. After Walter Callaway spotted a Boer at a considerable distance, his lieutenant encouraged him to shoot.[198] Callaway claimed he got 'great praise from all the boys' after he hit his target. Āhere Te Koari Hōhepa of the Hawke's Bay section of the Rough Riders was also reportedly one of the best riders and shots among the recruits for his contingent.[199] Twenty-one-year-old Harold Booth, who would be killed during fighting at Rensburg, was also a skilled marksman.[200] Described as daring to the point of recklessness, 'Happy-go-lucky Harold' won the Troopers' Champion Cup and reportedly held a commanding lead in the Troopers' Champion Belt when his contingent sailed.[201]

A number of contingent members wished to keep their weapons after their service ended. Shortly after the war, Geraldine MHR Frederick Flatman asked the acting minister of defence whether the government intended allowing Seventh Contingent veterans to retain their rifles and bandoliers in appreciation of their performance at 'the great Bothasberg fight'.[202] In reply, William Hall-Jones said that as all rifles and bandoliers remained the property of the imperial authorities the government's hands were tied. He added that other contingents who had performed well had relinquished their weapons. Hall-Jones noted, however, that contrary to orders some men had kept their rifles. Though Hall-Jones claimed the government lacked the authority to allow men to retain their weapons, this had already occurred on several occasions.

Eight months earlier former farrier-sergeant Sydney A'Court wrote to the under-secretary for defence.[203] A'Court, who had served in the Fifth Contingent,

Above: Gisborne members of the Fourth Contingent, including John Walker (Hōne Wāka) and Arthur Te Wawata Gannon, both troopers with Māori heritage.
Major (Retd) Noel W. Taylor ED** RNZIR

Right: Āhere Te Koari Hōhepa in his Volunteers uniform. Hōhepa served in the 1897 Jubilee Contingent, and in the Third Contingent in South Africa under the name Arthur Joseph. Whangarei Museum at Kiwi North, M. M. White Collection, 1967.5.353

informed Douglas that he intended enlisting in the Eighth Contingent and asked that his rifle be reissued to him.[204] The former trooper described the weapon in detail (including spur marks on the butt) and promised to return it to Defence Department stores if required.[205] A'Court received his rifle a month later, but he enlisted in the Tenth Contingent rather than the Eighth.[206] It was a decision that cost him his life as he contracted pneumonia on the voyage back home, died aboard the *Montrose* and was buried at sea.[207]

Frank Perham, John O'Reilly, Archibald Hutton and Frederick Bezar were among those reissued with their rifles after returning to New Zealand.[208] Hutton, who served in the Third Contingent, wrote directly to Seddon requesting the return of his weapon. In the letter, Hutton reminded Seddon of a promise to this effect which he claimed the premier had made at a banquet for returning soldiers in Christchurch.[209] In consenting to Hutton's request, the under-secretary of defence reminded Hutton that the rifle was not his personal property and that he was required to keep it in good order.[210] Following the end of his service in the Fifth Contingent, Perham signed the same agreement as A'Court.[211] Perham did not re-enlist in a later contingent and there is nothing to suggest he was ever required to surrender the weapon. Corporal Charles Nurse of the Third Contingent was another who received his service rifle after informing the Defence Department that his weapon had 'Nurse' branded into the butt and signing the same undertaking as the other men.[212]

In service of the empire

In 1900 Legislative Council member David Pinkerton claimed that the young men who had fought in the 1850s during the Crimean War and the Indian Mutiny were motivated by the same spirit that filled the soldiers fighting for the empire in South Africa.[213] The similarities were far closer than Pinkerton perhaps realised. The New Zealand contingents and irregular corps contained a leavening of men, from both New Zealand and the United Kingdom, who not only were animated with the same spirit as their forefathers, but also in some cases *were* soldiers who had participated in the empire's earlier wars.

Pinkerton's address would have struck a chord with Crimean War veteran Robert Bakewell, who at 70 was the oldest New Zealand contingent member.[214] Bakewell claimed that 24 hours before the Ninth Contingent was due to sail

Seddon had asked him to accompany it as surgeon-captain. Bakewell added that despite his age and infirmities (which he alleged were conveniently overlooked), he was 'fool enough to go'.[215] The doctor was hastily included after the death of Trooper Percy Leary at Te Papapa Camp led the government to dispense with the services of Captain-Surgeon A. L. Murray.[216] Despite suggestions in the press that Murray was the victim of 'a grave injustice' and had been made a scapegoat, Bakewell assumed his role, though apart from his duties aboard the *Devon* while en route for South Africa the aged surgeon proved largely ineffectual.

Suffering from emphysema and asthma while at sea, Bakewell took to smoking datura to relieve his symptoms.[217] The powerful hallucinogen suppressed the doctor's appetite, causing him to lose both weight and, occasionally, consciousness. On arrival in Natal, Bakewell was hospitalised, and although his condition improved he played no further part in the conflict, returning to New Zealand an invalid.[218] The doctor later criticised the premier for the Defence Department's refusal to award him a military pension after it found his condition was not due to military service. Following the war, Bakewell described Seddon as an 'irresponsible despot' and accused the premier of being indifferent to his financial difficulties.[219]

A New Zealand resident for 21 years, Hugh McDonagh was another seasoned veteran of Britain's earlier wars who had a lengthy military career prior to taking part in the South African War. The *Wanganui Herald* recorded his impressive service record: 'Royal Engineers, 17 years; South African Irregulars, 1 year; Indian Mutiny, China 1860, Abyssinia, South Africa 1900, and Distinguished Service medals'.[220] However, when the 64-year-old McDonagh attempted to enlist in the First Contingent, he was declined due to his age.[221] Adding the Crimean War and 'Maori war' medals to McDonagh's list of decorations, the *Otago Witness* reported that McDonagh had paid his own passage to South Africa and enlisted in the Duke of Edinburgh's Own Volunteer Rifles, but he suffered a back injury at Paardeberg that left him partially paralysed.[222] After he was invalided to England, Lloyd's Patriotic Fund contributed to McDonagh's care at the Soldiers' Home in London. In a letter to the New Zealand Agent General the fund's secretary stated that Sir Alfred Milner, the governor of Cape Colony and high commissioner of South Africa, was among those who had taken a personal interest in the aged soldier.[223]

The *Auckland Star* claimed there were troopers in the contingents who, having seen service in the Zulu War and Sudan, were more qualified to be NCOs than the men holding non-commissioned ranks above them.[224] Though he served as an officer in the Seventh Contingent, Lieutenant James Hamer already had extensive African experience, having enlisted in the Sixth Manchester Volunteers as a 15-year-old and served during the conflict with the indigenous population of the South African region of Griqualand in 1877, and in the 1879 Zulu War.[225] He was one of the few survivors of the resounding British defeat at Isandlwana where several hundred soldiers were overwhelmed by Zulu warriors. As well as mainland Africa and Madagascar, Hamer had spent time in North America.[226]

Despite spending 14 years in New Zealand, Hamer showed little inclination to return following the war.[227] After the conflict ended, he sailed to England for medical treatment for war-related injuries. There, he unsuccessfully petitioned the New Zealand Agent General and the War Office for half of his return fare to New Zealand in lieu, or a colonial posting in Uganda or West Africa.[228] He was told that as he had resigned his commission, securing a position in the West African Frontier Force or the King's West African Rifles was unlikely, and by March 1905 his postal address was the Regina Post Office in Canada's North-West Territories.[229]

Other men, such as Sergeant William Mahood, a 38-year-old Irish-born farmer from Tīrau, and Lieutenant Henry Browne, also had experience of Africa before joining New Zealand contingents.[230] Mahood spent four years in the Cape Mounted Rifles in the early 1880s, took part in fighting in Basutoland, and was sent to Tembuland in the Eastern Cape to eject Boers from land claimed by the imperial government. The 42-year-old Browne had joined the Cape Mounted Rifles in 1877, taking part in actions against the Galeka and Gaika tribes and serving in the Basuto Campaign of 1880–81.[231] He had also been involved in operations in Tembuland, Transkei, Cape Colony, Pondoland, Zululand, Bechuanaland and Basutoland, as well as serving in India.[232] In 1920 Browne was appointed sergeant-at-arms in the New Zealand House of Representatives, a role he held for 12 years before retiring to Paeroa.[233]

Before coming to New Zealand, Lieutenant Arthur Dewar of the Fifth Contingent had served in the 92nd Gordon Highlanders.[234] In 1895 he joined the Matabele Mounted Police as a corporal and took part in the Jameson Raid later that year, surrendering with the rest of Jameson's force at Krugersdorp.[235] On arrival in New Zealand in 1897, Dewar (reportedly a nephew of Lord Roberts) took up a position as an instructor in the Napier Volunteer district with the rank of sergeant-major, and later secured a position as a drill instructor in Whanganui.[236] During his comparatively short time in New Zealand, Dewar developed a close relationship with Wanganui Collegiate School. The school lamented losing his services when he left for South Africa, and in 1902 Dewar and his wife donated a competition cup to the school's cadet corps.[237]

Though the inflexibility of imperial officers occasionally irritated New Zealand soldiers, men like Fifth Contingent sergeant Edward Lascelles found the leadership of African old hands reassuring.[238] The 21-year-old Greenmeadows station cadet informed his father that his Coldstream Guards commanding officer was a strict disciplinarian, but added that he welcomed it as the New Zealanders were 'inclined to be a rather rough-and-ready crowd'.[239]

Thirty-one-year-old Captain Thomas Tanner from Napier, who served in the Fifth Contingent, was described in the *Otago Witness* as one of Cecil Rhodes' 'young men', having personally met the famed South African businessman.[240] After several years in Central Africa, Tanner had returned to New Zealand to recover from fever. He contracted the condition in an area of Africa where he and others like him endeavoured to colour the map with the red ink of British imperialism. In an episode reminiscent of a Conrad novel, Tanner had at first been carried, and then was transported by steamer down the Zambesi to the coast after falling ill in the interior.[241]

Sickness also ended Tanner's service in the Fifth Contingent and thwarted his subsequent attempts to re-enlist after recovering.[242] Following its arrival in Portuguese East Africa, the Fifth Contingent camped on a poorly chosen, swampy site in Beira, even though the town was known as an unhealthy locale.[243] Tanner was immediately laid low with fever and dysentery that required him to be invalided home on the *Gothic* in July 1900.[244] His certificate of discharge alluded to service in Rhodesia in 1891 and participation in the First Matabele War of 1893, where he acted as a 'galloper' in the Salisbury Horse.[245]

In one of those improbable situations that were surprisingly common in Africa, Tanner reportedly served in Africa with a friend and fellow Napier resident, George St Hill.[246] The son of a Napier minister, St Hill had helped establish the Hastings Football Club and served in the Volunteer artillery in Wellington prior to his departure for Rhodesia.[247] Another of Rhodes' disciples, St Hill had taken part in the 1893 Matabele War alongside Tanner as a corporal in the Salisbury Horse, and served during the 1896 Mashona Rebellion.[248]

Returning to England, St Hill joined the Devonshire Yeomanry following the outbreak of war in 1899 and returned to Africa with his regiment. Nonetheless, his colonial ties remained. Captain St Hill recommended Reginald Harper, another New Zealander serving in his regiment, for promotion.[249] 'Lance-Sergeant' Harper was recognised by the Royal Humane Society for saving a fellow soldier from drowning in the Vaal River, and on his return to England played a prominent role in the London branch of the Christ's College Old Boys' Association.[250]

An acquaintance of Harper who was also a New Zealand expatriate in Africa was former New Zealand Railways employee William Lancelot Miles. Born in 1863, Miles lived in Christchurch, where he started work at the Railways Department in 1878 before transferring to the general manager's office in Wellington in 1880.[251] He rose to the position of clerk first-class on a comfortable salary of £210 a year, but in 1890 he resigned and turned his back on both the railways and New Zealand.[252] Sailing first to England, Miles then travelled on to Zululand in 1894, where he held positions including district commissioner and police magistrate.[253]

During an 1899 visit to England, Miles extolled the virtues of his new home and stated his intention of returning to Zululand hopeful 'that Englishmen may be again respected in South Africa'.[254] Like countless New Zealanders who venture abroad, Miles retained a sense of connection with his homeland, and he planned to visit New Zealand on the way back to Africa; but events overtook him. Only days after war broke out, he boarded the *Umtata* and sailed directly to South Africa.[255] During the war he served in the Natal Corps of Guides, the Army Intelligence Department and as a trooper and corporal in the Colonial Scouts.[256] By late March 1900 he was based in Ladysmith, having taken part in the town's relief, and carried passes identifying him as a guide and interpreter in the Field Intelligence Department.[257] Even after falling ill and being invalided to England, Miles maintained his New Zealand identity and was among a large number of his fellow countrymen in the audience during a 1901 London lecture on the country.[258]

One of the more colourful New Zealanders in Africa when the war started was Arthur James Siggins. Siggins reportedly ran away to sea at the age of 14.[259] Making his way to Australia, he worked on sheep farms prior to departing for South Africa, where he became a sergeant in the Mashonaland Division of the British South African Police.[260] Siggins represented Rhodesia at the coronation of Edward VII and reportedly served as one of Cecil Rhodes' pallbearers.[261] In 1931, the *Auckland Star* claimed that Siggins had spent more than 30 years in African 'outposts of civilization' as a big game hunter, trapper, trader, planter, prospector and pearl fisher.[262] He married British actress Molly Adair, wrote several books, and was employed as the wild animal handler during the making of the 1929 Hollywood silent movie *The Four Feathers* featuring Fay Wray.

Significant numbers of New Zealanders serving in New Zealand contingents and irregular forces had little or no experience of warfare prior to their arrival in South Africa. However, there were also individuals such as Dewar, Mahood, Hamer and Browne serving alongside them who were familiar with both indigenous Africans and the Boers. These men knew the African climate and terrain and the difficulties it presented, but perhaps more importantly they had personal knowledge of their opponents.

The sons of others

Though politicians bore responsibility for the decision to send contingents to the war, several also had family members who fought in South Africa. With notable exceptions like Thomas Taylor, Henry Scotland and John Hutcheson, the country's politicians largely adopted a unified stance regarding the legitimacy of British actions in South Africa and New Zealand's involvement in the conflict. Nonetheless, when Richard John Spotswood Seddon received a commission in the Fourth Contingent, his father's rivals wasted no time in accusing the premier of favouritism. The *Manawatu Evening Standard*, which Seddon had referred to as 'an apology for a newspaper' shortly before the war, published an editorial titled 'The Premier's Nepotism'.[263] It accused Seddon of abusing his position as minister of defence, and claimed the most deplorable aspect of Seddon's lieutenancy was that his father had placed his son in a position where he might jeopardise not only his own life through his 'incompetence' but also the lives of those he commanded.[264] When Seddon later served in the Eighth Contingent, at least one of the soldiers under his command expressed doubts regarding his ability. Discussing Seddon in a letter to his father, Trooper Bert Stevens said he wished he had an officer in whom he and his fellow contingent members could place more confidence.[265]

Forced to defend his actions, the premier dismissed the criticism, claiming 'My son goes where I am sending the sons of others.'[266] Like Hugh Gourley and Seddon, a number of both serving and former politicians had a vested interest in the progress of the war. Henry Miller had a son captured at Koorn Spruit, and his fellow Legislative Council member Lieutenant-Colonel James Bonar had two sons in the contingents.[267] The eldest, Archibald Bonar, served in an irregular corps in South Africa, and was killed in action at Gallipoli in 1915.[268] Wanganui MHR Archibald Willis also had two sons in contingents.[269]

Lieutenant George Rolleston in the Fourth Contingent was the son of William Rolleston, who held several parliamentary seats.[270] George was mentioned in dispatches for meritorious service and, like Archibald Bonar, remained in South Africa for a time after the war, becoming the resident magistrate at Lichtenburg.[271] George Fisher, City of Wellington MHR, had one son in the Kaffrarian Rifles, while another, Francis Marion Fisher, was a captain in the Tenth Contingent.[272] For George Fisher senior and his wife, and the family of James Colvin, the mayor of Westport and Buller MHR, the war had particularly tragic consequences. Both parliamentarians lost sons to illness in South Africa either during the conflict or soon after.[273] When Seddon sought parliamentary support for the dispatch of the Second Contingent, Frederick Pirani agreed, stating that his eldest son was anxious to enrol.[274] It appears the younger Pirani's attempts to enrol in a New Zealand contingent failed, but he was not to be thwarted. Making his own way to South Africa, Percy Pirani served in Kitchener's Fighting Scouts before joining the Imperial Yeomanry branch of the Army Pay Department.[275]

At least one sitting parliamentarian joined up; Robert Heaton Rhodes junior, an Oxford-educated Canterbury farmer and Ellesmere MHR, served in the Eighth Contingent.[276] Born in 1861, Rhodes was an accomplished horseman and keen polo player who became a wealthy landowner following his father's death in 1884.[277] He spent 14 years in the Canterbury Yeomanry Cavalry prior to enlisting for service in South Africa and donated 10 horses to the Second Contingent.[278] At a farewell function at Leeston Town Hall, Captain Rhodes' supporters applauded his patriotism and selflessness, and were urged to prevent interlopers attempting to claim his parliamentary seat.[279] Prior to his departure, Māori from Banks Peninsula presented Rhodes with a whalebone mere.[280] Though declining to discuss his wartime experiences with the press following his return, he indicated that he intended returning to politics.[281]

The premier directly orchestrated the enrolment of at least one relative of a political crony. Robert Witheford, the son of Joseph Witheford, Birkenhead mayor and the Liberal City of Auckland MHR, sailed aboard the *Tropea* and on arrival at the Cape joined the First Contingent.[282] Witheford carried with him a letter, written on 'Premier's Office' stationery, from Seddon to Major Robin, stating that Witheford was 'accustomed to station life' and was understood to be a good rider and shot.[283] The letter instructed Robin to enrol him in the contingent despite Witheford indicating he had no military experience.[284] While in South Africa he was offered a commission in the Duke of Edinburgh's Wiltshire Regiment, but declined due to the expense.[285] Witheford served as a lieutenant in the Wellington Infantry Regiment during the First World War and died in 1936 after falling down a lift well in the capital.[286]

Seddon may also have assisted Frank Willis, whose father, Alexander Willis, worked in the premier's office as Cabinet secretary.[287] Willis's attestation form for the First Contingent was signed in Wellington on 8 May 1900 — several months *after* the contingent sailed and almost two months prior to him arriving in South Africa aboard the *Tropea* with Witheford.[288] Regardless of the circumstances surrounding his selection, Willis apparently proved a competent soldier who went on to serve in the Seventh Contingent before receiving a lieutenant's commission in the Tenth.[289]

If the premier was willing to oil the wheels of bureaucracy to ensure favoured individuals such as Witheford served in contingents, he was markedly less inclined to do so for those with connections to his political adversaries. Though Pātea MHR George Hutchison had supported Seddon's 1899 resolution, he described as 'niggardly' the premier's expectation that First Contingent men would supply their own horses, questioned the adequacy of contingent pension arrangements and, as we have seen, criticised the outdated First Contingent weapons.[290] Hutchison's visit to South Africa in early 1900, where he met with politicians, senior officers and First Contingent troops, was unlikely to have endeared him to Seddon, though the *Press* claimed Hutchison had benefited from the experience.[291]

Both George Hutchison and his father, William Hutchison senior, were Seddon's long-standing and implacable political rivals.[292] Without elaborating, in 1900 the premier told Parliament that a member of the Hutchison family was responsible for 'one of the greatest injuries ever attempted to be perpetrated on me'.[293] This mutual animosity was exacerbated when William Hutchison junior attempted to join the Fourth Contingent. During subsequent litigation, Seddon claimed that Hutchison had earlier been denied a place in the Third Contingent due to shortsightedness.[294] Hutchison's attempts to join the Fourth after resigning his job at the *Press* became the basis of a protracted and rancorous argument with the premier that played out in Parliament, the High Court and the nation's press.

At issue was whether Hutchison had actually attested as a contingent recruit. While Hutchison vehemently claimed he had, Seddon was equally adamant that there was no evidence supporting the claim. A *Press* article stated that after Awarua MHR Joseph Ward had helped Hutchison secure a place in the Fourth Contingent as orderly to its commanding officer, Hutchison had been sworn in at Forbury Camp.[295] However, when he attempted to board the troopship *Monowai* he was reportedly recognised by Louisa Seddon, who notified her husband that 'one of those Hutchisons' was trying to embark with the contingent.[296]

Seddon informed Hutchison that as his name did not appear on the contingent roll he would be allowed to travel with the contingent only if he went as a war correspondent and arranged payment for his passage.[297] Hutchison would later dismiss the war correspondent story as a 'fable' concocted by friends in the press to ensure he got to South Africa if he missed selection in the Fourth.[298] After Hutchison senior took up his son's case, the *Press* printed extracts from a letter William Hutchison had sent from South Africa giving his version of events.[299] In response to this, Seddon assumed the role of the aggrieved party when offering a personal explanation of the situation in the House.

The premier implausibly claimed that, despite previous clashes with the

William Hutchison, the son of Premier Richard Seddon's implacable political rival William Hutchison senior. Hutchison junior's attempts to travel to South Africa as a member of the Fourth Contingent were thwarted by Seddon, who insisted Hutchison could only accompany the contingent as a war reporter. THE COLLEGIATE SCHOOL, WANGANUI, IN SOUTH AFRICA, 1899–1900, WANGANUI: A.D. WILLIS, PRINTER, [1901], N.P., WANGANUI COLLEGIATE SCHOOL MUSEUM

Hutchisons, he bore them 'no ill-will' and stated that he had been approached by the editor of the *Lyttelton Times*, Samuel Saunders, who sought permission to send Hutchison to South Africa as a correspondent for a group of New Zealand newspapers.[300] According to Seddon, Saunders assured him that the newspapers would defray any associated expenses.[301] The premier added that he consented to the request and that it was only when Hutchison attempted to board the vessel in uniform that he became aware of Hutchison's intention to go as a trooper. After informing Hutchison that this was impossible, Seddon allowed him to proceed aboard the *Monowai* as far as the Australian port of Albany, though Hutchison was instructed by Major Frederick Francis to remove his uniform and wear civilian clothes.[302] Following the government's receipt of £25 from Saunders to cover his expenses, Hutchison sailed for Beira in Portuguese East Africa.[303]

But the protracted saga was far from over; when Hutchison requested his discharge, Major Francis informed him that as there was no evidence of him ever being in the Fourth Contingent he could not be discharged from it. Hutchison responded that as he had attested he was duty-bound to serve until he was either discharged or dismissed, adding that Fourth Contingent captain Harry Fulton supported his position.[304] Hutchison later claimed that as he was not 'of the right colour' he had been 'persecuted with outrageous vindictiveness'.[305] On reaching Marandellas in Rhodesia, an officer advised Hutchison to join the Rhodesian Field Force to avoid 'unpleasantness' with Major Francis. Heeding this advice, Hutchison served as a gunner with the corps and received his Queen's South Africa Medal for service in the Rhodesian Field Force, *not* the Fourth Contingent.[306]

Hutchison was nothing if not persistent, and it appears that neither Seddon's death nor the passage of years diminished his sense of injustice. After completing his service, Hutchison came back to New Zealand, although when his attempts to verify his Fourth Contingent attestation proved unsuccessful he returned to South Africa to work.[307] Following an approach by Hutchison in 1915, Defence Minister James Allen looked into Hutchison's claim that his attestation papers had been wilfully destroyed, but Allen indicated that without compelling evidence his hands were tied.[308] As intriguing as the idea of an incumbent New Zealand premier having official documents destroyed for political advantage may seem, there is no evidence supporting Hutchison's claims. Allen was no ally of Seddon and had no apparent reason to conceal irregularities if proof of them existed.

Soldiers of the queen

Although the term 'ANZAC spirit' was coined following the First World War Gallipoli debacle, the camaraderie underpinning the Australian and New Zealand military relationship first coalesced more than a decade earlier on the battlefields of South Africa. Just as New Zealanders and Australians fought side by side in the Dardanelles, so too had they fought at Langverwacht and in numerous other South African engagements. Corporal Francis Ryan of Christchurch, who later became a lieutenant and was killed in action at Paardeplaats, wrote of spending 14 hours pinned down by gunfire under a burning sun at Rhenosterkop with a corporal from New South Wales who was attached to the Third Contingent.[309] To the New Zealanders' right, Queenslanders endured similar conditions. A war correspondent with a penchant for alliteration reportedly said of the engagement at Rhenosterkop that 'the behaviour of the New Zealanders and Australians was one of the finest features of the war. Finer fighters never faced a foe.'[310]

Though he arrived in New Zealand as a child, Major Alfred Robin was born in Australia, and DCM recipient Farrier-Sergeant William Rouse of the Sixth Contingent was from Melbourne.[311] Another Australian serving alongside the New Zealanders was Archibald Wookey of the Fifth Contingent, who was killed at Doornbutt while attempting to escape his Boer captors.[312] Wookey was reportedly shot after killing one Boer and wounding another with a revolver.[313] Though Wookey gave his address as Tinwald in Canterbury when he attested, he was a Tasmanian.[314] Scouting ahead of his convoy, Corporal Angus McCartney (who was among those wounded at Ottoshoop) assumed men spotted on a kopje were Boers. Cautiously approaching, he discovered they were New South Wales Bushmen and, while conversing over a billy of tea, McCartney realised that one of the heavily bearded Australians was an old family friend.[315]

Several Australian units also included New Zealanders who were either Australian residents or travelled there to enrol. When Duncan Sinclair's attempts to join the Second Contingent were stymied by his lack of Volunteer experience, he enrolled in the New South Wales Citizens' Bushmen's Contingent and was subsequently mentioned in Baden-Powell's dispatches for his actions at Zeerust.[316] Lamenting the inflexible 'redtapism' that had denied Sinclair a place in a New Zealand contingent, the *Manawatu Evening Standard* described Sinclair as 'a worthy son of the Empire' and expressed pride that the gallantry of one of the region's own had been recognised.[317] Despite serving in an Australian force, Sinclair maintained his connections to the Manawatū region through letters.[318] Describing the South African landscape, he likened the vegetation to taramea (speargrass) growing in the Ruahine Range and the sand to that on Rangitīkei Beach.[319]

Trooper T. J. Reid, whose father lived in Mākino, served as a trooper in the 2nd New South Wales Mounted Infantry, while Edwin Gatland of Coromandel was living in Australia when he joined the Queensland Mounted Rifles.[320] Like Gatland, 20-year-old Wanganui Collegiate School Old Boy Bertram Finn was also resident in Australia when he joined the Fifth Victorian Mounted Rifles.[321] Finn's force suffered one of the worst Australian defeats of the war after Boers surrounded their camp under cover of darkness and opened fire at point-blank range. Relaxing around their fires, the men were easy targets. Finn described the pandemonium that followed and the sombre sight that greeted the survivors at sunrise: 'what an awful sight met our eyes, our dead and horribly wounded comrades, stretched on that frozen scene of horror'.[322]

New Zealanders were also present in the forces of other former British colonies. After leaving New Zealand for Canada, Henry Andrews of Greymouth enrolled in the Royal Canadian Regiment and was wounded during the Battle of Paardeberg. In an article titled 'Plucky Act by a New Zealander', the *Otago Witness* reported that Andrews and two companions moved a wounded officer to safety under heavy fire, though Andrews was himself shot in the leg.[323] Individuals with New Zealand connections could also be found among the Boer forces. Several New Zealand newspapers reported that Boer leader Commandant J. H. Olivier had lived in Māngere for a time.[324]

Significant numbers of New Zealanders such as Frederick Hughes served in irregular corps raised in South Africa. Among the 23 grooms selected by the Agriculture Department to travel aboard the *Tropea* were Walter Glass, Walter Keddell and Walter Bayley, as well as Alfred Emery and Alfred Collins.[325] For the majority of the grooms, the job was a convenient means of getting to South Africa where they could join military units. Of the 23, at least 15 joined Brabant's Horse.[326] Some, like Glass and Corporal Kerr Maxwell, joined other New Zealand contingents after serving their time in Brabant's Horse, while Trooper Archibald Hutton joined the Third Contingent soon after his arrival.[327]

An incongruous figure among the *Tropea* grooms was 59-year-old John Taylor Marshall, who became a sergeant in Brabant's Horse shortly after the vessel docked.[328] In the late 1850s, Marshall had joined the 2nd Dragoon Guards and served with the regiment in India.[329] On arrival in New Zealand, he served under Major-General Trevor Chute in 1866 before joining the New Zealand Armed Constabulary in 1870, and in 1874 commanded the Militia and Volunteer District of Nelson.[330] Injuries Marshall sustained when a Cobb & Co. coach overturned did not end his military career, and in 1886 he became the commanding officer of the Te Aro (Wellington) Rifle Volunteers.[331] It is tempting to imagine Marshall regaling the younger grooms with tales of the Raj and Waikato bush fighting as the *Tropea* slowly steamed towards Africa.

Of the 48 grooms who sailed for South Africa aboard the *Ormazan* with her cargo of horses, several went on to join irregular forces and New Zealand contingents.[332] On arrival in Africa, Albert Bezar enrolled in Kitchener's Horse, and later served both in the Fifth Contingent and as a sergeant in Scott's Railway Guards.[333] Otto Cook also joined Kitchener's Horse soon after the *Ormazan* docked at Port Elizabeth, and then served as a sergeant in the Eighth Contingent.[334] Twenty-three-year-old Frank Jackson accompanied his friends Fred Carter and Stanley Scott to South Africa with similar intentions. The trio made their way to Sydney, where they secured passages aboard the *Hyson* as grooms, but on arrival in Port Elizabeth all three enrolled in Brabant's Horse.[335] Scott also served in the Utrecht and Vryheid Mounted Police and as a lance-corporal in the 1st Scottish Light Horse, but he died of disease in Durban the month after hostilities ended.[336] In late 1902 Minister of Native Affairs James Carroll unveiled a drinking fountain in Ponsonby in Scott's honour.[337]

For some who failed to enrol in contingents, stowing away aboard vessels bound for South Africa became an attractive option. The captain of the troopship *Waiwera* landed three stowaways when his ship reached Albany, though one, 22-year-old William Henry White of Pleasant Point, was sworn in as a member of the Second Contingent while at sea.³³⁸ White had reportedly paid a crew member £5 to conceal him, but after being discovered and working in the ship's engine room he was permitted to carry on to South Africa.³³⁹ Fellow Pleasant Point resident Leonard Worthington of the Temuka Rifle Volunteers was another *Waiwera* stowaway.³⁴⁰ Worthington, a South Canterbury Acclimatisation Society ranger 'of the tough stamp', was a good shot, but his attempt to enrol for active service failed.³⁴¹ Though the press admired his 'valiant efforts', he did not attest at sea but joined Brabant's Horse on reaching South Africa.³⁴² Worthington later joined the Eighth Contingent and returned to Africa, where he was sentenced to 10 days' pack drill after being caught gambling.³⁴³ In 1914 he enlisted in the New Zealand Veterinary Corps but in 1916 returned to New Zealand a 'prisoner incorrigible' following repeated alcohol-related incidents in Egypt.³⁴⁴ Discharged in New Zealand, he re-enlisted in the Wellington Regiment and served in France, receiving the Military Medal for gallantry in the field.³⁴⁵

Although some admired stowaways' resourcefulness and determination, the editor of the *Otago Daily Times* considered they undermined military discipline, set a bad precedent for others and made a mockery of regulations. He contrasted men like Worthington with those who bore their disappointment with good grace after unsuccessfully attempting to enrol: 'The military stowaway is an undisciplined egotist, and no sentimental leniency should be extended to him.'³⁴⁶ The *Press* reported that every effort was made to prevent men concealing themselves aboard contingent vessels and described the 'stowaway nuisance' as serious.³⁴⁷ It expressed concern for the fate of young New Zealanders left in Sydney after being discovered aboard the *Norfolk* and claimed that stranding stowaways in foreign ports without support would force them into a life of crime.³⁴⁸ But public admonishments failed to dampen the enthusiasm of those determined to get to South Africa, and stowaways continued to be discovered aboard contingent vessels.³⁴⁹ When two Kaiapoi youths slipped aboard the *Monowai* in their Volunteer uniforms, the *Star* expressed none of the reservations exhibited by its Otago counterpart. The newspaper admired their eagerness to get to the front and claimed it would be a pity to dampen the ardour of such enthusiastic volunteers.³⁵⁰

Attempts to gain free passage to South Africa were not limited to men. Shortly before the *Maori* weighed anchor in Worser Bay, a girl discovered on board dressed as a trooper resisted attempts to remove her.[351] She was reportedly an orphan who had spent time at the Fifth Contingent's camp. On reaching Beira, the *Star*'s correspondent described the voyage with the 'Fighting Fifth' as uneventful and claimed the trip would have been 'more romantic if the stowaway of the gentler sex had not been so promptly discovered and put ashore'.[352]

The age of consent

Although Seddon originally indicated that volunteers aged between 23 and 40 would be considered for the First Contingent, this had been relaxed by early 1900.[353] In a newspaper advertisement, Lieutenant-Colonel Banks sought unmarried recruits aged between 21 and 40 who were skilled riders, good shots, at least five feet six inches tall and no more than 168 pounds in weight. In practice, age restrictions were only loosely applied, and when the final contingent left New Zealand shores it was a poorly kept secret that underage recruits were gaining admission. By the end of the war, instead of the 17 years that Seddon originally envisaged, a 54-year gap separated the youngest troopers and the septuagenarian Bakewell.

In a light-hearted moment during a Tenth Contingent luncheon, the premier noted that contingent members were supposed to be over 20, but he added that if he instructed everyone in the room under 20 to stand, many recruits would have to rise to their feet. When there was fighting to be done, Seddon claimed, 'the boys of New Zealand were like the ladies — they would not tell their age'.[354] The premier told an underage recruit he could not sail with the Tenth Contingent, but then enrolled him in the Ninth Contingent, where the minimum age was 19.[355]

Seddon was also willing to bend the rules regarding height restrictions. At a Heretaunga Mounted Rifles send-off for Eighth Contingent recruits, a soldier stood while Major Loveday explained his predicament to Seddon. After hearing that the soldier was slightly shorter than the minimum height, Seddon reportedly informed Loveday, 'that young man has grown half an inch while you were speaking' and then indicated that if the recruit visited him the following day he would personally enrol him.[356]

Due to the common practice of recruits inflating their ages when enrolling, identifying the youngest New Zealander to serve in either a contingent or an irregular force is problematic. Possibly the youngest to serve in a contingent was 16-year-old Alan Saunders, who made no attempt to disguise his age when he enlisted in the Sixth.[357] Though not enrolled as a contingent member at the time, Saunders was only 15 years old when he sailed with the Second Contingent as Surgeon-Captain Percival Fenwick's assistant.[358] Despite his age, Saunders apparently went with the blessing of at least one of his parents, Samuel Saunders of the *Lyttelton Times*. Though his initial attempts to join a contingent were rebuffed due to his youth, Saunders was appointed a hospital orderly, and after being invalided home he enrolled in the Sixth Contingent, giving his correct age of 16, and returned to South Africa.[359] When he was discharged in 1902 with an 'exemplary' character assessment, Saunders was still only 17 years old.[360] Newspapers referred to him as 'A Plucky Boy' and claimed he was 'the youngest colonial soldier' and the youngest soldier serving in the entire British Army in South Africa.[361] Possibly in response to newspaper reports that Seddon was considering dispatching a contingent to assist in suppressing the Boxer Uprising, on his return Saunders expressed a desire to travel to China.[362]

Although New Zealand newspapers referred to 'Trooper Louis Peat, from Kimberley', and following the death of his father in March 1900, 'Louis Peat' received government financial assistance for passage to New Zealand, Harcourt Eugene Louis Peat's name does not appear on the medal rolls of Kimberley's defence units.[363] Nonetheless, Peat was born in the same year as Saunders and was 15 years and eight months old when the siege of Kimberley began.[364] If he did attest in one of Kimberley's defence corps, Peat was possibly the youngest New Zealand soldier to serve in an irregular unit in South Africa.

John Duigan and Henry Vercoe, a trooper with Māori ancestry, were also only 16 when they joined either irregular corps in South Africa or New Zealand contingents.[365] Although Vercoe claimed to be 21 when he enrolled in the Seventh Contingent, if a 1946 military file is correct, he was 16 years and seven months old (three months younger than Saunders when he enrolled in the Sixth).[366] Vercoe, who later served in the Ninth Contingent, appears to have just celebrated his eighteenth birthday when he was discharged from the regiment in 1902.[367]

Duigan worked his passage to South Africa and enlisted in Brabant's

John Evelyn Duigan, who as a 16-year-old worked his passage to South Africa, where he enlisted in Brabant's Horse. After being seriously wounded in action at Wepener, he was invalided back to New Zealand but returned to South Africa where he joined the 2nd Battalion of Kitchener's Fighting Scouts. After being wounded a second time he again returned home, but then sailed for South Africa as a lieutenant in the Tenth Contingent. Duigan served as a staff officer during the First World War and in 1937 was appointed general officer commanding and chief of the general staff in the New Zealand Army. THE COLLEGIATE SCHOOL, WANGANUI, IN SOUTH AFRICA, 1899–1900, WANGANUI: A.D. WILLIS, PRINTER, [1901], N.P., WANGANUI COLLEGIATE SCHOOL MUSEUM

Horse together with Jack Jensen of Whanganui, and James Mowlem and Alexander Peddie of Palmerston North.[368] Although hit by a shell fragment and receiving two bullet wounds that were initially considered life-threatening, Duigan recovered sufficiently to return to Brabant's Horse.[369] Nonetheless, he appears to have grown weary of the war and said he would be 'glad to get out of it' and return to New Zealand — a wish that came true when he was again incapacitated and invalided home.[370] Although he had tired of soldiering, life in Whanganui apparently held even less appeal. After being denied a place in the Sixth Contingent due to his youth, Duigan once again worked his passage back to Africa aboard the *Kinclune*.[371] After the vessel docked in Cape Town and unloaded its cargo of oats, wheat and barley, Duigan enrolled in the 2nd Battalion of Kitchener's Fighting Scouts and soon found himself again under fire.[372] While scouting, Duigan's party encountered a much larger force of Boers near Heilbron and was compelled to surrender.[373] After being released by his Boer captors, Duigan again returned to New Zealand, where he received a lieutenant's commission in the Tenth Contingent.[374] The young officer, who later became chief of general staff of the New Zealand Army, had just turned 19 when the contingent set sail for Africa.

'I don't care if I do get the damned thing'

Even before the First Contingent returned to New Zealand, politicians were discussing how to reward soldiers for their service, but when Joseph Witheford mooted the idea of the government calling for competitive designs of a medal or clasp, Seddon's response was initially cautious.[375] Concerned that the government issuing medals could set a 'dangerous precedent', the premier recommended the matter remain in abeyance.[376] Yet within a month his position had changed. While rejecting a proposal that contingent members receive a medal entitling them to free rail travel for life, Seddon announced the government's intention to have a New Zealand medal struck for contingent officers and men.[377] When asked whether he would seek permission for the proposed award to be worn by New Zealanders serving in the British Army, he replied that it was not required. It was, he said, intended as a 'New Zealand medal'.[378] The premier indicated that New Zealanders in the British Army would receive two medals: one imperial and the other from the New Zealand government. After Robert Rhodes pointed out that if a New Zealand medal was issued, *all* New Zealanders in the contingents would be entitled to two medals, Seddon concurred, but he assured Rhodes that the New Zealand medal was for New Zealand contingent members and for them alone.

While a number of unofficial medals were presented to veterans, the proposal to issue an official New Zealand medal went no further.[379] For the vast majority of contingent members, South African service was rewarded with the balance of their pay, a £5 colonial gratuity and either one or two imperial campaign medals. Acts of gallantry were, however, sometimes recognised with higher awards. In 1901 Farrier-Sergeant William Hardham became the only New Zealand soldier to receive the Victoria Cross (VC) for bravery in South Africa. When Hardham's section encountered Boers near Naauwpoort, he rode to the assistance of his friend Trooper Jack McRae after McRae was wounded and had his horse shot from under him. Under heavy fire, Hardham placed McRae on his own horse and then led him to safety while running alongside.[380]

Sergeant-Major Ernest Lockett, Trooper Ivanhoe Baigent, Lance-Corporal

Francis Ryan and Trooper Duncan McLaren were all involved in similar incidents, though none were considered to have met the criteria for the VC. This was also true of Sergeant-Major Walter Callaway and Lance-Corporal Henry Vercoe, men with Māori heritage who rescued two dismounted troopers under heavy fire at Witkop but were only mentioned in dispatches.[381] In a skirmish at Patriotsfontein, Lieutenant George Leece and Sergeant-Major Daniel Love were killed, while Trooper John Helm and Lockett were seriously wounded.[382] Lockett received the DCM for rescuing a New Zealand prisoner during the engagement.[383] Baigent also came to the assistance of a comrade under fire and was sufficiently irked that his 'little adventure' only earned him the DCM, rather than the VC, that he petitioned Seddon expressing dissatisfaction.[384]

During operations near Wepener corporals Frank Hemphill and John Page and troopers John Law, Horace Strange-Mure, Henry Strawbridge, Ralph Letts and Baigent volunteered to carry dispatches to an adjacent column.[385] After riding for several hours, the party was challenged by Boers at 'Bastard's Drift' on the Caledon River. When Hemphill reportedly opened fire, the Boers fatally shot Strawbridge, grazed Page with a bullet that passed through his hat, and captured Letts.[386] Although Baigent escaped unscathed, after Hemphill's horse was shot the trooper came to his corporal's rescue 'under a murderous fire'. According to Baigent, he had been recommended for the VC by General Plumer (a claim supported by Trooper Leonard Law in a letter to his mother).[387]

A year earlier, Trooper Hugh Grahame praised the bravery of Lance-Corporal Francis Ryan during the Boer ambush at Roode Kopjes where Luke Perham was killed and Lieutenant Bradburne was mortally wounded.[388] According to Grahame, when Ryan saw that Grahame's horse had been hit, he galloped in front of the enemy, seized a wounded horse and attempted to lead it to Grahame. When the scared animal refused to face the gunfire, Ryan held it within range of the Boers until Grahame reached him. Grahame claimed that Ryan immediately rode back across the Boer firing line a second time and rescued Trooper Charles Lusk, whose horse had also been hit. With both Ryan (who was later killed in action at Paardeplaats) and Lusk on the lance-corporal's horse and Grahame on the wounded animal the trio retired out of range.[389]

Though similar to Grahame's account, Ryan's version, in which he claimed Grahame and Lusk ran to him while he held two horses, illustrates how perceptions could differ in the confusion of battle.[390] While retreating, the trio attempted to rescue Bradburne, but as they were drawing enemy fire he told them to leave him. Grahame later said, 'No man ever earned the V.C. more honestly or gamely than Ryan, but just because the only people to see these actions were troopers he gets nothing.'[391] Though Ryan was mentioned in dispatches, he noted that Surgeon-Captain Sidney Godfray also braved the intense fire to treat Bradburne.[392]

Even the support of an officer was no guarantee that a soldier would receive official recognition of valour. In a memorandum to Lieutenant-Colonel Porter, Major Arthur Bauchop drew attention to the bravery of Trooper Duncan McLaren.[393] During operations in Ermelo district McLaren's regiment was forced to retreat under fire, but seeing a comrade in difficulty the trooper immediately rode to his aid. Bauchop stated that the passage of time had not diminished his admiration for McLaren's actions, which he ranked among 'the most courageous'.[394] Though McLaren was promoted to corporal, he does not appear to have received an award.

More aggressive actions also received official recognition, and in 1901 Ryan praised the conduct of Private Mark Pickett at Rhenosterkop after the Poverty Bay trooper charged a Boer position concealed among rocks.[395] Pickett, who was Australian, was later promoted to sergeant-major and received the DCM.[396] Corporal O'Dowd and privates Rumble and Drinnan were mentioned in dispatches by General Babington and promoted after charging a Boer gun under heavy fire.[397] During the same action at Vaal Bank, Private Frederick Wylie, who was later killed in action at Klipfontein, was promoted to corporal after charging another gun and killing two of the four Boers who were defending it.[398]

New Zealand nurse Mary Warmington was clearly proud of her wartime service and wrote to the authorities enquiring about her entitlement to the Queen's South Africa Medal.[399] Unlike the majority of New Zealanders who served in contingents or irregular corps, the medals awarded to nurses such as Warmington, Mabel Brooke-Smith, Geraldine Jeffreys and Nellie Redstone were issued without clasps, though Redstone and Janet Williamson received their awards directly from King Edward VII.[400] In Williamson's case, the king presented her with the prestigious Royal Red Cross for exceptional devotion to duty in South Africa.[401]

One of the more unusual forms of recognition received by a New Zealander for courageous conduct was the 'Queen's Scarf' awarded to Trooper Henry Donald Coutts of the First Contingent. Amid the confusion of the Koorn Spruit withdrawal, Coutts hoisted a seriously injured British lance-corporal onto his saddle while under fire and carried him to safety.[402] Coutts was selected as the sole New Zealand recipient of one of several scarves knitted by Queen Victoria and awarded for 'conspicuous gallantry'.[403]

Although Baigent received a £20 gratuity as a DCM recipient, with the exception of author, journalist and soldier Sergeant Arthur Vogan there is little evidence that financial reward was a motivation for men who carried out acts of bravery.[404] Like Baigent, Vogan felt short-changed by the imperial authorities after they downplayed his role in the capture of Boer commandant Gideon Scheepers. When war broke out, the London-born Vogan was in Australia, but after making his way to South Africa he joined the New Zealanders' squadron of the Prince of Wales' Light Horse.[405] While scouting ahead of a party of 10th Royal Hussars, Vogan reportedly entered a farmhouse where he found the seriously wounded Scheepers.[406] Following Vogan's return to New Zealand, where he had reportedly lived since the 1880s, he wrote to Joseph Ward complaining that he had received no reward, despite a senior officer promising to pay £100 for Scheepers' capture and Vogan reportedly being recommended for a gallantry award.[407]

Raising Vogan's case in Parliament, William Herries asked whether the government intended taking action.[408] Acting Defence Minister William Hall-Jones replied that where possible he would assist New Zealanders who served in irregular corps but stressed the government bore no responsibility for them. Vogan's complaints were dismissed by a senior officer in South Africa, who claimed Scheepers had intentionally been left by his men due to his condition.[409] Undeterred, Vogan railed against what he described as the 'gross injustice' of the authorities' refusal to acknowledge his actions, and the 'base ingratitude' of the English public for their treatment of men who, despite serving in irregular corps, had not received their medals.[410]

Vogan's claims were not entirely unfounded, as his own campaign medal was not issued until 1904.[411] For others the wait was even longer; Corporal Allan Aislabie of Brabant's Horse received his medal in 1908, while Harold Nunneley (one of Hughes' recruits) and James Christie of the Bushveldt Carbineers did not receive their awards until 1909, though the enormity of the task made delays inevitable.[412] Most individuals who served in British, colonial and irregular forces in South Africa (as well as non-combatants such as war correspondents, nurses and doctors) received the Queen's South Africa Medal, while some were eligible for the King's South Africa Medal following Queen Victoria's death. Locating former soldiers and correctly issuing their medals was a daunting task that taxed Defence Department resources. Captain James Clark, who was ordered to compile rolls for the King's South Africa Medal as well as the clasps for all 10 contingents, faced the laborious job of assessing the eligibility of over six thousand officers and men, many of whom had served in more than one contingent, changed addresses or emigrated.[413]

While men like Vogan waited years to receive campaign medals, the authorities pursued Baigent after he received his Queen's South Africa Medal twice. When he was awarded his DCM in London he also received his campaign medal, but a duplicate medal was mistakenly issued two months later. Perhaps resentful over his failed attempts to receive the VC that he believed would enhance his prospects in New Zealand, Baigent surrendered the second medal only after repeated Defence Department requests and being visited by police in 1909.[414] He could not resist a parting shot at Seddon's son at the Defence Department: 'As you seem to attach so much value to this small article you will kindly acknowledge receipt.'[415] Despite the dispute, Baigent received an officer's commission during the First World War; he died fighting in Palestine.[416]

The importance of campaign medals increased when the men who earned them died during the war. After his son's accidental death, Thomas Byrne unsuccessfully sought permission to wear William Byrne's medal, while Catherine Bruce (whose son David was among the Langverwacht dead) thanked the governor for sending her son's medal, assuring Ranfurly it would be 'greatly treasured'.[417] Following the conflict, Sir John Stewart asked in Parliament what steps were being taken to ensure that families of men who had been killed in action, died of their wounds or died of disease received their relatives' medals.[418]

Some contingent members expressed resentment that men from the Ninth and Tenth contingents received medals despite never having been in action. Vogan claimed that a trooper who had served with distinction had expressed frustration about delays in receiving his medal: 'I don't care if I do get the d[amned] thing now. They can keep it. Medals given to men who arrived too late to see a shot fired are of no value to me.'[419] Frank Hemphill felt he was more entitled to the King's South Africa Medal than men who had arrived in South Africa 'as Peace was proclaimed'.[420] Though Hemphill did not receive the medal he sought, he did receive an unofficial award for his actions at Bastard's Drift. A Cape Town newspaper, the *Veld Weekly*, sent an engraved gold Veld Cross of Honour that was forwarded to the mayor of Whangārei for presentation to Hemphill in 1903.[421]

Though medals were a source of pride, some former troopers also valued the character assessments that appeared on their discharge certificates. The issue was raised in the House by Waipawa MHR Charles Hall, who said that he had been approached by former contingent members wishing to return to South Africa.[422] The men were concerned that their discharge papers gave no indication of their character and felt this could affect their chances of re-enlisting. Seddon replied that giving character details had been impossible in situations where men had returned from South Africa while their officers remained overseas. He assured Hall that as this was no longer the case, soldiers would receive discharge papers with character assessments. After returning to New Zealand, Bugler John Murray disputed his 'fair' character assessment, suggesting that a mistake had occurred resulting in his complexion being recorded instead of his character rating.[423]

Regardless of these post-war issues, New Zealanders generally performed well in action, and although many lacked experience, they adapted well to the challenges of their mounted infantry and scouting roles. By travelling light, often moving at night and supplementing their supplies with livestock and produce looted from Boer farms, they often acted with a degree of independence from the slow-moving British infantry and its ponderous supply trains. Yet if the New Zealanders' adaptability was a strength, their limited training was not. The deaths of Corporal Byrne and Private Raynes may well have been avoided had New Zealand soldiers received better weapons training.

Nonetheless, most New Zealanders had little interest in reports of their soldiers' shortcomings. Though Raynes' death was reported as accidental, in at least three government documents Byrne was listed as 'killed in action' — a fiction repeated in the press.[424] Though it didn't elaborate, the *Manawatu Evening Standard* was one of the few papers to record Byrne's death as accidental.[425] The same newspaper had earlier adopted a supercilious tone when reporting the expected arrival of several imperial sergeant-majors and their wives in 1901: 'These gentlemen are being brought out to the colony to teach New Zealanders methods of warfare, which the New Zealanders are teaching the British regulars to unlearn in South Africa.'[426]

Yet at New Zealand Hill the New Zealanders fought in the same manner as the Yorkshire Regiment regulars alongside them. There was nothing revolutionary about soldiers repelling an enemy attack at bayonet point; British Army regiments had been doing so for generations and Major Robin noted it was the Yorkshiremen who had first fixed bayonets and forced the Boers to seek cover. Robin said that it was only when they saw their officer wounded and their sergeant killed that they wavered.[427]

A feature of the New Zealanders' style of fighting was their willingness to make use of terrain when attacking or defending a position. Robin touched on the importance he placed on this aspect of modern warfare: 'I have at all times tried to ingrain into our men the necessity of taking or making intelligent cover.'[428] Regardless of their failings, the New Zealand contingents and the majority of New Zealanders in other corps displayed a level of competence in the field that more than satisfied the British officers under whom they served. The praise New Zealanders in South Africa received for their actions would remain one of their most enduring legacies.

CHAPTER FOUR

'LOYALTY TO THE BRITISH EMPIRE'

MĀORI RESPONSES TO THE SOUTH AFRICAN WAR

When the South African War began it had been almost 60 years since the Treaty of Waitangi gave Māori the rights of British citizens, though the intervening period had hardly been harmonious. Following the more localised Northern War in New Zealand in the 1840s, Crown–Māori relations were further tested by sovereignty and land disputes, as well as the rise of the Kīngitanga (the Māori King movement) and Māori spiritual groups such as Pai Mārire. During the 1860s full-scale war erupted, resulting in the Crown confiscating sizeable tracts of land from Māori deemed to be in rebellion. Despite having their own grievances, members of iwi such as Te Arawa, Ngāti Kahungunu and Ngāti Rangitihi allied themselves with British and colonial forces and provided military assistance. To facilitate this, the government issued approximately 3500 firearms to 'Friendly Natives' between 1865 and 1868.[1]

Nonetheless, Māori who supported the Crown did not always receive equitable treatment. The 1866 Military Pensions Act established pensions for officers and men of the Colonial Forces enrolled by and serving under the government who had sustained wounds and injuries while on active service. Although conditional, the legislation also provided for widows and dependants of men who were either killed in action or died of their wounds or illnesses contracted in the field. The act defined 'Colonial Forces' as all men, whether Māori or Pākehā, serving under the New Zealand government, but in most cases it provided lower pension entitlements for Māori.[2]

In addition, despite the military support the government received from

Māori, and a significant number of New Zealanders sharing both Māori and European ancestry by 1899, many Pākehā displayed racist attitudes towards Māori and continued to treat them with suspicion and disdain.

Post-1860s Māori–Pākehā conflict

The end of the fighting that played such a significant role in shaping Māori–Pākehā relations did not signal an end to conflict. On more than one occasion in the decades that followed, the government used military force to suppress what it considered to be Māori civil disobedience. Events such as the 1881 invasion of the peaceful settlement of Parihaka (described in the press at the time as 'a stronghold of fanaticism and disaffection'), the arrest of Māori prophets Te Whiti-o-Rongomai and Tohu Kākahi, and the detention of significant numbers of Māori without trial heightened tensions.[3] Eleven years later, police again raided Parihaka and arrested dog tax defaulters, with a New Plymouth magistrate stating, 'It would never do to let the Maories [sic] think that the arm of the law cannot reach Parihaka.'[4]

The 1880 Dog Registration Act was extremely unpopular with Māori, who in many cases owned multiple dogs and had been specifically excluded from the provisions of the earlier 1844 Dog Nuisance Act. Māori, whose dogs were often used for hunting, felt that the controversial tax levied per animal placed an unfair financial burden on them. The 1880 legislation removed the exemption for Māori and required most dog owners to register their animals and pay a registration fee.

In 1889 armed members of the Permanent Artillery sailed to Wharekauri Chatham Islands to support local officers collecting the dog tax from Māori.[5] The Commissioner of Police, Walter Gudgeon, advised that 'powerful' men should be selected for the trip as he expected there might be 'a rough & tumble'.[6] Gudgeon also recommended keeping the men in the dark regarding their destination to avoid the trip being 'noised about'. Wharekauri Māori continued to oppose the dog tax, and during the South African War the government again dispatched an armed force to the islands.[7]

In the 1890s the military were again used to enforce the law. Members of the Permanent Artillery, Crown officials and a police constable sailed to Te Hauturu-o-Toi Little Barrier Island to remove the residents, including the Ngāti Manuhiri chief Tenetahi,[8] after the Crown acquired the island through the 1894 Little Barrier Purchase Act. Although Tenetahi and other Māori had initially agreed to sell Te Hauturu-o-Toi to the Crown, some, including Tenetahi, had a change of heart.[9] Although the 1894 legislation vested Te Hauturu-o-Toi in Crown ownership, Tenetahi continued to reside peacefully on the island despite being instructed to leave.[10] Two years after the act's passage the under-secretary for Crown lands sought Defence Department assistance in evicting the Te Hauturu-o-Toi residents.[11] Landing at dawn, the force surprised Tenetahi, his wife and a handful of others, who were transported to the mainland after hastily gathering their possessions.

The dog tax continued to be a divisive issue in 1897 when several Northland Māori were imprisoned after refusing to pay.[12] The local constable described the defaulters as 'fanatics' and alleged they had been collecting arms and ammunition.[13] The following year matters came to a head at Waima in Hokianga when armed Māori again refused to pay taxes.[14] In a show of force in a region where the Crown's exercise of sovereignty remained superficial, the government dispatched over one hundred Armed Constabulary officers and Permanent Force men armed with machine guns and artillery. In an example of gunboat diplomacy, the Royal Navy sloop HMS *Torch* anchored in Hokianga Harbour.[15] Fighting was narrowly avoided after Hōne Heke Ngāpua, MHR for Northern Maori, persuaded 16 Māori to surrender their weapons. After being charged with 'intending by conspiracy to levy war against the Queen' and using force to prevent the collection of taxes, four were sentenced to 18 months' imprisonment with hard labour, while the remainder received fines.[16]

Shortly before the South African War, the minister of native affairs noted that Māori who had been hostile during the disturbances at Waima were again beginning to take an interest in the Native School at the nearby settlement of Ōmanaia.[17] This incremental thaw in relations continued during the war when Seddon received a letter from Hōne Tōia, one of the main protagonists in the Waima confrontation. Following the Duke of York's 1901 visit to New Zealand, Tōia reportedly sought 'atonement' for his actions at Waima by offering a force to 'go to any part of the world to fight for the King'.[18]

Nonetheless, the government remained suspicious and attempted to regulate weapon sales to Northland Māori by taxing storekeepers who sold Māori firearms. The minister of justice described the tax as the only viable means of controlling the sale of arms and ammunition to Māori.[19] Following a wartime visit to Hokianga, Lord Ranfurly reflected the Crown's lingering mistrust of Northland Māori when he notified the secretary of state for the colonies that he had encountered Boer sympathisers among Māori in the area:

> *I found a Pro-Boer element which had been fostered by some local Dutch and German Priests working chiefly among the Maoris. As this was the seat of the disturbance about two years ago, and the Natives were not too friendly at any time with the Europeans, I deemed it advisable to inform my Government, who are now enquiring into the matter.*[20]

Two years after the Hokianga confrontation, Father Smiers, a Dutch Catholic priest in Whangārei, indignantly denied accusations that he was pro-Boer.[21] The *Northern Advocate* reported that it had received two letters accusing a 'minister of religion' of stirring up pro-Boer feeling in the region. The newspaper noted that if the accusations were true, steps could be taken to stop the expression of anti-British sentiments.[22] In response, Smiers asserted his rights under British law and offered five guineas to anybody who could substantiate the claims made against him.[23]

'Superiority to racial prejudice'

Shortly before the South African War, City of Wellington MHR John Duthie publicly opposed a government proposal to restrict Māori land transactions to leaseholds. Addressing a Westport meeting, the politician said, 'Fancy a free-born Britisher having a nigger for a landlord.'[24] Following criticism, Duthie conceded that his statement was 'inaccurate' as Māori were 'not negroes and to so speak of them was a regrettable use of a common vulgarism'.[25] Duthie was correct in his assertion that the use of racist language was commonplace. Almost four decades earlier, Native Secretary Donald McLean drew attention to European attitudes toward Māori in the North Island:

> *The threats, curses, and opprobrious epithets used by Europeans towards [Māori] confirm their worst suspicions. The offensive terms 'bloody Maori,' 'black nigger,' [and] 'treacherous savage,'*

are frequently applied to them, and though uniformly kind and hospitable to all strangers, they are themselves often treated with cold indifference, and sometimes with contempt when they visit the English towns.[26]

By 1891 little had changed, with an Auckland newspaper referring to Whatiwhatihoe, the Waikato pā of King Tāwhiao, as a 'nigger metropolis'.[27] Once the war started, there were men in the contingents who harboured equally racist views. Apparently excluding soldiers with Māori ancestry who fought alongside him, Trooper Hugh Ross wrote: 'The Boers make the niggers go in all the warm corners. We won't allow any coloured men to fight for us, as it is a white man's war . . .'[28] The author of a 1900 letter to the *Evening Post* also opposed the use of 'savages' to fight in South Africa: 'No slight is intended to the Maoris [sic], who are a grand race, but it is difficult to draw the colour line in our piebald Empire.'[29]

Only days earlier the newspaper's editor had predicted that if Europeans were to employ 'yellow and black troops in their wars with one another, the end of European civilisation would be within measurable distance'.[30] New Zealand newspapers hypocritically accused the Boers of racial prejudice, but focused on their dislike of the British rather than their treatment of black Africans.[31] Discussing the African American 'problem' in the United States, the press applauded President Roosevelt for inviting 'a negro' to dine at the White House.[32] An *Auckland Star* editorial claimed it was difficult for New Zealanders to understand the significance of Roosevelt's actions because they had 'long since proved [their] superiority to racial prejudice'.[33]

But New Zealand had no cause for sanctimony. Many New Zealand soldiers considered themselves superior to the racial groups they encountered in South Africa, with black Africans often seen as little more than objects of amusement.[34] The offensive term 'kaffir' regularly appeared in New Zealand newspapers before and after the war, with the New Zealand government using the expression in a 1904 official report listing prohibited immigrants.[35] In 1900 the *Otago Witness* published a photo of Fourth Contingent sergeants with 'Kaffir Joe', a black African, sitting at their feet.[36] As contingent members often employed indigenous Africans to do menial tasks, 'Joe' was in all likelihood a servant.[37] Even the Eighth Contingent's commanding officer was granted temporary permission for a young Zulu orphan aboard the *Britannic* to land in New Zealand as his 'personal servant'.[38]

In a letter published in the *Wanganui Collegian*, a Fourth Contingent sergeant related how sentries had shot and killed a cow that had strayed close to their lines at night. He noted that his men had previously 'flattened a nigger' under similar circumstances, adding that the troopers 'reckoned they were quite right as native game shooting starts on that day'.[39] Even Edward Renata Broughton, a Ninth Contingent member with Māori heritage, appears to have been unimpressed by some of the ethnic groups in South Africa, describing Ladysmith as the dirtiest town he had ever seen, 'chiefly populated by Jews, Persians, Hindoos [sic] and Kaffirs, with a sprinkling of Dutch'.[40] Lieutenant-Colonel Newell of the Fifth Contingent described the African Matabele he encountered as 'a small people, not nearly so fine as the Maoris [sic]'.[41] Unlike Newell, most references in diaries, letters and newspaper articles did not differentiate between African groups. In a letter home, one man described black Africans as 'thick-headed darkies', while a Masterton trooper who did make a distinction between 'kaffirs and blacks' described the former as 'a filthy people' before noting there were plenty of opportunities in South Africa for 'a white man "driving niggers"'.[42]

'The service so freely offered': Māori and military service

Māori politicians were among the first to support New Zealand's involvement in the war, but their support was not unequivocal. Speaking during the 1899 First Contingent debate, Eastern Māori MHR Wī Pere claimed the imperial authorities had forced the Boers into a corner, leaving them little option other than to fight. He then obliquely referred to the possibility of Britain's enemies casting covetous eyes on New Zealand if England were defeated in South Africa. Solely to avert this, he supported sending the contingent and offered to personally lead a force of 500 Māori.[43] Hōne Heke Ngāpua noted that differences had periodically arisen between the imperial government and Māori since the signing of the Treaty of Waitangi, but he considered Māori support for the British Crown to be their duty if assistance was required.[44] While Southern Maori MHR Tame Parata and Western Maori MHR Hēnare Kaihau did not address Parliament during the debate, they, too, voted in favour of sending the First Contingent.[45]

Minister of Native Affairs James Carroll also advocated Māori participation and pointed to Māori serving in the New Zealand contingent sent to England to celebrate Queen Victoria's 1897 Diamond Jubilee. During the jubilee, Private Taranaki Te Ua had been one of four New Zealanders who served in the colonial section of Victoria's bodyguard, and Captain Tunuiarangi and Private Hohiana Te Puni were introduced to the queen.[46] Victoria asked Lieutenant-Colonel Pitt to tell the Māori members of the contingent how pleased she was to see them, adding that despite the New Zealand wars she was sure that they were 'good loyal subjects'.[47] During the contingent debate Carroll referred to England as the 'Mother country' and claimed Māori were well suited to military service: 'I know there is a yearning in their hearts, induced by loyalty, to add whatever they can towards holding up the military glory of the Empire.'[48]

Yet as Māori soon discovered, there were limits to the rights of non-European 'loyal subjects', with the imperial authorities repeatedly rebuffing their offers of military assistance. In this, Māori were not alone; similar offers from Lagos and Hong Kong were also politely declined.[49] In response to an offer of 300 Malay States Guides, Secretary of State for the Colonies Joseph Chamberlain said he was 'compelled to demur', claiming their absence would weaken the Singapore garrison.[50] The imperial authorities had no such reservations about accepting contingents of Europeans from New Zealand, Queensland, Victoria, South Australia, New South Wales and Canada.[51] The language remained diplomatic, but there was no disguising the underlying message: colonial European forces would be gratefully accepted, but non-European troops need not apply.

Throughout the war Seddon remained a vocal supporter of Māori involvement. In early 1900 he advocated Māori either being accepted for service in South Africa or held in reserve: 'The Maori are among Her Majesty's most devoted subjects, and it is hard to make them understand that their loyalty must not be submitted to the supreme test.'[52] When asked whether he favoured the inclusion of Māori in contingents, Seddon noted that influential Māori were 'naturally a little aggrieved' that their young men were not permitted to fight for their queen; the 'colour line', he claimed, had been removed by the Treaty of Waitangi.[53] The following month the premier announced a Māori offer of 2000 warriors to serve in South Africa, adding that young Māori were as eager to fight as their Pākehā counterparts.[54]

Although these initial offers of assistance came to nothing, it appeared Māori wishes would finally be granted in December 1900 when Lord Ranfurly informed Chamberlain that the New Zealand government intended sending a 200-strong Sixth Contingent to South Africa including 100 Māori.[55] The governor cabled 'Natives most eager' and asked if Chamberlain had any objections. Official notices appeared in the press announcing that Māori who could speak English would be eligible to apply, with preference given to Volunteer corps members.[56] The *New Zealand Herald* quoted 'the Agent General' who predicted that most Māori who joined the contingent would be well-to-do 'free citizens' who could not be compared to 'subject races'. He condescendingly added that Māori were 'kindly, homely and humane people, equal in those respects to any white people'.[57]

While several New Zealand newspapers predicted the inclusion of Māori would receive widespread approval, others opposed the idea.[58] The *Press* had earlier described a Zulu and Basuto offer to fight with British forces as a dangerous experiment, and warned there would be 'no telling where the warlike spirit, once infused into a race of semi-savages, would stop'.[59] The editor of the *Evening Post* dismissed Māori participation in the Sixth Contingent as a grave mistake: 'New Zealand surely does not wish to have the blame of introducing the coloured element into the war.'[60] The *Timaru Herald* called the use of Māori troops 'simply ridiculous' and 'arrant nonsense',[61] while a Manawatū newspaper claimed the proposal had aroused a storm of indignation throughout the country:

> Not that there is any intention to belittle the Natives or to cast a slur upon the fact that they could not in any sense of the term be called white troops, but because it would be a highly improper thing to raise the question of the employment of aboriginals in a war between white races.[62]

Other New Zealanders were equally critical: a former trooper called for the war to be brought to a close as it had been begun — 'by white men against white men'.[63]

Joseph Chamberlain's hand was forced in the final days of 1900 when the *Times* in London also reported the inclusion of Māori, leaving the British public with the impression that Māori involvement had been formally sanctioned.[64] Apparently caught off guard, Chamberlain quickly moved to scotch the idea, repeating his earlier position that unspecified political considerations made Māori involvement impossible.[65] Nonetheless, Chamberlain was not implacably opposed to Māori participation providing it was discreet: 'I am really sorry not to give the Maoris [sic] a chance[.] If they had sent them without asking & mixed up with others no one would have known the difference.'[66] Although the members of New Zealand Volunteer corps such as the Papawai (Native) Rifles and the Ngāti Porou Rifles formed prior to the war were almost entirely Māori, the prospect of Māori fighting in South Africa caused consternation in the Colonial Office, with officials warning that angry protests could be expected from pro-Boers.[67]

Following receipt of Chamberlain's cable, there was a rapid reversal of the Defence Department's position regarding Māori service in New Zealand contingents. On the last day of 1900, an amended version of the Sixth Contingent recruitment notice appeared in the press with all reference to Māori expunged.[68] Nonetheless, at least one newspaper appeared slow to digest this change. Two weeks after Chamberlain had officially rejected the idea, the *Free Lance* continued to publish a recruiting advertisement stating that Māori would be considered.[69] Once it became apparent that this was not the case, newspapers accused Seddon of giving Māori false hope when the imperial government had clearly expressed its opposition.[70] The editor of the *Otago Daily Times* argued that Māori should have been spared the disappointment of refusal, and noted that the imperial government's decision not to employ the services of Māori was in keeping with the policy adopted from the outset.[71]

Although Chamberlain continued to block Māori involvement in the conflict, war-related initiatives had the potential to directly impact on New Zealand Māori. When the imperial authorities proposed confining Boer prisoners on Rakiura Stewart Island in 1901, the *Star* applauded Seddon's rejection of the idea. However, instead of completely opposing the imprisonment of Boers on New Zealand soil, the newspaper suggested interning them on Wharekauri Chatham Islands where, it claimed, the combined population of Māori and Moriori was so small that their land could easily be exchanged for property on the mainland.[72] The islands, the newspaper noted, had already served as a 'penal settlement' for the prophet and military leader Te Kooti and a number of his followers. Christchurch schoolmaster John Newlyn suggested an alternative plan involving the use of Māori as guards for Boer prisoners in Ceylon (Sri Lanka) or on Saint Helena.[73] The Wharekauri proposal resurfaced in 1902 when the imperial authorities again proposed interning Boer prisoners in New Zealand. Although Seddon again rejected the idea, Lyttelton MHR George Laurenson and a deputation of island residents met the premier to advocate the establishment of a 'depot' for 3000–4000 Boer prisoners on Wharekauri.[74]

By 1902, the imperial authorities' attitudes to non-European recruitment had not changed; the attempts by a sergeant of Samoan heritage to serve in South Africa were blocked despite Joseph Witheford championing his case.[75] In early February Ranfurly informed Chamberlain that as Māori were not permitted to serve in South Africa, they instead offered a contingent to serve anywhere in the British dominions. The governor envisaged the contingent being officered by New Zealanders and estimated it could involve in excess of 1000 men.[76] The draft of Ranfurly's telegram concluded, 'The acceptance would give satisfaction.' However, the message had been edited and a passage crossed out. Originally, it ended '. . . as [Māori] fail to see why they cannot be permitted in S.A.'.[77]

When the Colonial Office did not respond immediately, Ranfurly again cabled Chamberlain saying that Māori were anxiously awaiting his reply.[78] Adding his weight to the argument, Seddon claimed that if Māori were included they would quickly show their worth and make a significant contribution to the British cause.[79] Three weeks passed before Chamberlain finally declined the offer. He informed Ranfurly that while the imperial government received the offer with 'extreme gratification', and appreciated the loyalty and patriotism that inspired it, it was with much regret that Māori 'service so freely offered' could not at that time be accepted in South Africa or elsewhere.[80] Chamberlain acknowledged Māori military prowess but cited the 'the peculiar circumstances' that obliged the rejection.[81] Perhaps to assuage Māori indignation, Chamberlain assured Lord Ranfurly that offers of help from the Indian Army and Indian feudal princes had also been rejected.[82]

While this may have been Chamberlain's final word on the subject, Seddon was unwilling to let the issue drop. Shortly before leaving for England in 1902 to attend the colonial leaders' conference, the premier addressed a Māori gathering at Pāpāwai that included several iwi.[83] In a typically bellicose speech, Seddon assured his audience that he would ask the imperial authorities not to refuse again an offer by Māori volunteers to fight for the empire and implied that Māori exclusion breached the Treaty of Waitangi.[84] The Christchurch *Star* also suggested that the enrolment distinction drawn between Māori and European amounted to a virtual abrogation of the third article of the Treaty guaranteeing Māori the rights and privileges of British subjects.[85] If European New Zealanders had the right to fight for the empire, then surely the Treaty extended the same right to Māori.

At Pāpāwai, Seddon criticised the 'kid glove' treatment he alleged the Boers were receiving in South Africa. The month before the war ended, Seddon claimed that if a force of 5000 Māori were permitted to participate unconstrained by imperial regulations they would quickly put an end to the trouble.[86] A newspaper reported the premier's address: 'With the Maoris [sic], continued Mr Seddon, war was war, and fight was fight; they were never afraid of hurting their enemies.'[87] Seddon's remarks were widely reported in the Australian press. In Perth, the *Western Mail* said that while Māori were above such behaviour Seddon's meaning was clear; if given free rein against the Boers Māori would not take prisoners. The Tasmanian *Examiner* reported that the government had a 'new' Māori policy that involved trusting Māori and treating them with justice and equity.[88]

Seddon's Pāpāwai remarks were unlikely to have endeared him to authorities at the War Office in London. By the time Seddon arrived in Sydney he was already attempting to distance himself from the press reports by claiming that the comments of a Māori orator had been mistakenly attributed to him.[89]

Sections of the New Zealand press questioned Seddon's sincerity; a *Timaru Herald* editorial said his statements were intended to 'tickle the ears of the natives'.[90] The newspaper claimed that while some Māori were industrious, their numbers were in decline due to 'idle and dissipated habits, and a half-civilised half-savage manner of life'.[91] Seddon was certainly not beyond exaggerating Māori support for the war while downplaying Māori concerns about more pressing issues such as land loss. Shortly after his Pāpāwai address, he told a Christchurch gathering that throughout New Zealand the only Māori grievance was their exclusion from playing an active role in the defence of the empire.[92]

For disaffected Māori, Seddon's claim must have seemed particularly disingenuous. Despite a 1902 Taranaki Māori Council resolution expressing support for the Crown, Māori in the Taranaki region influenced by the Māori prophets Te Whiti-o-Rongomai and Tohu Kākahi remained suspicious of government intentions. The 1901 Māori census enumerator for Taranaki and Pātea was asked by Māori why the government wished to ascertain their numbers: was it to send them to South Africa to fight, or to deport them?[93] Māori occupying pā at Pīhama, Ōeo, Ōtākeho and Kaūpokonui in the shadow of Mount Taranaki shared these views and were equally guarded.

And yet a significant number of soldiers with European names and Māori/Pākehā ancestry did serve in South Africa.[94] These included Edward Renata Broughton, Walter Callaway, William Thomson and Henry Reiwhati Vercoe. Vercoe, whose mother was Te Arawa, served in the Tauranga Mounted Rifle Volunteers and the Seventh and Ninth contingents.[95] Following the 1895–96 Jameson Raid, Te Arawa leaders had cabled the governor offering to supply a guerrilla force to serve in Transvaal.[96] While commending the iwi for their patriotism, the Earl of Glasgow assured them the danger had passed.[97] Early in 1902 Te Arawa members passed a resolution regretting not officially being permitted to serve in South Africa. Although the claim was unrealistic, Te Arawa indicated they could have 5000 men ready to fight in two weeks.[98]

Other men with Māori heritage successfully overcame official opposition and joined contingents with both the knowledge and support of politicians. During an 1899 election speech Wanganui MHR Gilbert Carson told his supporters that James Thorpe had attempted to enrol in the First Contingent but was declined as he was 'half-caste'. Although Carson had opposed New Zealand sending the First Contingent, the politician claimed to have taken Thorpe's case up with Seddon, who agreed that colour should not prevent men from serving. According to Carson, Lieutenant-Colonel Penton was adamant that Thorpe was ineligible, but after meeting Seddon he relented and permitted Thorpe to sail with the contingent.[99]

In October 1899 Arthur Guinness, the MHR for Grey electorate, asked whether men who had served in the contingent sent to the UK in 1897 for Queen Victoria's jubilee would be given preference during the South African

Farrier-Sergeant William Thomson, a blacksmith in the Waikato town of Kihikihi. Thomson served in both the Fourth and Ninth contingents. NATALIE MCCONNELL

contingent selection process.[100] In response, Seddon outlined the contingent's eligibility criteria, indicating that men who were 'first-class shots' would be selected first, though 'all other qualifications being equal' Jubilee Contingent men would be favoured. Āhere Hōhepa qualified on both counts. His target shooting score was the highest in the Napier Guards Rifle Volunteers and he was reportedly one of the best riders and shots in the Hawke's Bay detachment of the Third Contingent.[101] Despite his proficiency, Hōhepa's initial attempt to accompany his contingent was blocked as he was a 'full Maori', with a Māori newspaper claiming the Pākehā press had criticised his selection.[102] However, after the intervention of some of Hastings' leading citizens, the native minister reportedly promised that Hōhepa would sail with his contingent.[103]

Hōhepa, who enrolled in both the Napier Guards and the Third Contingent under the name Arthur Joseph, had travelled to London with the Jubilee Contingent.[104] Whether his enrolment as Arthur Joseph was a necessary piece of subterfuge is unclear, but if Hōhepa's true identity was intended to be a secret, it was a poorly kept one. A 1900 newspaper article included excerpts of a letter written by 'Trooper Ahere Hohepa (Arthur Joseph), of the Third Contingent and the only real representative of the Maori race in the South African war'.[105] Hōhepa reportedly posted a sample of dirt he had collected in the South African republics, which he predicted would soon be British property.[106] The extent to which men such as Hōhepa, Callaway, and Kihikihi blacksmith William Thomson (who served in the Waikato Mounted Rifle Volunteers with at least one other Māori, Hēnare Te Karutu) embraced their Māori heritage appears to have varied.[107] Callaway penned the First Contingent haka and wore traditional Māori dress for a posed studio photo.[108] All three were fluent Māori speakers serving in predominantly Pākehā Volunteer corps.[109] Thomson, who was a Freemason and whose mother, Rea Te Kopa, was Ngāti Maniapoto, may have considered himself materially better off in Pākehā society. Grey Holden, who came from Rangiaowhia in Waikato and had Māori ancestry, also served in South Africa with his two brothers, Richard and George.[110] Like Thomson, Richard served in the Waikato Mounted Rifle Volunteers.

Unlike many Pākehā contingent members, Hōhepa had first-hand experience of the pomp and ceremony of Victoria's jubilee in London. The importance men such as Hōhepa, Callaway and Thomson placed on being members of the

British Empire is difficult to gauge, but whether they went to South Africa for adventure or out of a desire to support the imperial cause, they retained links to both the Māori and Pākehā worlds. For Hōhepa, Seddon's claims of racial equality in New Zealand would later ring hollow. In 1916 Hōhepa wrote to the *Hastings Standard* claiming that, despite serving in South Africa and being wounded at Gallipoli, he had been refused entry to the dress circle of a Hastings picture theatre because he was Māori.[111] Echoing Seddon's own words, the 1916 article was titled 'The Colour Line'.

'The mana of our Queen': Māori support for the war

If the Dutch and German priests in Northland were attempting to encourage Boer sympathies among Māori in the region, their efforts appear to have had limited effect. A number of Māori from the tiny settlement of Parapara south of Doubtless Bay reportedly donated £1 1s to the Rough Riders Contingent Fund after hearing that Pākehā in Kaitaia were collecting money.[112] Following the 1901 Māori census the enumerator for the Northern District noted that events in South Africa were being watched with great interest by Northland Māori and claimed many would volunteer if required.[113] At least one northern Māori, Henry Tau, was a member of the Kawakawa Rifle Volunteers, and further west David Moetara served in the Hokianga Mounted Rifle Volunteers.[114]

The 'khaki fever' that swept much of New Zealand was also evident among a number of Māori women in Northland. In 1901 New Zealand newspapers featured an article about 'half-caste' Ngāpuhi women who dressed in matching khaki uniforms (based on those of New Zealand mounted corps) and called themselves the 'Ngapuhi Sisters of Mercy'. The women were reportedly training to become nurses who could serve in the field if volunteers were required for active service.[115] An *Auckland Weekly News* photo of five members of the Whangārei-based 'Ngapuhi Nursing Sisters' (presumably the 'Ngapuhi Sisters of Mercy') showed Māori women dressed in khaki tunics with bullet bandoliers and water bottles.[116] The women, who were reportedly all relatives of the renowned Ngāpuhi rangatira Hongi Hika, were prepared to tend to the wounded 'should a strange foe attack their native land'.[117]

Māori support attracted interest in other parts of the empire. The British *Globe* praised Māori contributions to the War Relief Fund and claimed there was no more loyal race in the empire. The *Westminster Gazette* noted that even though Māori were prevented from serving in South Africa they had contributed £1000 to the New Zealand Patriotic Fund.[118] Closer to home, the story of the Ngapuhi Nursing Sisters appeared in the Tasmanian press.[119]

In Rangitīkei, Ngāti Apa Māori subscribed £30 to the More Men Fund, and several Whanganui Māori serving on the Fund's Native Committee contributed money to equip soldiers for South Africa.[120] These included Takarangi Mete Kingi, Te Aohau Nekitini and Wāta Wiremu Hipango, all members of the Aotea Maori Council.[121] The following year Kingi sent Robert Baden-Powell a Māori taonga — a 'relic of historic value' in recognition of his defence of Mafeking.[122] The *Weekly Press* reported that although New Zealand was only to supply soldiers of 'British blood', the Native Committee had collected £80 for the fund and supplied two horses.[123] Horses were also gifted by influential Ngāti Kahungunu figures Hamuera Tamahau Mahupuku and Airini Donnelly, together with the sons of Māori MHRs Tame Parata and Hēnare Kaihau.[124] Not to be outdone, wealthy Ngāi Tūmapuhiārangi leader Taiāwhio Te Tau, who attended the 1897 jubilee and named his son Richard John Seddon, provided two horses from his personal stud.[125]

On the Jubilee Contingent's return to New Zealand, its members attended a large gathering of Māori held in their honour at Pāpāwai in Wairarapa.[126] Mahupuku and his wife arranged the event, which was attended by MHRs, regional dignitaries, Seddon's wife and daughter and Captain Robin (prior to his promotion). At the time, the *Evening Post* commented on the camaraderie of the Māori and Pākehā sections of the contingent. It claimed that if there were any doubts about the possibility of Māori and Pākehā fighting side by side 'when their country called', they had been dispelled.[127]

Like their Pākehā counterparts at Wanganui Collegiate School and Christ's College, many Māori students at Te Aute College in Hawke's Bay followed the war's progress with interest. The Anglican boarding school aimed to transform its students into 'educated, model Christian Māori citizens' and its cadet corps was an extension of this goal.[128] Te Aute Native College Rifle Cadets was established in 1897, and the following year it was the only predominantly Māori cadet corps in New Zealand.[129] While Waimana Native School, which opened in 1899 in 'Tuhoe or Urewera Country', received 36 wooden 'model rifles' for its

Above: The 'Wanganui Native Committee of the "More Men" Fund'. Back row from left: Takarangi Mete Kingi, Hōri Pukehika, N. E. Goffe, Captain McDonnell, Waata Wiremu Hipango. Seated: E. Rangirihau, Te Kerei, Hōri Kerei, Weraroa Kingi, Te Aohau Nekitini. WEEKLY PRESS, 14 MARCH 1900, P. 62, NEWSPAPER HISTORY OF THE BOER WAR IN SOUTH AFRICA, 1899–1902, COMPILED BY GEORGE FANNIN, P F968 FAN, ALEXANDER TURNBULL LIBRARY

Below: The 'Ngapuhi Nursing Sisters', a group of Ngāpuhi women who were descendants of the renowned Ngāpuhi rangatira Hongi Hika. The Whangārei-based group were reportedly 'willing to devote their time, and lives, if necessary, to the humane work of tending the wounded should a strange foe attack their native land'. They travelled in the Northland region providing medical assistance. Back row from left: Sergeant A. Calkin, Bugler M. Kaire. Seated: Sergeant-Major C. Calkin, Captain Kingi, Lieutenant G. Waetford. AUCKLAND WEEKLY NEWS, 5 JULY 1901, P. 8, AUCKLAND LIBRARIES HERITAGE COLLECTIONS, PHOTOGRAPHER J. T. COWDELL, 7-A12385

cadet corps in 1901, four years earlier Te Aute College headmaster John Thornton succeeded in getting the Defence Department to issue the school's cadet corps with 60 Snider carbines for drill purposes.[130] Among the boys who served in the Te Aute cadets was Sergeant Peter Buck, who was second in command of the Pioneer (Maori) Battalion during the First World War.[131]

One of the most vocal advocates of Māori involvement in the war was Ngāti Porou rangatira Tuta (Matutaera) Nihoniho. In 1886 Nihoniho had been appointed commanding officer of the newly formed Māori 'Ngatiporou Rifles', which by 1887 had a membership of 88.[132] The following year Nihoniho responded to reports of increased international tensions in Europe by urging Seddon to raise and train a force of three to four thousand Māori to provide assistance should the 'mana of our Queen' be challenged.[133] Nihoniho's loyalty, which was lauded in the British press, was again displayed in 1899 when he offered 500 Ngāti Porou warriors to fight alongside English troops in South Africa.[134] Though Nihiniho was deeply disappointed that his offer was declined, the rejection failed to dampen his patriotic fervour.

When the son of Lord Roberts, the British commander-in-chief, was killed during the Battle of Colenso in South Africa, Nihoniho sent Roberts a greenstone mere that had been in his mother's family for 19 generations (see page 204). Named Porourangi, the taonga had originally belonged to the warrior chief Kahukuranui, whom the *Sheffield Daily Telegraph* described as the 'native Duke of Wellington'.[135] On receipt of Porourangi, Roberts wrote thanking Nihoniho for his willingness to 'fight the enemies of our country, if called upon to do so'.[136] The gift was reported in several British newspapers, with the *Times* also claiming that Roberts' letter would be treasured by Ngāti Porou 'as was the *mere*'.[137]

British plaudits notwithstanding, the imperial government's rejection of Ngāti Porou offers of assistance irked Nihoniho, who tactfully expressed his sense of frustration and injustice:

> *Great was the love for you and the grief for ourselves the Maoris [sic], because the native people were not permitted to go with you to the assistance of our Mother England[.] Now, O my children, this is a new thing to me under the sun — namely, having two children, one white and the other brown, that when trouble overtakes me, their parent,*

Left: Airini Donnelly, the influential Ngāti Kahungunu who was one of the first women to actively support New Zealand's military involvement in South Africa in 1899 by offering two horses for the First Contingent. ALEXANDER TURNBULL LIBRARY, 1/4-022134-G

Right: Ngāti Porou rangatira Tuta Matutaera Nihoniho in the uniform of the Ngāti Porou Rifles. Nihoniho was a strong advocate of the inclusion of Māori in the contingents New Zealand sent to South Africa and gifted the mere Porourangi to Lord Roberts. *NARRATIVE OF THE FIGHTING ON THE EAST COAST (NGA PAKANGA KI TE TAI RAWHITI) 1865–71: WITH A MONOGRAPH ON BUSH FIGHTING (ME NGA KORERO MO UENUKU)*, BY TUTA NIHONIHO, OF NGATI-POROU, WELLINGTON: DOMINION MUSEUM, 1913: MUSEUM OF NEW ZEALAND TE PAPA TONGAREWA

that I should forbid my brown child to come to my assistance and invite my white child to die with me.[138]

Nihoniho's speech was published in several New Zealand newspapers and in the Exeter *Western Times* in England.[139] Nihoniho noted that the imperial government's response could be interpreted in two ways: '(1) a loving regard lest harm should befall a much-loved child; or (2) a feeling that the child is despised as being incapable of accomplishing any great deed'.[140] He did not indicate which option he felt was true of Ngāti Porou.

Thwarted in his attempts to have Ngāti Porou serve in South Africa, Nihoniho instead turned to fundraising. He was chairman of the 1900 Grand Maori Carnival in Gisborne's Theatre Royal, which raised money for both Indians affected by the famine and the South African War funds and included Māori performing haka, poi dances and patriotic songs.[141] Commenting on a performance by young Māori girls from Tikitiki during the carnival, the *Poverty Bay Herald* noted they were 'animated by patriotic spirit'.[142] When Mafeking was relieved, Nihoniho led Māori in a parade through Gisborne dressed in his Ngāti Porou Rifles uniform with sword in hand.[143]

In 1900 Seddon canvassed both branches of the New Zealand legislature regarding a second South African contingent. Hēnare Tomoana, the Ngāti Kahungunu and Ngāti Te Whātu-i-apiti politician who had fought with the Crown against Te Kooti, cabled a response accepting the proposition. He expressed sympathy for New Zealanders fighting in Transvaal and offered to immediately take a Māori contingent to Africa to assist the imperial forces.[144]

Even before the war began Kuku Karaitiana, who had also served in the 1897 Jubilee Contingent, expressed a willingness to serve in South Africa, while in 1901 Mahupuku offered to raise and fund a Māori force.[145] Deflecting blame from the government, James Carroll told Māori participating in a patriotic event that they should not complain as it was a directive from England that prevented Māori inclusion.[146]

Possibly the largest display of Māori support for the war occurred during the Maori Carnival held at Wellington's Basin Reserve in March 1900, which raised over £525 for the Transvaal War Fund.[147] The carnival involved Māori groups including Ngāti Kahungunu, Ngāti Raukawa, Ngāti Toa and Ngāti Kauwhata, while an estimated 10,000 spectators watched performances including a haka titled 'Kiki te Poa' (Kick the Boer).[148] Māori women were heavily involved, including Katherine Te Rongokahira Parata of Ngāti Pūkenga and Ngāti Pikiao, who helped organise the event.[149]

Smaller patriotic displays occurred elsewhere; at a Gisborne fundraiser a Māori spokesman noted the desire to fight 'for the old flag' and pointed out that Pākehā could not appreciate the strength of Māori feeling because they were unable to read Māori newspapers which warned against any Māori 'pro-Boerism'.[150] When another 'native festival' held at Pāpāwai raised £500, the money helped fund the Fifth Contingent, part of which sailed to South Africa aboard the *Maori*.[151] It is possible that Māori were aboard the vessel, but few (if any) travelled with official sanction.

'Our only distinctive feature': Māori culture and the war

Māori military service may have been rejected, but Pākehā had little compunction about appropriating Māori language and cultural imagery during the conflict. Although this was certainly not a new phenomenon, it was especially prevalent during the South African War. This was particularly true of the gifts presented to Baden-Powell. James Carroll was consulted about 'a suitable Māori curio' to be sent by the citizens of Wellington, and an illuminated address that accompanied the sword of honour that Dunedin sent to 'the Hero of Mafeking' was mounted on a rod featuring silver Māori heads.[152] A writing desk sent by the grateful citizens of Auckland included Māori designs and depictions of traditional Māori life, while a gold salver presented by the Public Service was etched with 'the emblems of chieftainship — the mere, the taiaha and the hatchet'.[153] Three model cannons presented by the people of Tīmaru to Baden-Powell and the Ladysmith and Kimberley commanders also featured Māori inscriptions.[154] Another example of cultural appropriation involved a drawing by Horatio Robley, the British general who had served in New Zealand during the New Zealand Wars. The drawing, which was reproduced in magazines and on postcards, depicted President Kruger 'decorated like the late Maori King' with tā moko. Robley's picture included a caption that appeared to compare Kruger's 'cheek' in challenging British supremacy to King Tawhiao's desire to maintain Māori sovereignty in the King Country.[155]

In 1901, the *New Zealand Herald* noted the national reliance on Māori

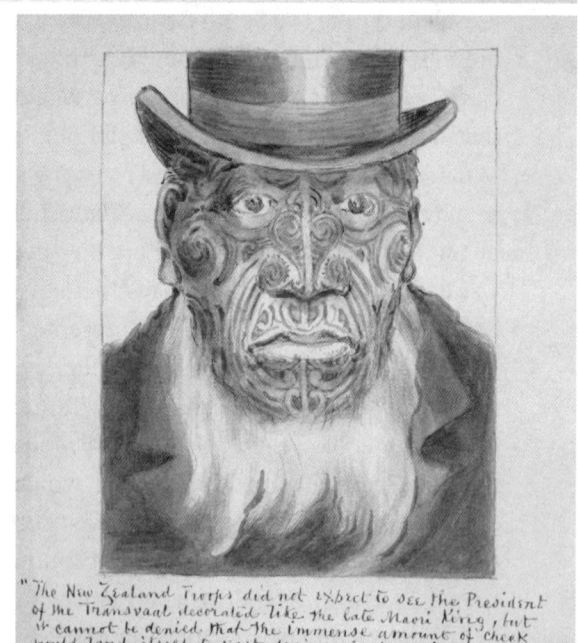

Above and right: Caricatures of South African Republic president Paul Kruger with tā moko. The 1900 postcard was sent from New Zealand to William Pember Reeves, the New Zealand Agent General in London. Both are versions of a cartoon originally produced by Major-General Horatio Gordon Robley, an English officer who served in New Zealand during the New Zealand Wars and amassed an extensive collection of toi moko. ALAMY; ALEXANDER TURNBULL LIBRARY, A-080025

culture when it suggested arranging a large Māori gathering in honour of the visit of the Duke and Duchess of York. New Zealand, the newspaper claimed, could not compete with the patriotic displays in England and Australia; Māori culture 'is our only distinctive feature. It is the only thing peculiar to the colony that we can offer.'[156] Although the use of the Māori language by Pākehā MHRs was not confined to the war period, on several occasions parliamentarians used Māori words when discussing the conflict. Seddon claimed that Māori wanted to uphold the Queen's 'mana' and 'were *pouri* (sorrowful) at not being allowed to go with our sons in the contingents to South Africa'.[157] Napier MHR Alfred Fraser wrote of volunteers for the contingents being eager to fight for 'the old flag and have *utu* for their brother's blood'.[158]

Together with the version Walter Callaway produced for the First Contingent, both the Third and Fifth contingents had their own Māori war cries.[159] Early in 1900 a woman from the Southland town of Gore sent a cake to a contingent in South Africa decorated with its war cry alongside the Union Jack.[160] In 1903 a New Zealand contingent member who had settled in Transvaal attended a concert at the New Zealand Club where a Māori war cry concluded the evening's entertainment.[161] Other Māori influences also appeared in South Africa; while serving in the Kaffrarian Mounted Rifles, Wanganui Collegiate School Old Boy Norman Palmerston described buying 'tiki' beer in Cape Town.[162]

'Free, loyal and peaceful citizens of the British Empire'

In 1900 Seddon had accompanied various Cabinet ministers and the governors of both New Zealand and New South Wales on a visit to Waikato where their party was greeted by the Māori king's band playing the National Anthem. When Seddon read a telegram reporting British military successes in South Africa the news was reportedly received with 'an outburst of enthusiasm, and cheers'.[163] Although Waikato Māori were markedly less supportive of British actions in South Africa than groups such as Ngāti Porou and Ngāti Kahungunu, the visit was indicative of a gradual thaw in their relations with the Crown.

Māori swiftly responded to the 1902 German accusations of British soldiers committing atrocities in South Africa.[164] Te Aute College students passed a resolution condemning the German claims and expressing the willingness 'of the young Maori party to maintain at all hazards the "mana" of England'.[165]

Māori from the Wairarapa, Hauraki, Rotorua, Tauranga, Hawke's Bay, East Coast and Taranaki regions, as well as Ngāti Maniapoto from the King Country, passed their own resolutions expressing confidence in the British government.[166]

As an 1899 incident showed, such support was not unanimous. When a group of Māori gathered in Tauranga's Strand the main topic of conversation was the war. The debate reportedly ended in blows after some expressed British sympathies, with the remainder voicing support for the Boers.[167] Despite these divisions, Hōri Ngātai of Ngāi Te Rangi, the Tauranga iwi that had defeated British forces at the 1864 Battle of Gate Pā, sent Carroll a telegram objecting to the 'untruthful slanders' and 'German insults'.[168] The Ngāi Te Rangi 'free, loyal and peaceful citizens of the British Empire' were reportedly indignant at being refused places in the Eighth Contingent but nonetheless offered a force of trained and armed Māori 'at the King's immediate command'.[169]

The inclusion of Māori in the contingent sent to the 1902 coronation of Edward VII may have gone some way towards reducing Māori disappointment. The Māori Contingent, led by Captain Taranaki Te Ua of Ngāti Kahungunu, also included men from a number of iwi, ranging from Ngāi Tahu in the south to Ngāpuhi in the north.[170] At least one Māori enrolled in the Tenth Contingent appears to have joined the Coronation Contingent. In May 1902, the Tenth Contingent's commanding officer informed the under-secretary of defence that after his contingent left the West Australian port of Albany, Trooper Robert Cameron was found to have been left behind in Wellington. Requesting the removal of Cameron's name from the contingent roll, William Messenger, a British-born Tenth Contingent officer who had served during the New Zealand Wars and taken part in the Armed Constabulary invasion of Parihaka in 1881, said he believed Cameron had been left behind 'for the purpose . . . of joining the Maori Coronation Contingent'.[171] Te Whatu and Tirua Tukiri of Waikato, and Hare Wahanui of Ngāti Maniapoto, were among Māori selected to attend the coronation.[172]

Yet Waikato Māori passing pro-British resolutions and cheering British military successes in South Africa, coupled with the service of a few of their number in the South African and coronation contingents, did not reflect a general rapprochement between Waikato Māori and the Crown. Nor did it represent widespread support in the Waikato region for the South African

War. During the war the Māori King movement was viewed with suspicion by many Pākehā, with a newspaper accusing the Kingites of being hostile to the advancement of New Zealand colonisation.[173] As late as 1907 a government report noted that largely due to the Waikato and Taranaki wars Māori in the region were divided into 'factions' and remained wary of European law and justice.[174] Only days after the South African War ended, a newspaper reported that an unnamed Aucklander had circulated pro-Boer literature in the King Country, though it claimed the chiefs to whom the parcels had been sent 'took very little notice of them'.[175] Regardless of their opinion of the war, there was a degree of interest in its progress among influential supporters of the Māori King movement. During a visit to Maungatapu, Ngāti Hauā leader Tupu Atanatiu Taingākawa Te Waharoa specifically requested an update about the South African conflict, with the 'Maori Premier' seeking information about the 'Boer War', 'liquid air' and wireless telegraphy.[176]

In 1902 the *Manawatu Evening Standard* reported that a Māori who had been instructed to leave Trentham Camp 'was classed amongst the smartest drill-sergeants'. The newspaper noted the astonishment of some camp visitors upon seeing a 'coloured man in charge of a squad of Europeans'.[177] But as press reports showed, the sergeant was not alone in gaining admission to a contingent only to be later ordered to leave:

> *Several Maoris and half-caste Maoris [sic] who got into the Eighth Contingent are very much mortified at being told at the last moment that they could not be allowed, in obedience to Imperial instructions, to go to Africa. It is a delicate question, and the Maoris [sic] are a high race. The leading chiefs are sure to protest.*[178]

Only a day before this report, the *New Zealand Herald* claimed that at no time during the selection process had the imperial authorities objected to the inclusion of 'half-castes' in the contingents.[179] The newspaper gave the examples of Callaway and Sergeant-Major William Tutepuaki Pitt as contingent members with Māori heritage. Pitt, who was Ngāti Porou, initially served as a trooper in the First Contingent and was promoted to lieutenant and paymaster in the Eighth Contingent.[180] When Pitt enrolled in the Eighth Contingent he gave his mother, Māta Te Owai, as his next of kin. Bernard Reed, who served in the Fifth and Eighth contingents before enlisting during the First World War, recorded his Māori mother, Titiripa Tūrei, on his First World War enrolment form.[181] However, when enlisting in the Eighth Contingent 15 years earlier he had made no mention of her, and instead gave an aunt with a European name as his next of kin.[182]

Regardless of Seddon's real intentions regarding Māori involvement, the issue was soon rendered academic by events in South Africa. By 1902 the Boers were facing increased military pressure in the field and the war's final outcome was less uncertain. The inclusion of Māori forces was guaranteed to be politically divisive while at the same time unlikely to hasten the conflict's end. The extent to which Seddon's advocacy of Māori participation was driven by domestic political considerations, rather than a genuine belief that he could reverse imperial policy, may never be known.

A list of 'general principles' formulated during the 1902 colonial leaders' conference (which Seddon attended) suggests his proposal would have met with little support:

> (a) The main burden of a great struggle between the British Empire and one or more States of European race or descent must be borne by the white subjects of the King. (b) Military contingents, therefore, of other than men of European descent need not be considered with regard to this particular problem, although the great value of the Indian army and the usefulness of the African and other Native forces are fully recognised.[183]

Though the war ended without any significant change in the imperial authorities' position regarding Māori troops, two years earlier an *Ashburton Guardian* editorial accurately predicted the future. Criticising the British creation of the 'colour line', the newspaper noted the willingness of 'Maoris, Ghoorkas [sic], Sikhs and other coloured subjects of the Queen' to fight in defence of the empire. The newspaper prophesied that '[a] time may come, and that before very long, when necessity will override sentiment, and Britain will be only too glad to accept the services of all her sons — be their colour what it may'.[184]

CHAPTER FIVE

'YELLING YAHOOS IN YELLOW'

THE BEHAVIOUR OF NEW ZEALAND SOLDIERS DURING THE SOUTH AFRICAN WAR

At a 1901 Timaru church parade Archdeacon Harper claimed the South African War's freedom from looting, cruelty and individual crime made it unique, and that the British Army's record in South Africa would have been impossible during the previous century. Adding that New Zealand soldiers downplayed their wartime exploits, Harper described them as 'Striplings come back men'.[1] Yet Harper's appraisal did not tell the whole story. While there were documented cases of men exhibiting bravery under fire, claims that New Zealand soldiers' conduct was 'exemplary' do not withstand closer scrutiny. Months before the war began, Lieutenant-Colonel Arthur Penton gave a stern lecture on the need for discipline in the ranks. Both officer and non-commissioned officer (NCO) were, he said, to erase all traces of civilian life while in uniform and enforce strict obedience of orders, as a reluctance to 'compel obedience' to commands was evident throughout the New Zealand force.[2] As later events would show, Penton's admonishment had limited effect.

Had Legislative Council member Henry Scotland perused the *Evening Post* in June 1901 he could have been excused a wry smile. Scotland had earlier been accused of senility for opposing New Zealand's involvement in the war and suggesting a South African contingent would attract 'larrikins and loafers'.[3] He must have felt vindicated by accounts of ill-discipline among Volunteers as they initially represented one of the larger contingent recruitment pools. The newspaper reported that a group of Volunteers had taken to the streets of New Zealand's capital city to protest substandard food at Newtown Park military camp. A parliamentary inquiry heard evidence that 20 or 30 men in uniform paraded down Wellington's Lambton Quay carrying rotten meat suspended from a pole with a sign attached that read 'Newtown Park Rations'.[4] Seddon was furious when informed of the protest; as minister of defence it was bad enough that the military's dirty laundry was being publicly aired, but the protest was even more galling as it coincided with the Duke and Duchess of Cornwall's visit to the capital to award medals to returned contingent members.[5]

When no soldiers admitted to participating in the protest during a parade at Newtown Camp the following day, Penton stated that those involved were a disgrace to their uniforms.[6] He referred to the culprits as 'infernal curs and cowards' for sullying their comrades' reputation.[7] Others seemed less concerned; witnesses claimed that two officers observing the protest seemed to enjoy it and did not intervene.[8] The incident was raised in the House of Representatives by Richard Monk, who demanded a parliamentary investigation.[9] But instead of criticising the Volunteers' behaviour, Monk took exception to the language Penton used when addressing them, claiming that in England either a profuse apology or a court martial would have resulted. While conceding that rain had left Newtown Park a muddy quagmire, Seddon claimed to have heard Penton ordering the camp's evacuation and blamed junior officers for not carrying out the instructions. Though the premier was prepared to absolve Penton of blame, he was less forgiving of the protestors. During a parliamentary debate on the 'Newtown Park Scandal', Seddon claimed that the Volunteers' invasion of Government House grounds constituted a military offence that must be dealt with under the Defence Act.[10]

Drunk and disorderly

The following year troopers' behaviour remained an issue. While addressing Sixth Contingent recruits, Tuta Nihoniho cautioned that 'drinking and such-like pastimes must be left behind you in New Zealand'.[11] In a 1902 article titled 'Roughs in Uniform' newspapers reported the invasion of a non-smoking carriage at Wellington's Thorndon Station by 'yelling yahoos in yellow' from the Eighth Contingent returning to Trentham Camp.[12] Many troopers were reportedly smoking and in various states of intoxication, with a Petone resident claiming to have been spat on and assaulted.[13] Newspapers criticised the 'cowardly supineness' of the authorities and warned their inertia encouraged similar behaviour.[14]

Two months later, Ninth Contingent members were also involved in drunken violence on trains. In court, a railway guard testified that civilian defendants had been fighting with troopers in a carriage containing female passengers, with the judge saying he had personally witnessed inebriated troopers on trains.[15] An article titled 'Our New Militarism' in the *Observer* reported that before dismissing the charges the judge had condemned the lenient treatment troopers received. He also noted the frequency with which troopers took control of carriages following the establishment of Te Papapa Camp. 'Cowardly troopers' reportedly broke carriage windows, assaulted railway guards, frightened passengers and used 'blackguardly language'. The *Observer* claimed that other incidents involving the contingent had been hushed up.[16]

At the time of the Wellington fracas a returned contingent member was fined £5 for a drunken assault on a restaurateur, while five months after the war a 27-year-old former trooper was fined £25 for assaulting a Mangamāhū publican.[17] In court the defendant sought leniency, saying he had recently returned with the Sixth Contingent. In another incident, an Eighth Contingent trooper cited alcohol as a mitigating factor in a sexual assault case. He was fined 20s after being found guilty of drunkenly molesting women on Auckland's Queen Street. The accused claimed to have no recollection of the offences, adding he had just returned from South Africa and had been induced to drink by his friends.[18]

While the war occurred against the backdrop of the Boxer Uprising in China, Dunedin's Chinese community contributed £25 to the Fourth Contingent Fund to purchase a horse called 'Canton', and Chinese in Auckland gave £20 for the Rough Riders contingent.[19] Although the extent to which the war influenced Chinese children is unclear, the *Otago Witness* published a photo of four-year-old Henry Sew Hoy dressed in an immaculate khaki uniform and armed with a toy rifle.[20] On the same day the photo appeared, the newspaper also reported 'outrages' allegedly committed by Chinese on missionaries in Beijing, and only days earlier Whitcombe & Tombs publishers had distributed 'China of To-day: the Yellow Peril' in Dunedin.[21]

Despite their contributions to patriotic funds, Chinese laboured under the burden of racial prejudice, negative press portrayals and discriminatory legislation. The stated purpose of the 1896 Asiatic Restriction Act was to 'safeguard racial-purity' and prevent an influx of 'Persons of Alien Race who are likely to be hurtful to the Public Welfare'.[22] While Jews and Indians were specifically excluded from the act's provisions, the legislation obliged all Chinese entering New Zealand to pay a £100 poll tax.

Newspapers also reported the prosecution of Chinese living in Wellington's Haining Street 'slum' for their involvement in illegal gaming, prostitution and opium use.[23] Concerns about the drug led to the passage of the 1901 Opium Prohibition Act, which made opium importation illegal without a permit.[24] The act prohibited the issuing of permits to Chinese. It also permitted police armed with warrants to search premises where they believed opium was being used, unless occupied by Chinese (in which case no warrant was required).

Their support for the war effort did not prevent Chinese in New Zealand also becoming targets of troopers' violence. Only days after returning from South Africa, First Contingent trooper Theodore Casey was arrested for an unprovoked and vicious attack on Young She Sue in Haining Street.[25] At his trial, Casey denied kicking his victim, but admitted the assault. Though his prowess as a brawler hardly needed reinforcing, Captain Henry Coutts (who had received a commission in the Seventh Contingent) and Sergeant-Major Wilkinson, who had both served with Casey in South Africa, attested to his 'excellent fighting record' and blamed the assault on his drinking since his return.[26] Newspapers reported that Casey's war record meant he was 'let off' with a fine of £2 and £3 court costs.[27]

Shortly before he was due to sail with the Eighth Contingent 22-year-old Bugler Leonard Noly Jacobs assaulted Sue Har and Len Shing.[28] Once again, the incident occurred in Haining Street, which newspapers described as an 'unsavoury thoroughfare'.[29] Following an earlier altercation with Sue Har, Jacobs, who had served in the Fifth Contingent, returned with two other troopers and a civilian and attacked Len Shing.[30] Describing the incident as a 'murderous assault', newspapers reported that Len Shing was still hospitalised a week after the incident.[31] Though the press accused Jacobs of disgracing his uniform, he appeared in court wearing it and his Queen's South Africa Medal ribbon.[32] Though Jacobs admitted he had been drinking at the time, he denied

Four-year-old Henry Sew Hoy pictured in his khaki uniform in 1900. Henry was the son of Dunedin merchant Choie Sew Hoy and Choie's second wife, Eliza Prescott. PETER AND JON JUDSON

Len Shing explained to Mr. Gully that he was holding the axe over his head merely in an "absent-minded way."
To Mr. Wilford Shing said he was holding it up and waving it a little, "just for fun."

A HAINING-STREET EPISODE.

Above: A cartoon depiction of the 1902 Haining Street incident involving Bugler Leonard Noly Jacobs and Len Shing.
FREE LANCE, 10 MAY 1902, P. 13,
NATIONAL LIBRARY OF NEW ZEALAND
TE PUNA MĀTAURANGA O AOTEAROA

Right: Bugler Leonard Noly Jacobs (left) of the Fifth Contingent who was charged with viciously assaulting Len Shing in Wellington's Haining Street in 1902. OTAGO WITNESS, 16 JANUARY. 1901, P. 49,
ALEXANDER TURNBULL LIBRARY, N-P-2173-49

being drunk and alleged that Sue Har had provoked him. Making no mention of the civilian, Jacobs claimed to have returned with fellow contingent members Sergeant Thomas Page and Taranaki school teacher Trooper Francis Raikes to 'fix the matter up if possible'.[33]

Though conceding he might have broken his victim's ribs during the struggle, Jacobs claimed to have been threatened by Len Shing, who admitted waving an axe 'just for fun'.[34] When a farmer who witnessed the assault and criticised the troopers' behaviour gave evidence, Jacobs' lawyer accused him of being a pro-Boer, leading to the judge saying he hoped New Zealanders' freedom would not be undermined by punishing individuals for their views regarding the war.[35] Despite Len Shing's injuries and the farmer's evidence, the jury found Jacobs had acted in self-defence and returned a verdict of not guilty.[36] A *Free Lance* cartoon of the incident illustrated prevailing racial stereotypes.[37] It showed Jacobs (in uniform) cowering before an enraged, axe-wielding Len Shing.

As the judge in Jacobs' case observed, although violence against Chinese was neither uncommon nor confined to contingent members, it was no less serious than violence against Europeans.[38] The judge suggested that Chinese were entitled to 'the greater protection of the law', but a disparity nonetheless remained between the sentences soldiers and former soldiers received for assaulting Chinese and the fine handed down in the case of the Mangamāhū publican. Newspaper accounts of atrocities committed during the 1900 Boxer Uprising, including the murder of Christchurch missionary Edith Searell, may have exacerbated existing prejudices, but judges also appear to have taken defendants' military service into account when passing sentence.[39]

On active service, alcohol came as a welcome respite from the rigours of military operations but also contributed to ill-discipline. Trooper Frank Perham described 'rum flying around pretty freely' during a campfire concert, and another trooper said he had never seen men more drunk than after 16 cases of whisky were mistakenly delivered to the Ninth Contingent.[40] Sixth Contingent bugler Percy Mowlem was sentenced to 14 days' imprisonment with hard labour after removing a case of whisky from a railway wagon.[41] Other Sixth Contingent men were also charged for alcohol-related offences, though punishments varied. Lance-Corporal William Laurie was 'admonished' for being drunk on guard duty aboard the *Cornwall*, while Ernest Hart was sentenced to 96 hours' imprisonment with hard labour for being 'drunk when parading for the front'.[42]

Though martial law limited troopers' access to liquor in urban centres, alcohol was a recurring factor in violent incidents involving both British and colonial forces in South African towns.[43] One of the more serious cases occurred while the Fifth Contingent awaited repatriation in the Cape Colony town of Worcester. Corporal Arthur Luck and Trooper Herbert Dixie of the Sussex Imperial Yeomanry and Fifth Contingent saddler Robert Reid became involved in a fracas with a group of Malays.[44] The confrontation occurred after the drunken soldiers caused a disturbance in the town's Malay quarter, prompting irate residents to respond. After entering a cul-de-sac where they found their way barred, the trio attempted to escape through the window of a Malay home. The startled occupant, Abdol Salih, struck Dixie a fatal blow with an axe, seriously injured Luck and repeatedly stabbed the 28-year-old Reid.[45]

Nearly two months after the incident, New Zealand newspapers gave a very different account of events. The *West Coast Times* claimed the soldiers came upon a 'howling mob' of Malays and that the men — outnumbered, unarmed and not wishing to cause a disturbance — unsuccessfully attempted to escape.[46] Even before Salih's trial, New Zealand newspapers reported that the trio's comrades had burnt down 'the murderer's' house and a Malay mosque in the vicinity after unsuccessfully attempting to remove Salih from jail.[47] While English newspapers featured a single-paragraph Reuter's account of the incident, few details were included. The articles implied Dixie (who was not named) was killed quelling a Malay 'riot', rather than actually causing the disturbance.[48] The *Times* mentioned the incident twice, first reporting that Dixie died from a head wound, and then saying that he was killed during a riot between the military and the Malays.[49]

Frank Perham, who was in camp at Worcester, recorded 180 soldiers descending on the town to exact retribution.[50] They apparently included New Zealanders whose officers assisted the commanding officer at Worcester to eventually persuade the men to desist.[51] The day after the mosque was burned, camp orders instructed that 'in future, and until further notice' an additional picket of an officer and 25 mounted NCOs and men would be divided between the Worcester prison and the 'Native Location' from 5.30 p.m. until reveille.[52]

Though Salih was charged with murder and assault with intent to murder, he was found not guilty (a fact that went unreported in the New Zealand press).[53] During the trial, Salih's stepson testified that soldiers had been kicking doors in the Malay quarter. Some South African and New Zealand newspapers mentioned Luck's admission that the three had been out 'on the spree' and that he had no clear recollection of what was said immediately prior to entering Salih's home. Others, including the *Nelson Evening Mail*, did not.[54] When the Worcester Malay community requested compensation, the British government sought tenders to rebuild the mosque and replace destroyed furnishings.[55] But Reid, following a period of convalescence, sailed home, where a Dunedin Medical Board noted he was still suffering from his 'active service' injuries.[56]

While troopers were involved in a raft of alcohol-related incidents, NCOs and officers also faced censure for excessive drinking. In a confidential letter, three lieutenants criticised the inclusion in the Eighth Contingent of several 'utter & absolute wasters' they claimed had earlier been the worst troopers in the Second Contingent.[57] Their captain agreed, describing the men as 'utterly bad', but added a sergeant to the list who he claimed became 'a dangerous lunatic' when drunk.[58]

Alcohol also contributed to the premature demise of the New Zealand military career of Captain John Rose. Seen as a popular and experienced officer, Rose was given command of No. 5 Company in the Third Contingent before its departure for South Africa.[59] In a letter home, he praised the conduct and discipline of his men during the voyage to Western Australia and gave no hint of what was to follow.[60] But after his vessel took on coal in Albany, Rose was forced to make the ignominious choice between resignation and court martial on arrival in Durban.[61] On the voyage from New Zealand, Rose reportedly acted in such a 'ridiculous and objectionable' manner that he was censured by his commanding officer, Major Jowsey.[62] He exhibited similar behaviour after leaving Albany, whereupon Jowsey presented him with the ultimatum.[63] The press blamed Rose's erratic behaviour on alcohol, claiming that after being confined to his cabin the officer reportedly threatened to commit suicide and was placed under guard.[64] After reluctantly resigning his commission, Rose attempted to circulate a petition among the contingent NCOs seeking his reinstatement, but they had already indicated they would resign if forced to serve under him.[65]

On reaching South Africa, Jowsey reported seeing Rose intoxicated in a Durban bar.[66] The termination of Rose's New Zealand service did not, however, end his involvement in the war. He enlisted in Roberts' Horse and, according to the *Auckland Star*, was promoted to the rank of sergeant-major for gallantry in the field.[67]

Crime and punishment

For some, the war offered a way of evading responsibility. Chapter 2 has already detailed the case of the Eighth Contingent recruit who appeared in Paeroa Court charged with attempting to leave the country without providing for his unborn, illegitimate child, and that of the trooper caught embarking at Lyttelton without making adequate maintenance arrangements for his wife.[68]

Others sought to benefit from the largely favourable public sentiment towards war veterans. For a young Auckland man, the temptation to capitalise on the fame of VC recipient Farrier Hardham proved too great. After claiming to be Hardham in a Parnell bar and producing what he claimed to be a Boer explosive bullet, the imposter aroused the publican's suspicions when he was unable to locate sites in Africa connected with Hardham's exploits on a map. Others, it seems, were more gullible; 'Hardham' reportedly received a guided tour of an Auckland brewery, was fêted at the Orphans' Club, received a suit of clothes from a Parnell tailor and drew cash on a fraudulent cheque.[69]

Straitened circumstances may have also contributed to soldiers' criminal acts. Shortly before the war ended a former trooper was found guilty of stealing from his Wellington employer.[70] In an article titled 'Trooper in Disgrace', the *Wanganui Herald* reported that Isidore Cohen had been sentenced to six months' imprisonment with hard labour.[71] The previous October, Cohen, an Australian who had served in the South African Light Horse, wrote to Seddon claiming to be unemployed and impoverished and seeking financial assistance or temporary employment.[72] In 1903 Cohen, a repeat offender, was sentenced to 18 months in prison for forgery and false pretences.[73]

As well as the alcohol incidents, the offence reports of the Sixth Contingent document other activities that attracted disciplinary action in South Africa. Harvey Shields was imprisoned for 10 days for 'riotous behaviour' in Cape Town, and then sentenced to a further two weeks for assaulting an NCO in Durban.[74]

His service record features two discharge certificates bearing Lieutenant-Colonel Banks's signature but written in different hands. One rated Shields' character as 'fair' while the other recorded his conduct as 'exemplary', even though it was Banks who ordered Shields' punishment for assaulting the NCO.[75] In contrast, Sergeant-Major Ernest Lockett, who was awarded the DCM for gallantry and lost an arm due to a bullet wound, received only a 'good' character rating. Even soldiers captured by the Boers were not immune to military justice. After managing to elude his captors, Lance-Corporal William McDonald was sentenced to two weeks' imprisonment with hard labour for being absent without leave, not taking adequate steps to avoid capture and losing his horse and equipment to the Boers.[76]

Military justice could be fickle and if some soldiers got off lightly others felt its full weight. In a case that bore strong similarities to that of Shields, Lieutenant-Colonel Banks was less forgiving. After 29-year-old Private Charles Morris was charged with striking an NCO, creating a disturbance in Albany, drunkenness and disgraceful conduct, Banks found him guilty of gross insubordination. He was ejected from the Sixth Contingent and left in Australia.[77] Despite being 'discharged with ignominy', Morris made his way to South Africa and joined Kitchener's Fighting Scouts.[78] Like Morris, William Hewson, who was sentenced to 84 days' imprisonment with hard labour and forfeited his Queen's South Africa Medal after being found guilty of stealing from fellow soldiers, was discharged with ignominy.[79]

Trooper John Butler of the Third Contingent was another court-martialled and dishonourably discharged for insubordination, using obscene language, threatening another soldier and refusing to help pitch tents.[80] After the Third's commanding officer, Major Jowsey, read the court martial verdict on parade, Butler was paid his outstanding wages, stripped of all government property and evicted from the contingent's camp near Rouxville.[81] Seemingly unrepentant, Butler joined Brabant's Horse and was later killed in action at Winburg.[82]

While many New Zealanders serving in contingents and irregular corps expressed admiration for British soldiers, others were less effusive.[83] Following the arrest of Sixth Contingent Sergeant-Major Nathaniel Davis for insubordination and defying a lawful command, Davis conceded that his men had verbally abused Second-Lieutenant Walker, a young British transport officer who had demanded they surrender a cart they were using. When the New Zealanders refused and 'hooted' Walker, the officer reportedly said 'damn you, and damn your men'. Davis claimed that without his intervention the officer would have received a 'rough handling'.[84] Davis was reduced to the ranks, forfeited his medal entitlement and was initially sentenced to five years' penal servitude for insubordination.[85] Though Lord Kitchener reportedly remitted the prison term, Davis claimed he was discharged in disgrace and prohibited from attending a Wellington luncheon for returned soldiers.[86]

Other cases of insubordination involved troopers openly disobeying orders and banding together to air grievances. Farrier-Sergeant Samuel Haslett, Farrier George Norris, and troopers Onesimus Howe, Pressney Farrow and Henry Harvey were court-martialled and each sentenced to three years' penal servitude (commuted to six months' hard labour) for disobeying orders after refusing to walk to their column because they had enlisted as mounted troops.[87] Howe claimed that despite being hospitalised with rheumatism prior to the incident, he was ordered to walk 25 miles.[88] While conceding the seriousness of their offence, Seddon sought clemency, claiming the men had no previous military training, which was only partially true as Howe had served in the Onehunga Navals.[89] Though the men received no pay while imprisoned, forfeited their campaign medals and lost their £5 imperial war gratuities, Kitchener ordered their release so they could return to New Zealand with the Sixth Contingent.[90]

Reporting another incident, newspapers claimed Seventh Contingent men almost mutinied after two of them were court-martialled for abusing an officer.[91] Though Kitchener dismissed the account, the article titled 'How New Zealanders are Punished. Resentment in the Seventh Contingent' claimed the story came from a trustworthy source. The men were allegedly subjected to Field Punishment No. 1, which involved being tied spread-eagle to a wagon wheel (a practice newspapers labelled torture). According to a trooper who witnessed the incident, the New Zealanders cut the prisoners free and mutiny was only averted by a major promising to speak to the men's commanding officer. The trooper said that while the colonials did not wish to 'rule the roost', they particularly objected to the punishment.[92]

Frank Perham also claimed that Fifth Contingent men who were camped on the outskirts of Worcester were 'in mutiny' when the town was placed off-limits in 1901 due to the risk of typhoid. According to Perham, when the men refused all guard duty except for the railway station, they were given 10 minutes to fall in by an officer; after initially refusing, they finally relented under protest.[93] Perham claimed that Colonel Davis gave the New Zealanders a stern reprimand, though no charges appear to have resulted.[94] Collective action seems to have been more effective when Fifth Contingent men confronted their major about unpaid wages, with Perham noting that two days later the soldiers received their pay.[95]

The apparent leniency shown in these cases was notably absent when Sixth Contingent trooper Charles Tasker was court-martialled for sleeping while on sentry duty. Tasker, who in civilian life worked as a letter carrier at the Wellington Post and Telegraph Department, was the son of prominent Women's Democratic Union leader Marianne Tasker.[96] Yet being a former civil servant with a politically active mother was of little avail to the young trooper. He was sentenced to five years' penal servitude, commuted to one year's imprisonment with hard labour in England, despite being on duty for over 21 hours at the time of his arrest.[97] Tasker continued to take part in military operations for six weeks while awaiting confirmation of his court martial verdict, but was then, with little warning, sent to Cape Town and transported to Gosport Prison in England.[98]

An incensed Marianne Tasker accused the imperial authorities of exceeding their jurisdiction by sending a native-born New Zealander to the United Kingdom for imprisonment. New Zealand, she noted, was a self-governing colony that had offered troops for a specific purpose in a specific place. Beyond those limits, Marianne argued, New Zealand had the right to decide the fate of its troops. Tasker also criticised the 'dark ages' treatment her son had allegedly received en route to England. She claimed the prisoners had been confined below the waterline for the 15 days it took their vessel to pass through the tropics. The intense heat made eating and sleeping almost impossible and resulted in Charles Tasker losing so much weight that a ring Marianne had given him fell off his finger and was lost. On arrival in Southampton, the prisoners were subjected to the additional ignominy of being handcuffed in pairs and marched to the station.[99]

The imperial authorities' handling of Tasker's case was widely criticised in the New Zealand press. An article published in several newspapers titled 'Hardships of Outpost Duty' claimed Tasker's punishment was unlikely to inspire New Zealanders to serve the empire in South Africa. It added that colonial volunteers were increasingly 'out of place' in the British Army.[100] In a letter to the editor, 'A Trooper's Mother' also questioned the legality of Tasker's punishment, claiming that a section of the Army Act 1881, if still in force, prevented the imprisonment of members of colonial corps in the United Kingdom.[101] Seddon apparently agreed; he informed Ranfurly that he understood regulations required Tasker to have been incarcerated either in South Africa or New Zealand. While acknowledging the seriousness of Tasker's offence, the premier drew attention to the trooper's youth and lack of experience and training as a soldier.[102]

Joseph Chamberlain sought clarification from the War Office regarding the legality of Tasker's imprisonment and his continued service following his arrest. The Colonial Secretary was informed that when a column was on the march in hostile country it was impossible to make a distinction between a soldier doing his duty and a prisoner, but this in no way diminished the seriousness of the prisoner's offence.[103] The War Office also indicated that transfers to England of colonial contingent prisoners facing lengthy sentences were legal. Nonetheless, Chamberlain informed Ranfurly in mid-December that Tasker would be released in January 1902.[104] William Pember Reeves, the New Zealand Agent General in London, then asked the Adjutant General to reconsider Tasker's case but was met with a flat refusal. Not only was the Adjutant General totally inflexible, he was unwilling to even discuss Tasker. Reeves then asked the under-secretary of state to speak to Chamberlain about releasing Tasker before Christmas. Although some questioned Reeves' influence, Tasker was released on 21 December 1901 and sailed for New Zealand the same day.[105] When Edwin Jellicoe, the Wellington barrister the Taskers had engaged to plead their case in London, returned to New Zealand he stressed that Tasker's crime was 'punishable by death'. The lawyer, who petitioned Edward VII for Tasker's release, had also met with the Adjutant General while in London. Jellicoe described the New Zealand Agent General's office in London as 'worse than useless', and instead credited 'the King and Lord Roberts alone' with Tasker's early release.[106]

The Taskers' plight struck a chord with parents in New Zealand concerned about the treatment of their own sons serving in South Africa. A correspondent to the *Evening Post* spoke for many when she wrote:

> *We colonial mothers did not give our sons body and soul to the Imperial Army to do as they like with. Our boys may be brave and fearless as lions, but in the tigerish grip of the Imperial Army it shall avail them nothing in the day of trouble.*[107]

Despite his imprisonment Charles Tasker remained proud of his South African service and several years after the war was still attempting to receive the Queen's South Africa Medal he had forfeited due to his conviction.[108] In December 1919, almost exactly 18 years after his release, the imperial authorities finally relented and, after classifying Tasker 'a special case', awarded the Wellington Post Office employee his medal.[109]

The year before Tasker's arrest, a New Zealander serving in the Transvaal Provisional Constabulary was reportedly sentenced to 28 days' imprisonment for the same offence after Major Jowsey intervened. According to the press, Jowsey noted the offender's youth and inexperience, and claimed that severe punishment would mean lifelong disgrace as well as anxiety and suffering for his parents.[110] The following year, the *Tuapeka Times* claimed that the rigid imperial discipline that often turned a British officer into 'an arrogant and offensive martinet' was anathema to the average colonial. It also predicted that in future if colonial soldiers served under British officers there would be less official tolerance of the colonials' independent streak. The newspaper added that imperial military discipline stemmed from a state of society where 'caste and class privileges predominated, and when there was little, if anything, to distinguish between military service and slavery'.[111] During his farewell speech to the Ninth Contingent in 1902, the governor made an oblique reference to Tasker's case. Ranfurly stressed the importance of strict discipline in the contingents and the danger that a soldier sleeping on guard duty represented to his comrades.[112]

Added to these disciplinary transgressions, men serving in New Zealand contingents were also involved in looting, theft and violence in the South African towns they passed through, and during operations in the field. With their borders breached and the capital cities of Orange Free State and the South African Republic in British hands, the Boers were obliged to adopt new tactics. These changes heralded a shift away from the set-piece battles that had characterised the early stages of the war. Despite initial Boer successes, confronting the British in static positions had proven costly in terms of both manpower and resources. Instead, the Boers played to their strengths. Utilising their mobility and extensive rural support networks, they switched to guerrilla warfare, conducting raids into Cape Colony and Natal, targeting vital railway links and harrying the extended British lines of supply and communication.

Mounted troops became increasingly important in countering these threats. Colonial mounted infantry were often employed ahead of the ponderous British columns, conducting gruelling patrols in search of the elusive Boer kommandos, and engaging in the widespread destruction and pillaging of Boer farms suspected of sustaining the enemy. The result was a fluid and frustrating style of warfare. Roving imperial and colonial units often operated at a distance from their senior command structures and, with the tedium of patrols periodically punctuated by violent clashes with the enemy, each side accused the other of criminal behaviour.

A New Zealander claimed that British regiments had scores to settle after Boers allegedly killed British wounded at Vlakfontein, while Leonard Armstrong of the Seventh Contingent recorded finding the bodies of two black African scouts he claimed had been murdered.[113] Though allegations of Boer excesses represented a recurring theme in New Zealand newspapers, the imperial forces were not beyond reproach.[114] Recording a skirmish he had witnessed in his diary, New Zealander John Burnett wrote: 'There were a lot of women and children with the Boers who tried to retreat thinking the British would not fire on them. But they were mistaken for the infantry gave them a hot fire.'[115]

A bad business

As William Hewson's imprisonment showed, men caught stealing from their fellow soldiers could face stiff punishment. The ramifications of looting, however, were less clear. A newspaper article titled 'A Sample of Boer Loot' reported the arrival of items pillaged from the enemy, despite Lord Roberts expressly forbidding the molestation of Boer residents or the looting of their homes.[116] The *Press* claimed that the 'first' looted Boer possessions to reach New Zealand shores consisted of a bundle of music taken from a farmhouse by Trooper Arthur Batchelor.[117] As the war progressed, Boer homes and property were routinely targeted. A newspaper article titled 'Our Polite Army' praised

Above: Sixth Contingent men burning a Boer farmhouse after slaughtering the livestock. ALEXANDER TURNBULL LIBRARY, QMS-1676-59C

Right: A trooper with a pig, poultry and vegetables taken from a Boer farm. The photo is one of several in the diary of Trooper William Frederick Raynes of the Sixth Contingent. Raynes, in civilian life a Waikato farmer, was fatally shot by his own men in an accident at Mokari in September 1901. ALEXANDER TURNBULL LIBRARY, QMS-1676-60D

the behaviour of British forces in South Africa, but such reports painted an unrealistic picture.[118]

While attached to the South Rhodesian Volunteers, Hew Montgomerie of Whanganui recounted entering the town of Schweizer Reneke after the Boers had withdrawn: 'we had no fighting there, but got a good deal of loot of all sorts and burned a good number of houses'.[119] The men's actions were at least in part borne out of frustration. Train wrecking by Boers was relatively common, as were stories of white flag abuses and the use of expanding bullets.[120] Some soldiers expressed scepticism about the accuracy of the white flag reports, but others, such as Sergeant Duncan Blair, claimed to have witnessed the practice first-hand.[121] James Moore of the Fourth Contingent said that at Lichtenburg two Boers had dropped their weapons and hoisted a white flag. When an Australian officer and a corporal rode forward to accept their surrender, Moore alleged both were shot dead by concealed Boers.[122] Dunedin City MHR Alfred Barclay remained unconvinced, dismissing claims of Boer white flag abuse as 'an absolute calumny'.[123]

A number of activities considered illegal in New Zealand appear to have been either diplomatically overlooked or actively encouraged by some officers in South Africa. Trooper Wilfred Wilson, for instance, claimed that once his contingent was in the field, discipline was not strict except for serious offences.[124] The practice of soldiers supplementing their rations with whatever can be found through foraging is as old as warfare itself. This was especially true in South Africa, where operations against a highly mobile enemy inevitably stretched supply lines. The difficulties of accessing adequate supplies were illustrated by Whanganui trooper Norman Palmerston's description of his rations in the Kaffrarian Mounted Rifles: 'we get dry bread and tea for breakfast, a tin of stew for dinner, and dry bread and tea for tea'.[125]

With such monotonous fare, officers knew the salutary effect a good meal could have on flagging morale, and Lord Kitchener's policy of denying the Boers their support networks by burning farms, destroying crops and either removing or slaughtering livestock provided ample opportunity for looting. Trooper Donald Henderson told his brother that '[o]ur officers told us not to be caught foraging, but they did not tell us not to forage'.[126] As well as being a means of obtaining supplementary rations, looting had the added attraction of providing light relief and a brief respite from the rigours of trekking. Accounts of troopers' attempts to catch livestock appear in many wartime diaries, with Frank Perham frequently describing his contingent's pillaging activities: 'At the first farm house we reached the fun commenced. We were told we could take whatever we liked. It was great sport watching the blokes chasing the poultry.'[127] Perham secured vegetables, a turkey and a lamb, which quickly became the ingredients of the soldiers' meal.[128]

Livestock and vegetables were far from the only targets. On one occasion, Frank Perham was disappointed to find little of value in a Boer farmhouse.[129] His brother Luke jokingly claimed that the actions of Third Contingent men in the field were turning them into 'notorious thieves'; Private Leonard Armstrong obtained 'plenty of loot' from a farmhouse recently vacated by its Boer owners; and New Zealand nurse Bessie Teape sent her brother several 'mementoes' including a sovereign case reportedly taken from the pocket of a dead Boer.[130] Officers, however, were not always prepared to either acquiesce or turn a blind eye to their men's depredations. Some of Frank Perham's comrades were forced to pay restitution after being apprehended 'snapping mutton' when they stopped for dinner near a farmhouse, and Te Puke farmer Private Arnold Harris was sentenced to 14 days' imprisonment with hard labour after being found guilty of both stealing and receiving goods at Pietersburg.[131]

Not all New Zealand soldiers were comfortable with looting, with some decrying the practice. Although Trooper Harold Dickinson was critical of stealing from civilians, he seemed oblivious to the apparent contradiction in one of his letters:

> *There is a Boer farm here which has been taken by the troops. You should have seen the things the fellows took. One fellow of ours got a gold watch and chain, another a silver one, and others also got valuables. I myself would not go near the place, as I reckoned it a [damned] shame. Some fellows in the regulars pulled up the floor to see if there was anything hidden there, and others broke the piano, organ, and things, for the sake of saying they did it. I am glad to say that our fellows behaved themselves decently.*[132]

Other New Zealanders did not share Dickinson's scruples about the wanton destruction of Boer property. James Clarke proudly sent his sister a key of a piano smashed by his contingent in a Dutch farmhouse.[133] With troops leaving a trail of destruction in their wake, Boer women and children separated from their menfolk on isolated farms must have felt trepidation at the sight of approaching enemy horsemen. Lieutenant John Montgomerie of the Second Contingent provided a frank account of what Boer families suspected of aiding their compatriots could expect:

> *We then proceeded to Rustenburg, driving the enemy before us in small parties, burning down the houses flying white flags, which sheltered them, after removing any goods that were in them, and the women and children were left weeping over the wrecks.*[134]

During these operations, Boer family Bibles became popular (and controversial) targets of looters. Already reeling from the slaughter of their livestock and the destruction of their homes, the disappearance of these treasured heirlooms was an added blow to the deeply pious Boers. Their loss was all the more keenly felt as the Bibles often included the only copies of family records going back generations. Seventh Contingent farrier Patrick Mulhern removed the family Bible of B. M. Janse Van Vuuren from the Boer's burning home in Witpoort.[135] Mulhern estimated that the Bible, which was published in the seventeenth century, weighed 25 pounds.[136] The Bible of Joseph Johannes Fritz was exhibited in the Parliamentary Library in Wellington, while another 'found' by First Contingent trooper George Powell in the home of a Boer magistrate at Arundel was displayed in Wanganui Museum.[137] At least one visitor to the museum felt uncomfortable with its collection being 'disfigured by a looted Boer Bible' and recommended Powell return it to its rightful owner.[138] Describing Powell's acquisition of the Bible as sacrilege, Masterton MHR Alexander Hogg raised the matter in the House, indignantly observing, 'What a trophy for one Christian to loot from another!'[139]

In 1903 the editor of the *Ohinemuri Gazette* reported that in the recent past Boer Bibles had been exhibited in many cities throughout New Zealand.[140] Some were reportedly auctioned, while Fritz's Bible containing family records was originally displayed in the window of a Wellington pawnbroker.[141] A contingent member's father asked Christchurch City MHR Thomas Taylor to arrange the return of Peter Vanderberg's family Bible. It contained the Boer's will, a lock of hair bound with a blue ribbon, Christmas cards and several 'pathetic little souvenirs'.[142] When Otago trooper James Melville sent his father a number of items he had 'picked up' in a schoolhouse, they including two illuminated vellum pages from a Bible that recorded the births of the Steyn family's children.[143]

The language used to describe the Bibles reflected differing perceptions of the morality of their acquisition. The *Bay of Plenty Times* referred to the Bibles as 'stolen', while the *Marlborough Express* described them as 'captured'.[144] Shortly after the war ended, the Wellington Peace and Humanity Society added its voice to those calling for the return of Boer family records and heirlooms.[145] The scale of the issue became apparent in 1903 when Lord Roberts issued a notice requesting that Boer Bibles that had 'disappeared' from Boer farms be returned to their rightful owners.[146] A correspondent in the *Evening Post* suggested that Roberts should have gone further and requested the return of other looted articles such as watches, rings and letters.[147]

By late 1903 the premier had reportedly arranged the repatriation of several Bibles.[148] Yet despite Roberts' plea, a number remained in the possession of former contingent members long after the war ended. In early 1906 Lieutenant-Colonel Edward Chaytor forwarded to London a Boer family register he claimed had been 'picked up' on the veldt so it could be returned to its rightful owner.[149] During the 1921 Springbok tour of New Zealand, team member J. S. Olivier received a Bible from a Seventh Contingent veteran who had looted it from the Olivier family during the war.[150] Five years later, a Bible belonging to the Louw family was returned, and almost 27 years after the war had ended Mulhern sought the assistance of the government in returning the Janse Van Vuuren Bible.[151] In 1936, the *Auckland Star* reported that former trooper James Brown had returned a Bible belonging to Mrs J. Van Rensburg. The newspaper claimed that there were 'many Boer Bibles still in the hands of New Zealanders'.[152]

In at least one case, this was true. In 1960, New Zealand Press Association correspondent Graeme Jenkins accompanied Wilson Whineray's All Black team on a tour to South Africa. When Jenkins' elderly neighbour, Ellen Wood, heard about the trip, she asked him to take a Bible that her husband, Trooper James Wood of the Sixth Contingent, had taken from a Boer farmhouse prior to it being burnt. As Wood, an Anglican shepherd from Otago, came from a pious family it is possible he took the Moolman family Bible to prevent its destruction. On his arrival in South Africa, Jenkins wrote an article about the Bible that was published in the newspaper *Die Vaderland*. Within 12 hours, he received a response from a grateful member of the Moolman family, and within days the Bible was returned to its rightful owners.

While the practice of looting Boer Bibles was comparatively commonplace, an attempt by a Sixth Contingent officer to acquire more substantial wartime souvenirs was markedly less successful. In 1901, Surgeon-Captain John Purdy of the Sixth Contingent tried to ship two church bells to New Zealand after they were taken by his contingent at Piet Retief. Following the withdrawal of Boer forces from the town, the bells, which Purdy described in a letter to Seddon as 'curios of this war', were found packed in crates ready to be hung in the town's church. Surgeon-Major Sidney Skerman claimed they were taken as they 'were at the mercy of Kaffirs and others' who would have used the cases for firewood and damaged the bells.[153] When Purdy suggested that the bells go to Wanganui and Wellington colleges, Walter Empson, the principal of Wanganui Collegiate School, contacted Seddon asking that his school receive one in recognition of its Old Boys' wartime contribution.[154] However, after enquiring whether the bells could be sent to New Zealand as 'spoils of war', the New Zealand agent in Cape Town, Herbert Pilcher, was ordered to surrender them to the Army Ordnance Department. When the bells failed to arrive in New Zealand, Seddon cabled Pilcher on 23 December asking what had become of them and claiming they should have been 'pealing New Zealand today Merry Xmas Happy New Year'.[155] Lord Ranfurly later informed Seddon that Lord Kitchener had declined the request as they were either private or church property that should never have been removed from Piet Retief.[156]

Cases of indiscriminate looting occurred in all 10 contingents, with Trooper Amos McKegg of the First Contingent openly admitting his involvement: 'We have no compunction in taking anything we can find.'[157] Fellow First Contingent member William Bunten was equally candid but complained that after contingent members had looted a Boer home his share of the booty only amounted to a packet of candles and a love letter.[158] During a 1903 speaking tour of New Zealand, Australian war correspondent and poet 'Banjo' Paterson was

Trooper James Wood of the Sixth Contingent who brought the Moolman family Bible back to New Zealand when he returned from the war. In 1960 the Boer Bible was returned to its rightful owners by Graeme Jenkins when he accompanied Wilson Whineray's 1960 All Blacks to South Africa. ALAN WOOD

deeply critical of both looting and soldiers' apparent willingness to admit their involvement. Paterson was disturbed by the apparent lack of remorse exhibited by both officers and men when recording their activities, and by newspaper editors who published their stories. According to Paterson, soldiers stole from prisoners and looted watches, money and jewellery from private houses — some of which were in British territory. The *New Zealand Tablet* noted the broad smiles on the faces of Paterson's audience when he described the tactics Australian officers and men employed when stealing property. These included issuing forged receipts and illegally requisitioning articles for military use; the audience's amusement reflected the tacit acceptance of such activities by many New Zealanders.[159]

'I had, however, some trouble with that officer'

The majority of disciplinary issues involved troopers and NCOs, but officers such as Captain John Rose were also periodically accused of transgressions. Though reportedly popular with his men, Captain Nigel Markham's 'intemperate habits' saw him relieved of his command in the Sixth Contingent and prevented from assuming command of any other force in South Africa.[160] While New Zealand newspapers remained largely silent on the issue, and Markham and his squadron reportedly received praise for their conduct in action, a confidential letter from his commanding officer described Markham's arrest for being drunk on parade.[161]

In the letter, Major Albert Andrew claimed that Markham had frequently offended in the same manner. Andrew stated that after bringing the case to the attention of General Plumer it was initially decided that Markham could either resign his commission or face a court martial. Apparently keen to protect his contingent's reputation, Andrew pointed out that there had been no other serious disciplinary infractions in the corps and that it had performed well in the field. Plumer relented and settled for Markham being removed from his command and denied any further leadership role.[162] Regardless of his alleged alcohol issues, Markham appears to have been a competent soldier and was hailed as a hero of the South African War in his *Taranaki Daily News* obituary.[163] Adopting a similar tone, the *Observer* described Markham as a 'sterling officer' who possessed military foresight.[164]

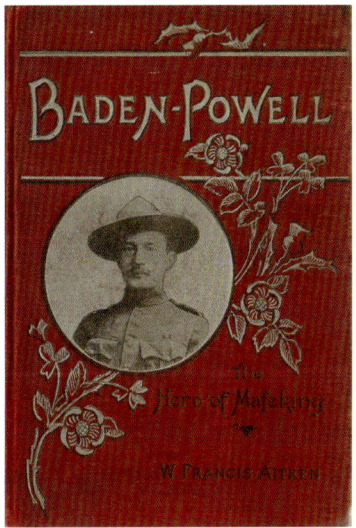

Above: One of the patriotic badges commemorating the relief of Mafeking that were widely available in New Zealand stores. RETTER COLLECTION

Below: William Francis Aitken's book *The Hero of Mafeking*. Published in 1900, it was sold in New Zealand bookstores during the war. It reinforced the image of Robert Baden-Powell as the 'gallant defender' of Mafeking, but contained inaccurate descriptions of conditions during the siege. NIGEL ROBSON

Right: A ribbon manufactured by McKee and Company of Wellington commemorating the relief of Mafeking on 18 May 1900. The ribbons cost sixpence, with the company recommending they be worn by 'every patriotic New Zealander' as a souvenir 'of a great event that will live in history'. NIGEL ROBSON

From top: A gift sent to Robert Baden-Powell by the citizens of the Otago town of Blue Spur following the relief of Mafeking. The medallion is engraved 'Our Trusty Knight of the Empire'.

Medallion sent to Robert Baden-Powell by the children of Campbell Street School in Palmerston North following the relief of Mafeking.

A presentation letter opener sent to Robert Baden-Powell by the citizens of the Southland town of Gore following the relief of Mafeking. THE SCOUT ASSOCIATION (UK) HERITAGE COLLECTION, GILWELL PARK

Above: Some of the Post and Telegraph Department postcards that were popular during the war. The postcards depicted a variety of contingent scenes. Retter Collection

Right: In 1900 the Post and Telegraph Department introduced a postage stamp commemorating the war. The stamp featured the words 'The Empire's Call' and a picture of contingent members framed by the New Zealand flag and a fern. Museum of New Zealand Te Papa Tongarewa, The New Zealand Post Collection, Gift of New Zealand Post Ltd

A postcard sent by Trooper Bert Stevens of the Eighth Contingent to his father in Hāwera. Posted in late April 1902, it did not arrive in New Zealand until after the war had ended. The postcard's reference to the 'Anglo-Boer War' was comparatively uncommon during the conflict. NATIONAL ARMY MUSEUM TE MATA TOA, 1999.3239

A member of the New Zealand Young Ladies' Contingent, a patriotic group formed by Lady Mary Douglas, the wife of the under-secretary for defence, Sir Arthur Douglas. ALEXANDER TURNBULL LIBRARY, A-091-050

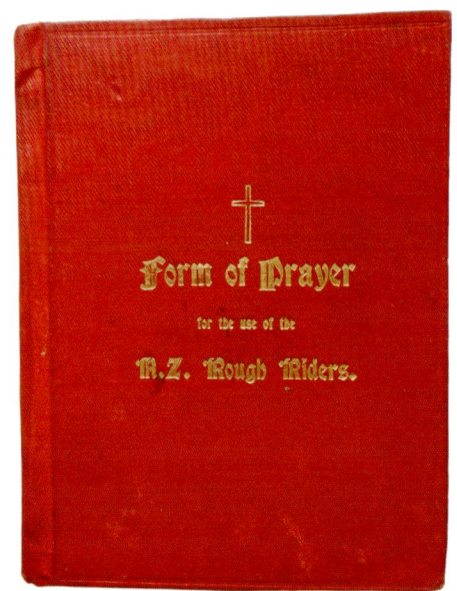

Above: One of the prayer books given to Major Jowsey by Bishop Julius of Christchurch in February 1900. Julius wrote each trooper's name in the books by hand. They were then distributed by Jowsey to the members of the Third 'Rough Riders' Contingent. In this book, given to Trooper Richardson, the bishop hoped that 'On the journey and in camp [the prayer book] may remind you of that only strength and protection which can support you in all dangers, and carry you through all temptations'.
(Retd) Noel W. Taylor ED** RNZIR

Below: A Soldier's New Testament distributed to contingent members in 1900 before their departure for South Africa. National Army Museum Te Mata Toa, 1996.1700

Opposite: A Presbyterian Church of New Zealand programme for a service at Dunedin's First Church on 11 March 1902 to commend Captain-Chaplain Daniel Dutton and the men of the Ninth Contingent 'to the safekeeping of Almighty God'.
Presbyterian Research Centre (Archives)

Presbyterian Church of New Zealand.

SERVICE

To set apart
the Rev. D. Dutton
as Chaplain to the Ninth Contingent,
and to commend him and the Men of
that Contingent to the safe-keeping of
Almighty God.

In First Church of Otago,
TUESDAY, 11th MARCH, 1902,
At 8 p.m.

**The Right Reverend the Moderator of
General Assembly (The Rev. J. Gibb)
will preside.**

S. N. BROWN CO. PRINT.

Above: A Sixth Contingent Christmas card. NATIONAL ARMY MUSEUM TE MATA TOA, 1996.1439

Opposite: Employees of New Zealand Railways Hillside Workshops who served in South Africa. ARCHIVES NEW ZEALAND, AAAA 21953 D579 BOX 1

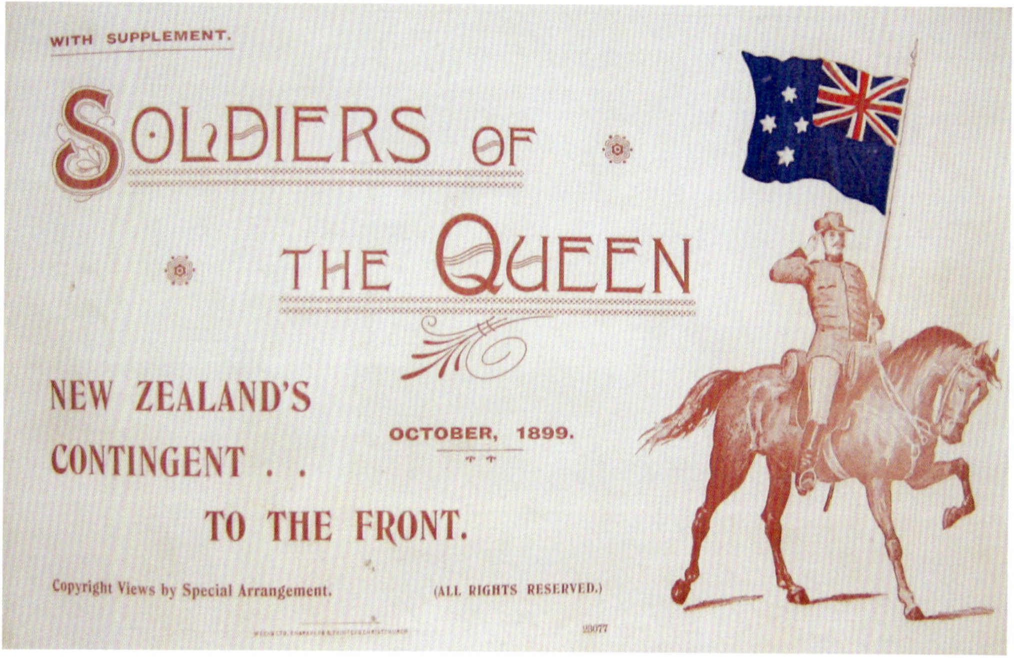

Above: 'Soldiers of the Queen', a souvenir booklet produced to mark the dispatch of the First Contingent to South Africa in October 1899. RETTER COLLECTION

Right: A jug featuring the image of Lord Roberts, the British commander-in-chief in South Africa. The jugs, which were sold for 1/6, were among other 'patriotic goods' sold in Nelson at G. Kidson's Trafalgar Street premises. The store also sold Lord Roberts plates, plaques of the 'Queen and generals', brooches of British generals, and 'sons of Britain' knives. RETTER COLLECTION

Clockwise from top left: The Veld Cross awarded to Sergeant Frank Hemphill of the Sixth Contingent by the Cape Town *Veld Weekly* newspaper in honour of Hemphill's actions at Bastard's Drift. Hemphill, of Maungataroto in Kaipara, also served in the Eighth Contingent.
CAROL AND MURRAY HEMPHILL

The Queen's South Africa Medal of Walter Armstrong, a Whanganui chemist who died of typhoid at Raadsaal Hospital in Bloemfontein while serving in the Royal Army Medical Corps. RETTER COLLECTION

Commemorative medallions sold to the New Zealand public as mementoes of the Second Contingent and the Rough Riders Contingent. RETTER COLLECTION

Porourangi, the mere gifted to Lord Roberts in February 1901 by Tuta Nihoniho following the death of Roberts' son at the Battle of Colenso. The mere originally belonged to the Ngāti Porou rangatira Kahukuranui. Porourangi had been handed down through 19 generations. NATIONAL ARMY MUSEUM, UK, 1971-01-25-4-1

The Cape Colony town of Worcester. On 2 June 1901 New Zealander Robert Samuel Reid of the Fifth Contingent was seriously injured after becoming involved in a fracas with Malay residents of the town. NATIONAL LIBRARY, CAPE TOWN, AG112

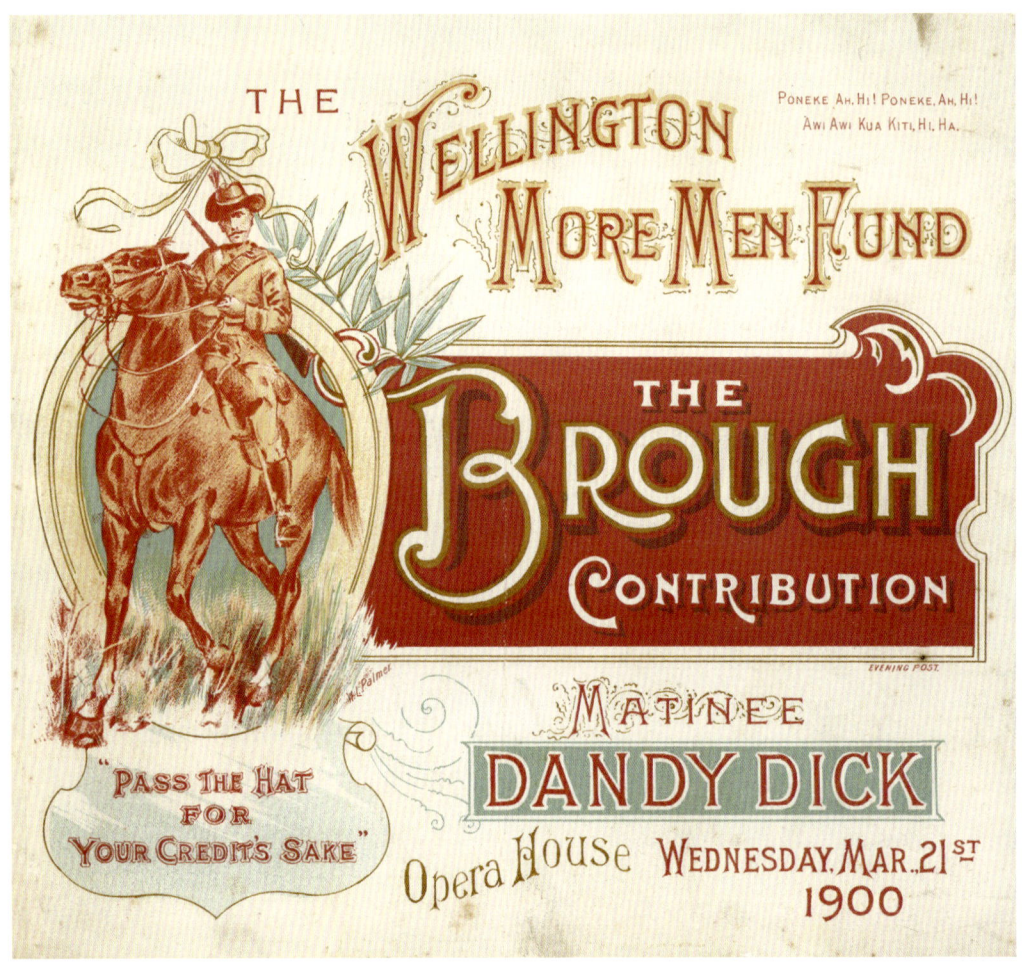

The programme cover of an event organised by the Wellington More Men Fund at the Wellington Opera House to raise funds for the South African War. ALEXANDER TURNBULL LIBRARY, EPH-A-WAR-SA-1900-01-COVER

South African War memorial window in the Whanganui Collegiate School chapel. Although originally in memory of Leslie Seton Melville, Walter Armstrong and Campbell Parkinson, it later commemorated all nine Old Boys who died in South Africa. The window was ordered from Powell & Sons in London at a cost of £40 and was unveiled by the Bishop of Wellington on 25 November 1901. NIGEL ROBSON

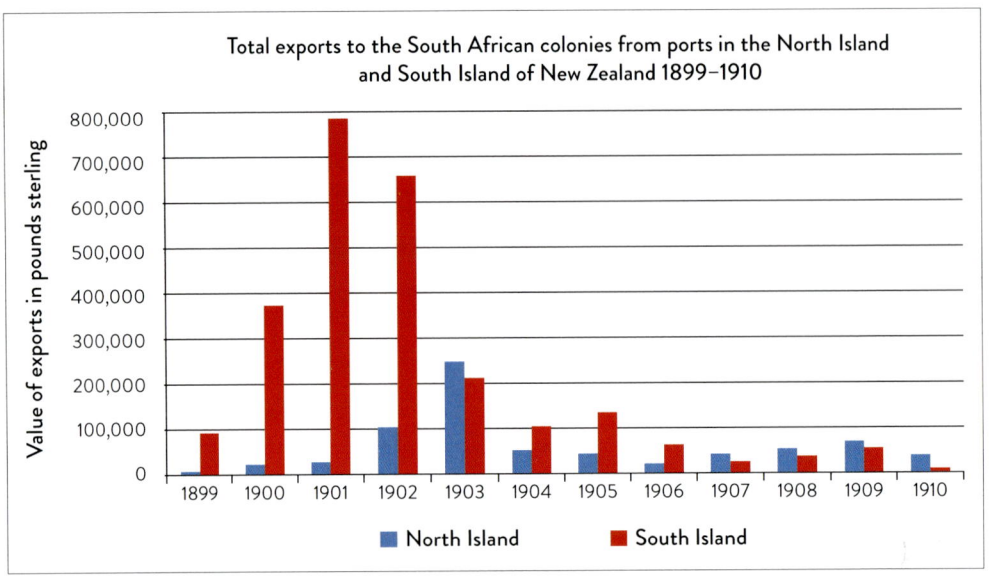

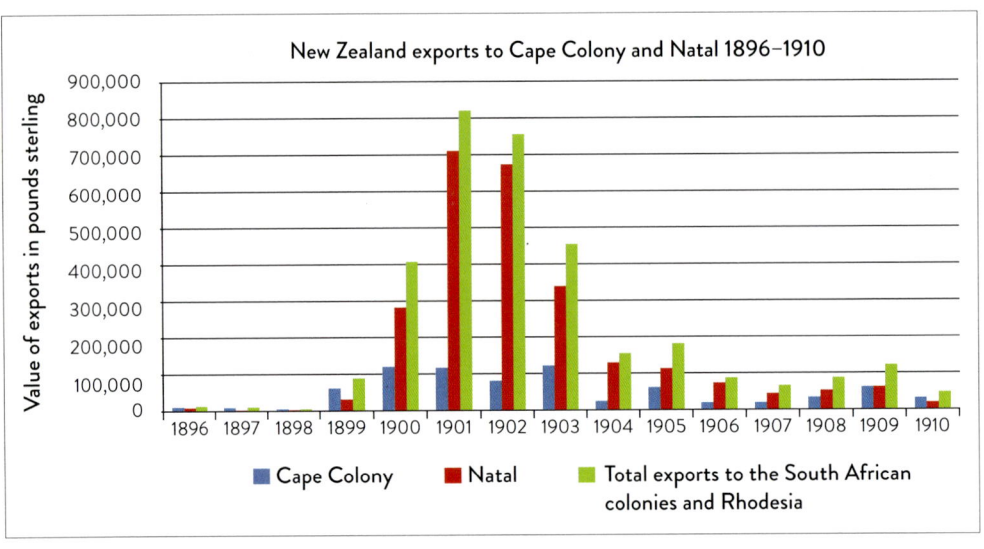

There was a significant increase in New Zealand exports to the South African colonies of Cape Colony and Natal during the war. The majority of this trade passed through New Zealand's southern ports. Following the annexation of Orange River Colony and Transvaal Colony in 1902, New Zealand also exported to these regions, but in the post-war years the African market proved unsustainable due to high transport costs and increased competition. NIGEL ROBSON

'Yelling yahoos in yellow'

Accusations of theft were also levelled against New Zealand officers, though in the case of one future MHR, Captain Francis Fisher, his accuser was neither a disgruntled Boer nor a South African shopkeeper. Following the war, the Commandant of New Zealand Forces asked Fisher's commanding officer in the Tenth Contingent, Major Andrew, whether Fisher had been placed under arrest while serving in South Africa. In a confidential 1903 letter, Andrew replied that he had not, but went on to say he had experienced 'some trouble' with Fisher and documented his subordinate's alleged offences while the contingent was stationed in the Natal town of Newcastle. Andrew claimed he had been approached by Fisher seeking permission to return to New Zealand via Canada aboard a Canadian troopship. As the war had ended, Andrew consented and instructed Fisher to pass command of his company to the next most senior officer. Andrew also said he agreed to Fisher receiving both pay due to him from the paymaster in Durban and his 40 days' 'passage advance'.[165]

The day after Fisher departed for the coast, Major Pennycook, the commanding officer of the Tenth Contingent's 2nd Regiment, accused Fisher of leaving without arranging for the care of his horse or repaying money owed to his brother officers. Pennycook also accused Fisher of not settling his mess bill and failing to return his saddle, bridle and government service revolver. On hearing this, Andrew cabled the Durban authorities, revoking Fisher's permission to embark and ordering the junior officer's immediate return to Newcastle (described by one trooper as 'a dirty little one-horse place') to answer the allegations. Andrew claimed that 'beyond a string of recriminations, as childish as they were impertinent [Fisher] had not the slightest defence to offer'.[166]

Andrew's irritation was increased by Fisher being the third case in a week of an officer being accused of stealing weapons and equipment belonging to the imperial government. As in Captain Markham's case, Andrew was apparently more concerned about the impact of a court martial on the image of New Zealand and its contingents than pressing charges. He noted that while Fisher's actions were court martial offences, the Tenth Contingent was due to leave for New Zealand and he had no desire to tarnish the reputation of New Zealand and its troops. Instead, he chose to order Fisher home with the Ninth Contingent following a stern reprimand.

Officers were also accused of embezzlement. Lieutenant William Tutepuaki Pitt, the paymaster of the Eighth Contingent, was asked to account for £314 12s 4d of missing imperial funds.[167] When the imperial authorities rejected Pitt's explanation for irregularities in his contingent's accounts, Seddon wrote to the New Zealand High Commissioner in London, informing him that Pitt had been chosen for the paymaster's role due to his trustworthiness and suitability. Though not entirely satisfied with Pitt's explanation, the premier noted that the lieutenant had cooperated fully with the investigation and there was no evidence of him misappropriating funds. He recommended that given Pitt's limited means, the debt should be written off, but after Pitt refunded a substantial amount of the funds, the Paymaster General proposed that Pitt repay the remaining £101 in £2 monthly instalments.[168]

Captain George Walker was another officer accused of criminal behaviour. In a 1902 letter, Seddon informed the commanding officer in Cape Town that Walker had left for South Africa without arranging financial support for his three daughters and his 'faithful industrious and deserving' wife, who had been forced into domestic service. After serving in the Fourth and Eighth contingents Walker, who received the Distinguished Service Order, elected to remain in South Africa, where he joined the Johannesburg Mounted Constabulary.[169] Although described as the perpetrator of 'vile cruelty' for his treatment of his family, it was accusations of theft that attracted press interest.[170] The *Wanganui Chronicle* devoted an entire editorial to Walker's imprisonment by the Cape civil authorities, and the *Taranaki Herald* reported that one of Taranaki's own had been arrested after being cleared months earlier by a military inquiry that found no grounds for a court martial.[171] Walker's case was raised in Parliament by Egmont MHR William Jennings, who described the captain's treatment as 'simply diabolical'.[172] The *Wanganui Chronicle* reported that Walker was accused of stealing £301 (Jennings put the sum at £410) from a Boer woman named Elizabeth Sieberhagen after he and fellow members of the 'Colonial Light Horse' searched her home during the war. After failing to get satisfaction from the military authorities, Sieberhagen lodged an affidavit again accusing Lieutenant Albert Barratt of the Cape Colonial Forces and Captain Walker of the theft, leading to their arrest and imprisonment.

In October 1903, Seddon wrote to the prime minister of the Cape Colony seeking details of Walker's case. In response, the premier received a copy of testimony taken during the case from the colonial prime minister's office.[173] Walker described 'Klipkraal', the Sieberhagen farm, as a 'rebel nest' in his evidence, and stated his patrol burned the family's mill and a number of carts. While he admitted to removing the money in gold and notes from a locked box after cartridges were found in the home, Walker denied stealing it and claimed he replaced it prior to his departure. Sieberhagen was equally adamant that the soldiers stole money from both her and her son, as well as taking a gold watch and two gold chains from their home. In February 1904 Seddon received a letter from the prime minister of Cape Colony's office informing him that, after examining the evidence, the Attorney General had declined to prosecute Walker, who then was in the employ of the Cape Colony government.[174]

Mayhem or murder? The Newcastle affray

Prior to the Tenth Contingent's 1902 departure for South Africa, an officer praised the troopers' behaviour, noting there had not been a single offence reported during the contingent's camp.[175] While this may have been true, his men would later be implicated in one of the more sensational discipline-related incidents involving New Zealand troops in South Africa. After their long voyage, contingent members were frustrated to discover that peace talks had effectively removed any prospect of them seeing active service. It also ensured they would return to New Zealand with, as the *Auckland Star* put it, 'fewer laurels than their more fortunate comrades'.[176]

Instead of heading to the front, the contingents made their way to camps near Newcastle, where they joined around 7000 other British, Canadian and Australian troops awaiting either their discharges or the troopships that would carry them home.[177] The 1068 Tenth Contingent officers and men who arrived shortly before hostilities ended narrowly missed seeing Seddon, who had passed through two days earlier and inspected the Seventh Contingent on the day it departed for Durban.[178] With the Eighth and Ninth contingents reaching Newcastle in late June, the number of New Zealanders encamped at Fort Hay swelled to over 3200.[179] While the battle-hardened Eighth Contingent had seen extensive combat, many in the Ninth and Tenth contingents had never faced the enemy, with Sergeant Arthur Stock claiming they were disappointed that they would 'not fire a shot'.[180] Ninth Contingent trooper Herbert Hart expressed similar views when he wrote of his frustration and annoyance at having given up a good job as a Carterton accountant, left friends and family behind and travelled all the way to South Africa only 'to be quietly packed off home again'.[181]

The extent to which this disappointment affected their subsequent behaviour is difficult to gauge, but a 1902 court of inquiry found New Zealand contingents were involved in looting Indian stores in Newcastle and the Johannesburg suburb of Jeppestown, with the Eighth Contingent also accused of looting a Natal hotel.[182] Indian storekeeper Nehora Sing testified that he had seen men wearing fern leaf badges among Canadian and Australian soldiers who looted his Jeppestown store.[183] The court heard that between two and three hundred troops created a 'great disturbance' in Newcastle's main street, where contingent members allegedly destroyed a bar.[184] A British provost sergeant noted that 'the most rowdy were Canadians and New Zealanders'.[185] Later the same month Hart recorded how New Zealanders attempted to steal the goods of Indian vendors, but 'did not get very much'.[186]

After a Natal hotel owner refused to sell Eighth Contingent men alcohol without a permit, he was allegedly assaulted by troopers who broke billiard cues and stole food and decorations prepared for an event commemorating Edward VII's coronation.[187] Although Eighth Contingent commanding officer Colonel Richard Davies denied the allegations, the court's opinion was that the Eighth, Ninth and Tenth contingents should each pay 11 per cent of the Newcastle restitution claims, with Canadian and Australian contingents liable for the remainder.[188] Public accounts from 1904 indicate the New Zealand government paid £20 16s 8d to the imperial government for 'claims for looting in South Africa by the Eighth and Tenth Contingents'.[189] Shortly after the inquiry, the commanding officer for Natal District informed Lord Ranfurly that 'extensive looting' had allegedly taken place in northern Natal while Canadian, Australian and New Zealand contingents were stationed in the area and indicated that 'many more claims' were anticipated.[190]

Although many details of the events that followed were obscured by official denials, varied and contradictory accounts, and allegations that the incident was hushed up, on 4 September 1902 several New Zealand newspapers featured 'sensational' accounts of a fight in Newcastle shortly before the Tenth

Contingent's departure.[191] The source appears to have been two unnamed officers of the recently returned Tenth Contingent. The officers reported riotous behaviour during a brawl between New Zealanders and imperial troops near the Newcastle town hall, with claims that the former had wielded stirrup irons attached to leather straps during the fight.[192] Newspapers printed claims that two of the New Zealanders' opponents had died in the brawl, which reportedly escalated to involve large numbers of Tenth Contingent men and a number of Royal Munster Fusiliers and British dragoons. If Herbert Hart's diary is accurate, the incident occurred on 11 July: 'Our men had a fight downtown today with the result that a Munster Fusilier died from wounds received.'[193]

In reply to a question in Parliament by Bruce MHR James Allen, Acting Premier Joseph Ward categorically denied the incident had taken place, dismissed the accounts as 'absolutely untrue' and admonished Allen for putting the matter on record.[194] Ward said he had sought verification from contingent officers and had been informed the accounts were 'incorrect', adding that it would be almost impossible for the fight to have taken place without the knowledge of New Zealand officers, who would have been duty-bound to inform the government. Yet Ward's hasty dismissal of the reports did nothing to suppress media interest. The Christchurch *Star* claimed that as accounts of the fight had been discussed by so many returning contingent members they must contain a degree of truth.[195] The newspaper reported that recurring elements were the death of two 'Tommies', and also the prominent role played by men from the North Island section of the Tenth Contingent. Wellington's *Evening Post* said a witness dismissed the claim that stirrup irons were used on the grounds that the troops had no horses.[196] While this may have been true at the time of the incident, Trooper Hugh Collison died after falling from his horse at Fort Hay on 16 June.[197]

One account claimed that two dragoons were found dead with their skulls fractured, and alleged that the incident had been hushed up, the deaths appearing as accidents in official reports.[198] The involvement of both Munster Fusiliers and dragoons is possible as in June 1902 several hundred officers and men of the 2nd Royal Munster Fusiliers were based at the Newcastle Defence Camp and Fort Amiel, while an equally large number of 6th Dragoon Guards were encamped at Windsor Castle Camp.[199] While admitting a fight between

Boers outside Newcastle Town Hall in Natal. Members of New Zealand's Tenth Contingent reportedly fought British and Irish soldiers in the vicinity of the hall shortly after the South African War ended. NATIONAL ARCHIVES OF SOUTH AFRICA, PRETORIA, TAB 30455

contingent members and the Munster Fusiliers had taken place, Captain Henry Coutts (who had joined the Ninth Contingent) claimed that only three New Zealanders were involved.[200] The *Evening Post* suggested there may have been some confusion with a prior incident in the Newcastle camp, where a drunken New Zealand trooper had reportedly run amok while wielding a stirrup iron.[201]

Yet another version had two New Zealanders breaking regulations by leaving Fort Hay without permission, entering Newcastle and getting drunk with two British soldiers. The English troops reportedly abused their colonial comrades using language 'no self-respecting colonial could submit to', with the ensuing fight quickly spreading to involve others.[202] Several newspapers claimed the incident began when the British provost marshal ordered Munster Fusiliers to administer summary justice after a New Zealander abused him for enforcing dress regulations.[203] Three hundred New Zealanders who allegedly came to their comrade's assistance were joined by five hundred other colonials in the brawl after the provost marshal ordered a squadron of mounted dragoons to charge the rioters using the flats of their swords.[204] Wesleyan clergyman Captain Chaplain Luxford claimed he saw the start of the fight and spoke to the superintendent of police the following morning, who assured him that while two dragoons suffered injuries, they were not serious. A conflicting account claimed Luxford had said that the two Munster men reported to be dead were only severely injured.[205]

There was an element of regionalism in coverage of the incident, with the *Otago Witness* devoting three columns to the story and printing an Otago soldier's claim that several North Island men had got completely out of hand and seemed to be looking for trouble. Though a Blenheim trooper claimed to have witnessed the incident, the Otago soldier asserted that the South Island men knew nothing of it until the following morning when they entered Newcastle and found the street looking 'as though a pitched battle had taken place', with windows and doors smashed and businesses looted.[206]

According to one Tasmanian newspaper, the behaviour of the troopers from the last three contingents had been poor at the end of the war.[207] Some of the men, it alleged, created disturbances while drunk in Cape Town, Durban and Newcastle. The newspaper also stated that in one town a party of New Zealanders had hired all the rickshaws and raced them through the main street while firing their weapons.[208] The *Western Australian* published a contradictory account; it praised the behaviour of men from the last three contingents but then added a trooper's claim that an imperial soldier had been killed at Newcastle, with several New Zealanders being court-martialled and fined.[209] Accepting that the brawl had occurred, the *Manawatu Daily Times* was less critical of the troopers' behaviour:

> *We know that our men are not made of the stuff that meekly submits to an insult, and we would be sorry if they were, while we also know that men placed in their circumstances, men just let loose from strict discipline, are never the orderly beings they are under more normal conditions.*[210]

The day after the incident, the *Natal Advertiser* accused colonial troops of having little respect for law and order.[211] According to the newspaper, a serious riot in Newcastle's Scott Street — caused, 'as usual', by colonial regiments — had resulted in several people being injured. It was high time, it added, that something was done to prevent a recurrence of similar incidents where rocks and other projectiles were hurled in the street. Claiming that officers of the regiments either could not or would not interfere, the newspaper suggested there were better ways of settling grievances between colonials and garrison police than rioting in the streets.[212] A contributing factor appears to have been British soldiers' resentment at colonial and imperial pay disparities.[213] Eighth Contingent captain Conrad Saxby reflected this imbalance when he doubted the New Zealanders would remain in South Africa following the end of hostilities as British troops could be kept more cheaply.[214]

Whether by plan or in an attempt at damage control, Tenth Contingent men departed for New Zealand within days of the Newcastle incident. The *Waimate Witness* referred to soldiers being 'bundled onto the transports before things were ready for them as a consequence of what took place at Newcastle'.[215] Herbert Hart also noted that most of the Tenth Contingent departed for home only days after the incident.[216] Questioned about the *Britannic*'s 1902 voyage from Durban to New Zealand, Major Lucius O'Brien said that the behaviour of returning troops was markedly worse than when they sailed to South Africa. When a commissioner asked whether O'Brien meant it was impossible to enforce proper discipline on colonial troops, the officer replied, 'Yes, sir, exactly.'[217]

Although the *New Zealand Herald* referred to the rest of the Tenth Contingent's hasty embarkation aboard the *Montrose* at Port Natal on 15 July, its departure was not entirely unexpected.[218] A telegram sent to New Zealand by Major Andrew in late June indicated that the contingent would probably embark for New Zealand early the following month.[219] Andrew did not refer to the *Britannic*, but in a second cable sent the day after the Newcastle incident Andrew indicated that the Tenth would embark from Durban aboard the *Montrose* on 15 July.[220] A passage in the 1902 Contingent Transport Commission report referred to the men aboard the *Britannic*, but it might just as well have applied to the Tenth Contingent while encamped at Fort Hay. The report warned of the need for close regulation when troops were only partially trained, and discipline was relaxed following the end of hostilities. It also alluded to the lack of *esprit de corps*, which, it claimed, comes only from 'long association coupled with traditions of an inherited glory'; the Ninth and Tenth contingents had neither.[221]

The exact details of what took place at Newcastle in July 1902 may never be known, but given the soldiers' accounts and reports in the Natal newspapers, it appears that neither Joseph Ward's blanket denial nor Coutts' claim that it was a minor fight involving a few New Zealanders accurately reflects what occurred. As implausible as it may seem, the possibility that details were intentionally concealed, as the *Evening Post* account claimed, cannot be entirely discounted. The conduct of the war had already been the subject of criticism in Europe, and the imperial government was unlikely to have welcomed evidence of discord between British and colonial forces.

Though briefly sensationalised in the press, incidents like the Newcastle brawl and the Worcester affray all but disappeared from the pages of New Zealand newspapers in the post-war years. A rare exception was a passing reference to the contingents' 'Worcester or Newcastle doings' in a 1903 letter to the *Wanganui Chronicle* editor by D. Fleming, the vociferous and outspoken critic of what he termed the 'disgraceful war'.[222] No mention of Abdol Salih's acquittal seems to have appeared in New Zealand, Australian or British newspapers, and the claim that Dixie died while attempting to quell a Malay riot was never retracted.

Some of the behaviour of contingent members, such as the violent attacks on Chinese, the looting of Indian stores in South Africa and the retribution exacted on the Worcester Malays, could be viewed as evidence of prevailing attitudes towards Asians among New Zealand's European population. Like attitudes towards black Africans, these negative associations both pre-dated the conflict and continued in its wake. In 1907, the same judge who fined Michael Sullivan £20 for his attack on the Mangamāhū publican fined a civilian £2 plus costs for assaulting an elderly Muslim.[223] Though the war may have exacerbated some contingent members' racial or religious intolerance, these views were already ingrained in New Zealand society. Newspapers like the *Observer* were instrumental in cultivating the impression that the overall behaviour of the New Zealand contingents was exemplary and largely beyond censure. The newspaper favourably compared the New Zealand soldiers with their Australian counterparts, and maintained they had behaved in a manner creditable to their country:

> *There may have been individual examples of pillage for the sake of gain, or misconduct in other directions, but collectively our contingents have carried themselves as gentlemen, and we have reason to be proud of them. We have no dirty linen to wash on their return, no stories of blackguardism to blush for, and no claims for depredations on private property in the towns they have passed through.*[224]

Disciplinary issues involved a comparatively small percentage of the more than 6500 New Zealand soldiers who served in South Africa but nonetheless tarnished their record. New Zealand soldiers were certainly not alone in engaging in illegal activities in South Africa, but the *Observer*'s sweeping rebuttal of 'dirty linen' is refuted by a substantial body of evidence suggesting otherwise. Sufficient cases of drunkenness, violence, theft, insubordination and ill-discipline occurred in New Zealand, on the troopships and when the men were in South Africa to suggest that New Zealand soldiers periodically chafed under the yoke of military authority.

CHAPTER SIX

'MAIMED, CRIPPLED AND COMPLETELY BROKEN'

THE HUMAN COST OF THE WAR

In an August 1900 letter published in the *Wanganui Collegian*, Sergeant Duncan Blair provided a brutally graphic account of the death of fellow Old Boy Luke Perham.[1] Whether Perham's family read the account is uncertain, but if they did it could only have added to their trauma. In addition to those who died in combat were men who sustained serious injuries, or contracted debilitating diseases either in South Africa or on the troopships. Though New Zealand casualties were considerably lower than the combined Māori and Pākehā death toll during the country's earlier internal conflicts, deaths attributable to the war in South Africa were not insignificant. A 1903 parliamentary return indicated that 182 contingent members died while on active service in Africa, including soldiers who lost their lives as a result of accidents, incidents, disease and various other medical conditions. Of these, 68 contingent members either were killed in action or died of wounds, while a further 74 died of typhoid (known as enteric fever at the time). The report also recorded 44 men who died after leaving South Africa either on board troopships or shortly after their return to New Zealand.[2] Yet the exact human toll of the war on New Zealand may never be known. A number of New Zealanders lost their lives while either serving in the armed forces of other countries or in irregular corps raised within South Africa without having recorded their New Zealand connections.

The New Zealand government's figures did not, for instance, include New Zealanders such as Leslie Melville (the eldest of the four Melville brothers), Charles Arkell and Walter Armstrong — all of whom succumbed to typhoid.

In 1899, Captain William Russell spoke in Parliament of contingent members returning 'covered with glory and with wounds', but it was typhoid rather than bullets that caused the largest number of deaths among New Zealanders.[3] Melville, who was born in Christchurch, served in the Natal Border Mounted Rifles and died of the disease at Ladysmith, while Arkell, an Auckland Grammar Old Boy, served in the Royal Army Medical Corps.[4] John Thomas was yet another whose life was cut short by typhoid; from Westport, he served in the 2nd Imperial Light Horse and died at Charleston.[5]

The 22-year-old Armstrong, who died at Raadsaal Hospital in Bloemfontein, was a skilled pharmacist, but his calling was no protection against the virulent disease.[6] After completing a six-year apprenticeship at a Whanganui chemist and scoring highly in his pharmaceutical examination, Armstrong left for London where he gained further experience at a Regent Street pharmacy.[7] The War Office selected him to fill one of 10 dispensers' positions, and two days after Christmas 1899 he enlisted.[8] While awaiting transport to South Africa in the garrison town of Aldershot, Armstrong wrote his will, leaving all his 'earthly belongings' to his mother, Louisa.[9] Once in South Africa, he served in the Royal Army Medical Corps as a 'compounder of medicines' at the 14th Brigade Field Hospital and the 19th Field Hospital, but he died in April.[10] Only days after learning of their son's death, Armstrong's grief-stricken parents received a letter from him saying he was in good health and including a photo showing Armstrong proudly posing in his Medical Corps uniform.[11]

Among other recorded causes of death for New Zealand soldiers were pneumonia, gangrene, malarial fever, dysentery, pernicious fever, chloroform poisoning, brain tumour and suicide.[12] Families deprived of husbands, sons, siblings and breadwinners often struggled to deal with their loss, as did those whose family members returned either sick, injured or traumatised. Even if the men and women who served in South Africa returned to New Zealand physically unscathed, wartime experiences were not easily forgotten.

In 1900, the brother of 31-year-old Captain John Harvey of the Fourth Contingent received an unexpected telegram from New Zealand's premier, informing him that John, in civilian life a Balclutha banker, had been killed during the fighting at Ottoshoop. Seddon expressed his 'sincerest sympathy' and 'deep regret' at the family's loss, and told Harvey he would leave it to him to break the unhappy news to the captain's mother.¹³ In a telegram sent in reply, Harvey said it was some consolation to know that his brother had died 'for the Flag' and that Captain Harvey and his men had 'upheld the honour' of New Zealand.¹⁴

Harvey's death was widely reported in the nation's press. Dunedin's *Evening Star* described the fighting at Ottoshoop as a severe disaster, and another newspaper noted that Harvey was killed six months to the day after being presented with a horse by the patriotic young women of Balclutha.¹⁵ Harvey, who had reportedly been a popular officer in the Clutha Mounted Rifles, served in the Volunteers corps for three years prior to joining his contingent.¹⁶ Earlier in the war, Queen Victoria had indicated she wished to have photographs of all colonial officers killed in South Africa, and following an approach from the Colonial Secretary a photo of Captain Harvey was duly dispatched to the aging monarch.

In one of the more unusual wartime commemorations, the Dunedin Horticultural Society marked Harvey's death by naming a variety of daffodil after him. The flower, newspapers reported, was named 'Captain Harvey' in honour of the Dunedin High School Old Boy who went to South Africa 'as a soldier of the Queen'.¹⁷ Harvey was held up as a role model for the boys at his former school. At a school parade a year after his death, the cadets turned out in full uniform and armed with rifles as the dead captain's revolver was ceremoniously presented to the corps by Lieutenant-Colonel De Latour.¹⁸ For even the least attentive student, the lesson was clear: dying on the battlefield in defence of flag and empire was admirable and those that did so would not be forgotten. Addressing the cadets, the mayor of Clutha said that Captain Harvey had 'lived an honourable and helpful life and died a hero's death'.¹⁹

Only days before Christmas 1899, news of the death of George Bradford, a Paeroa trooper in the First Contingent, reached New Zealand. The former Grenadier Guardsman had reportedly been killed during a skirmish with Boer forces at Arundel in northern Cape Colony. Bradford's death took on special significance as he was the first soldier to die overseas while serving in a New Zealand military force. For newspapers, that fact alone made Bradford good copy. The *Evening Star* claimed that he was likely to be remembered as the first colonial soldier to die in the imperial cause. At a time when many Europeans

Left: The photo of Captain John Harvey sent to Queen Victoria at her request. Together with Wairarapa sheep farmer Trooper Septimus McDougall, Harvey was killed in action on 16 August 1900 at Buffelshoek approximately 20 miles north-east of Mafeking. In Seddon's telegram of condolence, official records and most newspapers, the engagement in which the men lost their lives was referred to as Ottoshoop, the name of the nearby town. THE ROYAL COLLECTION © HM QUEEN ELIZABETH II, 2011. RCIN 2502305

Right: The Klee brothers Louis, Victor and George served together in the Seventh Contingent. Although they all survived the war, Louis was seriously injured after being accidentally shot in the leg while pig hunting in the Wairarapa in 1904. ALEXANDER TURNBULL LIBRARY, PAColl-5657-15

resident in New Zealand were not New Zealand-born, the public had few reservations about embracing the Englishman as one of its own.

By the following day, however, the story had changed. Bradford was reported to have survived his wounds and been taken captive by the enemy.[20] Further contradictory accounts emerged in the press: some newspapers claimed Bradford was alive and a prisoner, while others correctly reported that he had died in captivity. His death was portrayed as a distinction as he was the first 'to seal with his blood the new bond that was drawn when the contingent was despatched'.[21] Fatalities like Bradford's were presented to both New Zealand and the greater empire as an affirmation of the nation's commitment to the imperial cause. Australian newspapers featured articles about Bradford, variously claiming that he was wounded but likely to recover, or dead from meningitis.[22] Without elaborating, the Western Australian *Bunbury Herald* and *Kalgoorlie Miner* both described Bradford as the 'First Australian Killed'.[23]

Comparatively anonymous in life, in New Zealand at least, Bradford assumed celebrity status in death. A Masterton store sold handkerchiefs for a shilling featuring his image. The handkerchiefs, featuring the words 'God Speed Our New Zealand Contingent', depicted the trooper holding the New Zealand flag.[24] In Nelson, Starr's store sold pictures of Bradford, while Hounsell's booksellers sold Bradford patriotic badges.[25] In 1901 the idea was mooted of erecting a statue of the trooper to mark the nation's wartime sacrifice with the names of the dead inscribed on its base.[26] Though the idea was not adopted, in May 1903 Seddon unveiled a memorial commemorating Bradford on Kakaipo Hill in Paeroa.[27] Although he was of no particular significance to them, the Boers nonetheless erected a headstone on Bradford's grave.[28] Following the war, Bradford's bullet bandolier was adopted as the symbol of office of the dominion president of the South African War Veterans' Association of New Zealand.

Although wartime increases in cable traffic could slow proceedings, the deaths of men like Bradford and Harvey were usually recorded in succinct casualty reports sent by the imperial authorities in London to Lord Ranfurly. The governor then passed the sparse details on to the premier, who in turn notified the bereaved families by telegram. Conflicting accounts of casualties and the severity of their wounds did little to alleviate the anxiety of soldiers' families. Several newspapers reported that Corporal Bolton of the Fourth Contingent had been dangerously wounded in the groin at Ottoshoop — a claim Joseph Ward repeated in Parliament.[29] Bolton, a 35-year-old Nelson nurseryman, was, however, confused with Private Oscar Bottom of Denniston, also of the Fourth Contingent, who died in Kimberley Hospital after being shot in the thigh.[30] The *Marlborough Express* claimed that Bolton was seriously injured, but on the next page stated that Seddon had received confirmation that it was Bottom who had been gravely wounded.[31] A week later, at least one newspaper was still claiming that it was Bolton who had been shot at Ottoshoop.[32] Even Bolton's service record featured the handwritten annotation, 'Dangerously wounded in groin Ottoshoop 18.8.00', but a medical board report contradicted this, noting he had 'no wounds' but had contracted malaria at Klerksdorp and was 'only suffering from Alcoholism'.[33]

Other parents were also unnecessarily traumatised by false casualty reports. In July 1901 several New Zealand newspapers ran stories claiming George Fisher, the son of the City of Wellington MHR George Fisher senior and the brother of Captain Francis Fisher, had died at Waterval from a gunshot wound.[34] At the time, the younger Fisher was serving in the Kaffrarian Rifles, an irregular unit raised in South Africa.[35] Two weeks later, however, newspapers were forced to print retractions after George Fisher senior received a letter from his son mentioning the death of another unrelated Lance-Corporal George Fisher serving in the same corps.[36]

The Fisher family's relief was comparatively short-lived. George Fisher junior remained in Africa after the war, but died of typhoid in Lourenço Marques in Portuguese East Africa early in 1904, leaving a widow and child in Lower Hutt.[37] George Fisher senior had been a vocal supporter of the war and spoke of wiping out both President Kruger and the British defeat at Majuba Hill during the 1899 contingent debate.[38] He blamed the Boers' 'temporary victory' on what he perceived as the weakness of British prime minister William Gladstone's administration, which he contrasted with the 'determination and pluck' shown by the Marquess of Salisbury's government during the war. Whether his son's fate played a part in his own demise is unclear, but Fisher senior's death in 1905, at the age of 62, ended a 20-year parliamentary career and forced a by-election for his vacant Wellington seat — won by his other son, former contingent member Francis Fisher.[39]

In June 1900 Trooper James Fahey was also incorrectly reported to have died from injuries at Thaba 'Nchu in Orange Free State.[40] Though Fahey was gravely wounded, his injuries were not fatal.[41] A month later in Parliament, Seddon described as 'very gratifying' the news that Fahey had survived.[42] For Fahey's parents, who for a month had believed their son to be dead, the premier's words were an understatement. In July, they received a letter from Fahey informing them that he was alive and, if not well, at least recovering in hospital.[43]

The circumstances in which Trooper John Saxon of the Fourth Contingent lost his life led to criticism of the military authorities. Saxon died at Beira in Portuguese East Africa in June 1900 from a combination of pneumonia and dysentery. Although his death reportedly occurred 'despite every attention' being given, a very different picture of the young trooper's last days emerged in the press several weeks later.[44] In a report titled 'The Death of Trooper Saxon. A Pathetic Tale', Alf C. Morton gave his version of events.[45] The *New Zealand Times* 'special correspondent' claimed that after contracting malaria Saxon had been left behind by his contingent when it moved on to Bamboo Creek. Morton described discovering Saxon lying emaciated, alone and near death on the dirt floor of the squalid building that passed for a hospital.[46] The room, Morton claimed, was swarming with flies, while decomposing food and dirty utensils surrounded the helpless trooper. Newspapers immediately seized on the story, more than one claiming the trooper died of neglect.[47] Understandably upset, Saxon's father rejected the official explanation of his son's death, and in 1901 was still expressing his belief that it was the result of negligence.[48] He claimed to have received confirmation that instead of being hospitalised prior to the Fourth Contingent leaving Beira, his son was left in the care of a farrier who was himself too ill to travel.

Morton's version of events was contradicted by other accounts that reported Saxon had been left aboard a hospital hulk moored in Beira harbour.[49] Trooper Roland Houghton, who served with Saxon and was present at Beira, also dismissed Morton's account as wilful misrepresentation.[50] Houghton claimed that everything possible was done to assist the dying invalid, and the reason that only one officer visited Saxon was because other officers were also afflicted with fever. In an indication of the sometimes-strained relations between correspondents and contingent members, Houghton added that

Morton's presence was not missed when he did not accompany the Fourth Contingent to Rhodesia.

As in Walter Armstrong's case, slow postal traffic between New Zealand and South Africa compounded the Saxon family's anguish. After his death, a letter Saxon had written two months earlier that gave no hint of his illness arrived in New Zealand.[51] Shortly after Saxon died, Seddon raised in Parliament the issue of poor communications with South Africa and drew attention to the time it took for notification of deaths, injuries and sickness to reach New Zealand. He described the 15 days it had taken for news of an unnamed trooper's death to reach New Zealand as 'undesirable'.[52] Seddon informed Parliament that he had instructed Major Pilcher, the New Zealand Agent General in Cape Town, to report on the number of New Zealanders in hospital and supply details of their condition.[53]

When New Zealander William Smith died in Ladybrand Hospital after being accidentally shot by a member of his corps, it took nearly two months before news of his death reached New Zealand, and longer still before it was finally confirmed.[54] The Smith family of Havelock had already lost William's cousin, Gunner Cecil Smith of the Fifth Contingent, who died of dysentery in Mafeking, and the impact of William's death was exacerbated by inaccurate reports regarding the true identity of the dead man.[55] A newspaper reported that Smith had been a purser with the Australasian United Steam Navigation Company and as their son had never worked for the company the Smith family believed a mistake may have occurred and William was in fact still alive.[56] It cannot have helped that Smith failed to mention his injury in letters to his family.

Similar disturbing accounts of death, injury and disease did little to assuage concerns regarding the well-being of men serving in South Africa — concerns that increased exponentially when more than one family member was involved. There were several instances of brothers either serving together or in separate units during the war. The Retter, Donnelly, Hemphill, Klee and Gray families each contributed three sons to New Zealand contingents.[57] Hector Retter served in the Fourth Contingent, while Leonard and Darcy enlisted in the Seventh Contingent on the same day and received consecutive service numbers. The Retter family's worst fears were realised in 1902 when Leonard's name appeared on the roll of those killed at Langverwacht.[58]

Ann Donnelly's three sons, Robert, William and Peter, served in the Fourth, Eighth and Seventh contingents respectively. Peter, the youngest of the three, was discharged after being shot and contracting fever.[59] In a telegram to the Defence Department, a concerned James Hemphill sought information about his sons serving in South Africa after hearing rumours that one had been killed and another wounded.[60] Though Hemphill was assured, correctly, that this was not the case, one of his sons was less fortunate in 1916 when he sustained a gunshot wound to the head while on the Western Front in France.[61]

The Klee brothers, who all enlisted on the same day and served together in the Seventh Contingent, came from a French military family. Their parents, Emily and Victor, were both born in France. Although Louis Klee survived the war unscathed, in 1904 he was accidentally shot in the leg while pig hunting. As he was injured in an isolated location on Akitio Station east of Dannevirke, by the time Louis was stretchered to the farmhouse it was 10 hours after he had been shot and his wound was so serious his leg had to be amputated.[62]

The parents of Leslie Melville, who died at Ladysmith, contributed four sons to the conflict. David Melville (the manager of the Union Bank of Australia in Christchurch) and his Irish wife Louisa had one daughter and five sons born in New Zealand (though one son died in infancy).[63] The three elder boys — Leslie, and twins Hugh and Alex — attended Wanganui Collegiate School.[64] Following David Melville's retirement, the entire family immigrated to London in 1891, though their stay in the United Kingdom was brief and by December they were living in Dresden.[65] They intended staying in Saxony for two years for the children's education, but David's death meant the older boys became the family's breadwinners.[66]

After travelling to India, Leslie and Hugh Melville found work as tea planters.[67] From there, the New Zealand brothers sailed for Natal carrying letters of introduction and established themselves on farms at Lower Umzimkulu, intending to develop their land into orchards. The family was later reunited in Africa when Leslie and Hugh's mother and siblings joined them, though this was to prove short-lived. The brothers had erected a temporary dwelling and begun ploughing when war broke out in 1899.[68] All four of the surviving Melville boys — Leslie, Alex, Hugh and Hamilton — enlisted in the Natal Border Rifles to, as their mother put it, 'see some fun'. Louisa wrote that her sons had fought at the

Battle of Glencoe, and had also taken part in the Boer defeat at Elandslaagte where Hugh was shot in the lungs, injuries he confirmed in a letter to his old school, adding that his younger brother Hamilton was also an invalid.[69]

When the Boers encircled Ladysmith, the brothers were among Lieutenant-General Sir George White's forces besieged in the Natal town.[70] A 1900 edition of the Melville brothers' old school magazine asked, 'Why is a hard-boiled egg like the Siege of Ladysmith? Because there is White inside and shell all around.'[71] The siege, however, was no laughing matter, and it was disease, rather than Boer artillery, that took its toll on the Melvilles. Leslie succumbed to typhoid in the pestilent tent hospital of Intombi Spruit on the outskirts of Ladysmith shortly before British forces broke through the Boer cordon.[72] Hugh Melville claimed that for the last month of the siege patients in 'the "hospital", whose very name I loathe' were surviving on soaked dog biscuits, starch and water.[73]

At one stage during the siege, former Tīmaru resident Nurse Rose Shappere was reportedly caring for nine delirious patients at Intombi, all of whom were near death. It did not help that the tents were periodically flooded after heavy rain.[74] In a letter to her aunt in Wellington, Shappere claimed patients like Leslie Melville were 'dying like dogs', with the nurses receiving reduced daily rations that included 'a cup of horse soup' during the final days of the siege.[75] Following her eldest son's death, a letter to a Natal magistrate described Louisa Melville's precarious situation: 'Leslie was her guide and her future was bound up in him. At present the future looks dark and full of uncertainty for all the Melville family particularly the Mother and daughter.'[76]

Arkell, Armstrong and Leslie Melville shared more than dying of typhoid far from their country of birth. They, and many other New Zealanders like them, were part of the colonial diaspora: men and women who left their homelands in search of opportunities across the empire. Though their motivations and the routes they took may have varied, a common denominator was that all three of the New Zealanders found themselves in South Africa during the war.

It is likely that no New Zealand parents contributed more of their children to the conflict than Jane and Samuel Taplin, who had five sons fighting in South Africa. Originally from Pātea, troopers Samuel (known by his second name, Frederick) and William Taplin served in the Second and Sixth contingents respectively, while their younger brother, Burton, was a lieutenant in the

Clockwise from top left: Leslie Seton Melville, the eldest of the four Melville brothers who served in the Natal Border Rifles and were besieged in Ladysmith. Leslie died of typhoid at the tent hospital at Intombi on the outskirts of Ladysmith.

Patrick Fitzherbert, who as a teenager served in Roberts' Horse at the Battle of Paardeberg. Fitzherbert was deeply disturbed by the scenes he witnessed.

New Zealander Walter Armstrong, a Whanganui chemist who enlisted in London in the Royal Army Medical Corps as a 'compounder of medicines' and died of typhoid in a Bloemfontein hospital. THE COLLEGIATE SCHOOL, WANGANUI, IN SOUTH AFRICA, 1899–1900, WANGANUI: A.D. WILLIS, PRINTER, [1901], N.P., WANGANUI COLLEGIATE SCHOOL MUSEUM

Eighth.[77] Samuel junior went on to serve in the Imperial Light Horse, and Thomas Taplin, who joined the New Zealand Permanent Artillery in 1894 and then transferred to the Wellington Police Force before leaving for Australia, served as quartermaster sergeant in the New South Wales Mounted Infantry.[78] Frederick Taplin returned briefly to New Zealand in 1901 but then went back to Africa as a recruiting sergeant in Rimington's Guides.[79] Like the Melvilles, the Taplins also lost their eldest son to disease after Thomas died of typhoid at Elandsfontein in South Africa in 1902.[80] The family had a history of military involvement: Samuel Taplin senior served in the militia and Armed Constabulary during the New Zealand Wars (when he helped repel an 1868 attack by Pai Mārire Māori on Weraroa redoubt), and in 1882 he was elected captain of the Patea Rifle Volunteers.[81]

'The heat is very great'

For those who served in South Africa, a significant factor affecting their health was the extremes of the climate. In a letter home, Ronald Saxby of the Third Contingent described the water in his bottle freezing solid during field operations, while Patrick Fitzherbert of Roberts' Horse bemoaned sleeping on hard ground in the bitterly cold conditions.[82] Fitzherbert contracted mild rheumatic fever and dysentery after the men of Roberts' Horse slept on the veldt in saturated clothes during heavy rain without tents or blankets.[83] The African sun was of equal concern. Describing the heat as 'very great', Major Robin reported that Private Ernest Smith of the First Contingent was hospitalised after being 'mentally affected' by sunstroke, while in a letter home Trooper Thomas Hynes of the Sixth Contingent, who was hospitalised in Pretoria, wrote: 'I got sunstroke and went a bit off my "dot".'[84] After Fitzherbert's corps was pinned down by heavy rifle fire from entrenched Boers, one of the men in his unit was wounded. Fitzherbert claimed the man died of sunstroke after losing his helmet and not being able to recover it as it lay just beyond his reach.[85] The slouch hats that many New Zealand soldiers wore became a standard element of the contingents' uniform in much the same way as the lemon-squeezer hat would later symbolise New Zealand soldiers in the First World War. But despite its almost iconic status, the hat presented problems in South Africa.

The felt hats may have offered a degree of respite from the harsh African sun, but they afforded no protection from bullets and shell splinters. Despite this, it was to avoid being confused with the Boers, rather than a desire to provide adequate head protection, that saw at least some New Zealanders issued with alternative headwear. In December 1899, Major Robin reported that he would probably have to requisition 'helmets' for the contingent, noting that on several occasions colonial troops with wide-brimmed hats had been mistaken for Boers and fired upon by imperial troops.[86] The steel helmets almost universally adopted by soldiers in the First World War were yet to come into common use. British soldiers in South Africa were instead issued with the standard pith helmet worn in hot-climate campaigns like those in India, Egypt and the Sudan.

Robin described the First Contingent men as looking 'very ruffianly' after he ordered them to replace their brass buttons with bone versions and to remove the brass lion's-head hooks used to hold up one side of their hat brims.[87] To avoid the brass glinting in the sunlight and giving away their position to the enemy, he also instructed his men to remove the brass 'NZMR' shoulder titles from their epaulettes. In another report written at Springfield three months later, Robin noted that the helmets had reached the contingent and been issued to the men.[88]

Though at best the helmets might deflect a Mauser bullet or absorb some of the force of a shell fragment, some contingent members were nevertheless grateful to have them. Henry Duigan of the Third Contingent described an especially close shave when he 'felt something go crash' through his helmet, and 'the tears rushed to my eyes, and my head throbbed as if it would burst; I thought I was shot through my head but thank goodness I was wearing a helmet'.[89] Describing the Battle of Rhenosterkop, the *Otago Daily Times* 'special correspondent' noted that Private Robert Brown of the Second Contingent was wounded in the thigh by a piece of shrapnel, as well as getting two bullets through his helmet, two through his water bottle, and two through his bandolier.[90] In a letter discussing the death of Corporal William Byrne of the First Contingent, Trooper Charles Grahame indicated that Byrne was less fortunate and had been killed by a shell fragment that passed through his helmet.[91]

Weather extremes were sometimes of more concern than the Boer enemy. Lance-Corporal Percy Nation, who was later killed at Langverwacht, described a thunderstorm 'appalling in its intensity'.[92] Nation claimed that 10 men on a kopje were struck by lightning, with one soldier killed outright and the remainder left 'paralysed'. Sergeant Arthur Coleman of the Seventh Contingent and Trooper Henry Stephens of the Fourth sustained serious injuries when they, too, were struck by lightning on separate occasions.[93] Coleman reportedly remained unconscious for four days following the strike as well as suffering partial paralysis.[94] A medical board recommended that Stephens, who sustained nerve damage, undergo prolonged treatment on his return to New Zealand at the government sanatorium in Rotorua.[95] However, the 38-year-old trooper, who claimed to have served in the Murrumbidgee Light Cavalry in Wagga Wagga before the war, was obliged to return to Wellington due to a 'scarcity of funds'.[96]

In Parliament, Joseph Witheford asked whether money from Lloyd's Patriotic Fund could be distributed to an unnamed trooper who had been severely injured by lightning.[97] The parliamentarian explained that the Seventh Contingent soldier had a wife and two children to support. In response, Cabinet minister William Hall-Jones assured Witheford that the case was being 'dealt with'. Coleman, a former carpenter, was still receiving treatment in the early 1920s as an inpatient at both Queen Mary Hospital at Hanmer and Christchurch Hospital as a result of partial paralysis caused by the lightning strike.[98] Corporal Guy Ronalds and Private Peter Farquhar of the Sixth Contingent were also reportedly struck by lightning in 1901. The story, which was quickly picked up by newspapers, claimed that the force of the strike left the men's clothes and boots blackened and shredded.[99]

Some contingent members sustained serious injuries before their vessels had even dropped anchor in South Africa. William Hosking, the father of Rupert Hosking who was wounded during the siege of Mafeking, held a surgeon-major's commission in the Volunteers and dealt with a number of Fourth Contingent patients aboard the *Gymeric* en route to Durban.[100] Private Albert Samuel's toes were crushed by a cask, Trooper Armstrong suffered serious injuries when a horse rolled on him, and Trooper Vincent Hattaway's skull was fractured by a falling winch block.[101]

Even the end of hostilities was no guarantee of safety. Tenth Contingent trooper Hugh Collison never saw action but died following a fall from his mount while watering horses as he awaited repatriation at Fort Hay in Natal.[102] Lieutenant Robert McKeich, a Tuapeka butcher who had grown up in Victoria, was killed four days after the signing of the Vereeniging Treaty that ended the war. Despite Major Robin expressing reservations about the 46-year-old's age, McKeich obtained a commission after writing to his MHR seeking admission to the Ninth Contingent for both himself and his son, Walter (who was overlooked for the Ninth Contingent but sailed later with the Tenth).[103] When informed of Robert McKeich's death, Ranfurly passed the information on to Acting Premier Joseph Ward, who in turn notified McKeich's family in New Zealand.[104] Walter, however, had by then arrived in South Africa and learned of his father's demise from John Luxford, the Tenth Contingent's Wesleyan chaplain.[105]

According to Trooper Herbert Hart, McKeich and Lieutenant Henry Rayne had gone hunting after peace was declared and were confronted by three armed Boers. Despite the New Zealand officers' protestations that the war had ended, the Boers ordered Rayne and McKeich to dismount and remove their clothes. As the New Zealanders reluctantly complied, Rayne produced a revolver from his coat pocket, killed one of the Boers and gravely wounded another before making a dash for his horse. Hart claimed the wounded Boer rested his rifle on an anthill and shot Rayne in the thigh.[106] Unable to remount, Rayne initially took cover before limping several miles back to his camp. The following morning, patrols discovered McKeich's body lying near those of the Boers Rayne had killed.

Trooper William McFarlane took part in the firing party at McKeich's funeral. He told his mother that two Boers, one of whom was a parson, entered the New Zealanders' camp on the afternoon of the funeral to collect the bodies of their dead comrades. The New Zealanders unceremoniously tossed the bodies into the Boers' cart 'like a bag of oats'. McKeich's body was treated with more dignity; after being sewn up in a blanket and draped with the Union Jack, it was carried to the cemetery in a mule-drawn cart.[107]

'Paltry pensions'

Though politicians had displayed a degree of unity during the 1899 contingent debate, this temporary suspension of party rivalry was short-lived. When the question of providing pensions for soldiers disabled in South Africa first arose in Parliament in 1899, Seddon prevaricated, dismissing it as a matter 'for future consideration'.[108] Only days after the South African War began, the issue was again raised by Seddon's political adversary, Waitematā MHR Richard Monk.

Without elaborating, Seddon confirmed that the New Zealand government would pay pensions.[109]

The 1866 Military Pensions Act did not provide support for the dependants of contingent members who were killed or injured as a result of the South African War. Initially, there was also no official provision for those whose health suffered as a result of war-related disease or accidents. Another of Seddon's political rivals, Hawke's Bay MHR Captain William Russell, sought to address this in July 1900. Like Monk before him, in Parliament Russell questioned Seddon about the pension issue. Russell recommended that while Parliament's appreciation of the contingents' services was 'still keen' it should adequately provide for those who had suffered during the war. Seddon replied that he believed Parliament did not support the awarding of pensions, adding that injured soldiers were retained on full pay subject to medical board examinations. He assured his fellow parliamentarians that if soldiers' injuries were deemed permanent they could receive a lump sum or, subject to parliamentary approval, an annual payment.[110]

Changing tack, Russell then pressed Seddon on the issue of widows and orphans. The premier conceded that in some cases married troopers had managed to enlist despite their official exclusion. Seddon was nonetheless prepared to acknowledge that the government 'could not punish the widow in such cases, or visit the shortcomings of the father on the children'.[111] Instead, he proposed widows and orphans receive funds from the £50,000 raised through patriotic contributions. However, as casualties and the incidence of typhoid rose, the government finally relented and the Military Pensions Extension to Contingents Act 1900 was passed.[112] The act contained schedules of men covered by the legislation. It did not, however, extend to New Zealanders who had joined irregular corps in South Africa or those who had served in the armies of other nations. Pension legislation enacted in 1901, 1902 and 1903 either extended the provisions of the 1900 legislation to include later contingents or amended the earlier act.[113] None of the various acts made provision for New Zealand nurses who had served in South Africa, as they did so 'at the disposal of the Army Medical Department'.[114]

In Parliament, Seddon noted that a number of men invalided home from the war arrived penniless and, in some cases, wearing the same ragged uniforms they had on when admitted to hospital in South Africa. The premier said that despite these men being welcomed on arrival, 'what is everybody's business is nobody's business'. Despite substantial sums remaining from public subscriptions, parliamentarians on occasion personally advanced money to men who were essentially destitute on arrival. While stopping short of openly criticising the patriotic committees, in a thinly veiled threat Seddon stated that Parliament would pass legislation allowing it to take control of the contributions if the money was not distributed in the manner originally intended.[115]

On his way back to New Zealand after serving in South Africa in Brabant's Horse, John Taylor Marshall broke his wrist and his forearm in two places after being knocked down on the steps of Melbourne Post Office. A crowd of enthusiastic well-wishers had jostled Marshall and other returned soldiers and reportedly removed his sergeant's chevrons and a number of his buttons as souvenirs.[116] On reaching New Zealand shores, Marshall's circumstances did not improve. Lacking money to pay for accommodation, he sought help from Joseph Ward after arriving unannounced at the minister's home on a Sunday.[117] Ward gave Marshall £2 and an order for a week's board at the Panama Hotel. In a 1903 letter to the secretary of Lloyd's Patriotic Fund, the 62-year-old Marshall said his wrist injury prevented him doing anything apart from clerical work.[118] Claiming he was 'absolutely poor', Marshall added that his age worked against him.

Among the civilians worst affected by the conflict were the widows of men who had lost their lives. Charlotte Berry, the wife of Lieutenant William John Berry, was the first woman to receive a widow's pension under the Military Pensions Extension to Contingents Act 1900.[119] Her husband died of pneumonia in 1900, leaving her with the burden of raising the couple's young children alone.[120] Though Seddon indicated in 1899 that only single men would be selected for service, William Berry's attestation form listed his wife as next of kin and recorded the names of their two young children.[121] While marital status affected NCOs and men in the ranks, it was apparently not an issue for officers, as Berry and several other married men holding commissions were accepted for service.[122]

Charlotte Berry's straitened circumstances became national news after a letter written by her father was read at a Napier Borough Council meeting. William Berry claimed his daughter had received no financial aid from either the British War Office or the New Zealand government. He added that apart from £10 Charlotte received from the Napier committee of the Transvaal War Fund, no assistance had been offered by any of the patriotic funds, and she was still waiting for £250 from her husband's life insurance policy.[123] A press editorial described Charlotte's situation as a disgrace in an 'alleged liberal country'.[124] Napier MHR Alfred Fraser raised Berry's case in Parliament, claiming that although her husband had died two months earlier, there had been no expressions of sympathy or offers of support from the government.[125]

The premier vehemently denied the accusations and defended the government's actions. He claimed that George Swan (the mayor of Napier who had repeated the claims) was well aware that the government had taken steps to offer Berry financial assistance.[126] Joseph Ward, however, did not question the veracity of Fraser's accusations when he responded in Parliament. Instead, Ward stated that if the Defence Department had not already remedied the situation it should do so immediately.[127] Nonetheless, the following month a Napier newspaper criticised Swan for his 'unfounded attack' and expressed surprise that the mayor had not conceded he was wrong.[128]

Catherine Francis, wife of Lieutenant-Colonel Frederick Wyatt Francis, was another officer's widow left to provide for her children after her husband's death. Lieutenant-Colonel Francis, who served in the Fourth Contingent, was invalided first to England then to New Zealand, where he subsequently died.[129] His widow was awarded an annual pension of £120 for herself and £20 per annum for each of her four children.[130] As with the 1866 Military Pensions Act, the rank of deceased contingent members directly affected the amount of financial assistance their widows and dependants received. Emily McKeich, wife of Lieutenant McKeich, received a pension of £60 per annum with gratuities of £273 for herself and £91 each for her three children.[131] According to the schedule of payments included in the 1866 legislation, £60 was the maximum pension the widow of a lieutenant who had been killed in action could receive.[132]

For their lightning strike injuries, Trooper Stephens received a pension of 1s per day in 1902, and it appears Sergeant Coleman was finally awarded a daily payment of 2s 6d in 1904.[133] Newspapers were critical of 'paltry pensions' and drew attention to the disparity between the sums paid to officers and the entitlements of non-commissioned officers and men in the ranks.[134] The *Wanganui Chronicle* compared the £54 per annum received by Sergeant-Major

Left: Wellington baker Trooper Henry Alexander Stephens of the Fourth Contingent who on 27 October 1900 was injured by a lightning strike while on active service. Stephens also contracted typhoid.
OTAGO WITNESS, 6 MARCH 1901, P. 38, ALEXANDER TURNBULL LIBRARY, N-P-2173-38

Right: Lieutenant Robert Collins of the Fourth Contingent received an annual pension of £100 after he was shot in the wrist in August 1900. On his return to New Zealand, Collins, who prior to the war worked at the Department of Lands and Survey, named his Newtown home 'Ottoshoop' after the engagement in which he was wounded.
AJHR, 1901, C-1, N.P, NATIONAL LIBRARY OF NEW ZEALAND TE PUNA MĀTAURANGA O AOTEAROA

Lockett, whose arm was amputated due to a gunshot wound received in an incident where he was praised for his 'gallant conduct', and the 'very handsome' sum paid to a less-seriously wounded Wellington lieutenant.[135] The officer in question was probably Lieutenant Robert Collins of Kilbirnie, who served in the Fourth Rough Riders Contingent and was shot through the wrist at Ottoshoop in August 1900.[136]

Criticising the new Military Pensions Bill in Parliament in 1902, Palmerston North MHR Frederick Pirani claimed: 'This was class legislation. One thing for the "fat man" we hear so much of from the Government's supporters — that was, the officer — and another thing for the poor man — the trooper.'[137] Lieutenant Collins, whose father was Lieutenant-Colonel Robert Collins, received a £100 per annum pension under the 1900 act.[138] In an article titled 'A Shameful Contrast', the *Wanganui Chronicle* noted that pension inequality occurred despite officers and men of the New Zealand contingents frequently coming from the same socio-economic group. The newspaper claimed that Lockett was better qualified than the unnamed Wellington lieutenant, but as he was not an officer he was denied a comfortable life 'without fear of poverty'.[139] In Collins' and Lockett's cases, both men did come from a similar social level, as before the war both worked as surveyors.[140]

Twenty-one-year-old Lance-Corporal Albert Rosanowski's left arm was amputated after he sustained gunshot wounds to his head and arm at Langverwacht while serving in the Seventh Contingent. A miner in civilian life, Rosanowski's medical report noted that his disability would prevent him earning a full livelihood for the rest of his life.[141] He received £30 from the Otago Patriotic Committee and a pension of 2s 3d per day, though he found employment paying 7s a day as a messenger for the New Zealand Defence Forces in Wellington.[142] He required a further operation in 1907, when his doctor observed that he was in poor health, with the amputation making it difficult for him to look after himself away from home.[143] Lance-Corporal William Roddick, a Temuka labourer who served alongside Rosanowski, was killed during the same Boer attack.[144] For the loss of her 21-year-old son, Helena Roddick received a yearly pension of £26.[145] This was exactly half the amount awarded to Lieutenant-Colonel Francis's widowed mother after her son's death, though Mrs Francis received only £55 before she, too, died in 1901.[146]

The gunshot wound that Trooper Albert Beath received at Rhenosterkop also necessitated the amputation of his arm.[147] In 1902, Beath received £36 annually — £19 less per year than the pension awarded to Sergeant Lockett for what was essentially the same injury.[148] In a 1903 plea for government financial assistance, Beath told Colonel Robin that he was unable to earn a living and asked if he was eligible for the 'blood money' he understood was paid to those injured in the war.[149] The government was not entirely unsympathetic; Beath was awarded a pension of 2s per day subject to medical board review, but it was not until 1909 that the pension was confirmed as permanent.[150] In Parliament, Joseph Ward acknowledged the government's obligation to provide for families and dependants in such situations, but claimed a reasonable delay was necessary to enable the government to institute 'a thorough system'.[151] The money was small compensation for Beath, who in 1903 married Penelope Farquharson, the sister of his Second Contingent comrade William Farquharson.[152] Beath continued to suffer from ill health and in 1910 travelled to Scotland for medical reasons.[153] He died in his early forties in 1912.[154]

The granting of pensions to officers was not a foregone conclusion. After being invalided back to New Zealand, 70-year-old Surgeon-Captain Robert Bakewell was judged 'almost incapable' of earning a livelihood, with the medical board recommending he receive a pension of £150 per annum.[155] However, the Defence Department rejected the board's recommendation, with Bakewell deemed ineligible under the pension legislation as he had never been in action in South Africa.[156]

Post-war life was often dramatically affected as a result of injuries sustained in action. A head wound received at Ottoshoop left 27-year-old Trooper Michael Canavan permanently deaf in one ear and affected his spine.[157] Invalided home, he was awarded a pension of 2s a day.[158] In a letter to Seddon seeking employment as a parliamentary messenger, he explained that the effects of his injury prevented him from returning to his former occupation as a horse trainer.[159] Although Canavan received a letter of support from Joseph Witheford, Seddon indicated there was no suitable position for the former trooper.[160]

Early in 1902, Aucklanders William and Samuel 'Frederick' Finch sailed for Africa with the Eighth Contingent aboard the *Surrey*.[161] They were two of Samuel and Martha Finch's 10 children, with the couple's eldest son, Harry, already serving in the Seventh Contingent. A month to the day after the *Surrey* sailed, and while the brothers were still at sea, a telegram from the premier arrived at the Finches' Mount Albert home. Harry had been killed in action a week earlier at Langverwacht where, the family was assured, he had 'died the death of a soldier, fighting the battles of the Empire'.[162] It was not until the *Surrey* docked in Durban that William and Frederick learned of their brother's death several weeks earlier.[163] Though the under-secretary of defence forwarded Harry's knife, pipe and Bible in June, his bereaved father received little else and did not receive a pension.[164]

Pride appears to have been a factor in at least one instance where a South African War widow declined financial assistance despite the struggles she faced raising several children in rural Taranaki. Mary Patterson's 49-year-old husband, Private James Patterson, died of typhoid at Rondebosch.[165] Private Patterson's brother had been among the British dead at Majuba, and the farmer reportedly had spoken of avenging his sibling's death.[166] Mary was left in 'poor circumstances' with four children and a £240 mortgage on the couple's Inglewood farm, though Edward Dockrill, the mayor of New Plymouth, claimed that if a pension was offered, he was 'certain it would be refused'.[167] Dockrill was correct as, in a letter to Lieutenant-Colonel Ellis, Mary Patterson emphasised that she had never applied for assistance and had no intention of doing so.[168]

As Seddon noted, some men were prepared to bend the truth in order to secure a place in the contingents. Sergeant-Major Daniel Love of the Seventh Contingent was killed in action near Vasburg in August 1901.[169] When he had enlisted four months earlier Love, a Public Works Department painter, made no mention of his wife, Georgina, or the couple's three young children, and instead gave his mother as his next of kin.[170] A month after Love's death, Reverend Elliott of Kent Terrace Presbyterian Church in Wellington wrote to the Defence Department stating that Georgina Love was destitute.[171] Love subsequently received a gratuity of £164 (one year of her husband's wages), a New Zealand pension of £36 per annum, and £10 per annum for each of her children, though she supplemented this with employment.[172]

In 1908 Love again sought government financial assistance. She explained that she had managed on her pension until she contracted pleurisy but could no longer meet her financial commitments.[173] Though Love also received an

Ninth Contingent officers, including Lieutenant Henry Rayne (standing, right) and Lieutenant Robert McKeich (seated, left). McKeich was killed by Boers he and Rayne encountered while hunting after hostilities had officially ended. *Otago Witness*, 18 June 1902, p. 41, Alexander Turnbull Library, N-P-2174-41

imperial pension, the pension conditions had a direct impact on recipients' lifestyles. Widows could be denied financial assistance for a minimum of 12 months for 'unworthy conduct', including 'conduct of a character to create public scandal, such as [the] birth of an illegitimate child, co-habitation with a man to whom the widow is not married [and] disorderly habits leading to neglect of children'.[174] When Emily McKeich's husband was killed, the commanding officer of the Tuapeka Mounted Rifles (her late husband's Volunteer corps) recommended that any money due to her as a war widow should be distributed as an annual allowance as he suspected she would 'squander' a lump sum payment.[175] Although it is unlikely the government was acting on the officer's advice, in June 1902 Emily received an annual pension of £60 for herself and her children.[176] The 1903 Military Pensions Amendment Act provided for wives and children of men who died of injuries or illness attributable to the war. If the soldier was unmarried, the act provided for their mothers or sisters.[177] Female children, however, were disadvantaged as any financial assistance they received ceased if they married.[178]

Private Frederick Betcke was one of the more unusual individuals to receive a pension. The adopted son of German immigrants Otto and Antonia Betcke, Frederick was reportedly obsessed with the military life.[179] Though he worked as a carpenter, Antonia Betcke claimed her son never missed a Volunteer parade, leaving work early and walking 16 miles in order to attend. Betcke attested for the Seventh Contingent aboard the *Gulf of Taranto* several days after she sailed, suggesting he was yet another stowaway.[180] Although the nature of his injury is unclear, he was awarded a pension of a shilling a day in November 1903 under the Military Pensions Amendment Act.[181]

Betcke was discharged from the Seventh Contingent in September 1902 but was reluctant to bring down the curtain on his military career. The following year he was arrested in Feilding wearing the uniform of a Royal Horse Artillery sergeant-major.[182] When approached by police, Betcke attempted to flee but was tackled and subsequently sentenced to four months' imprisonment for bringing the king's uniform into disrepute and theft of a field service cap. 'Brigadier-General Betcke' resurfaced in 1917 in Sydney, where he was again arrested and sentenced to two months' imprisonment for impersonating a senior officer by wearing a general's uniform, complete with medals.[183] Betcke had reportedly kept up the charade for the previous six years while continuing to draw his shilling from the Imperial Pensions Office in Sydney. The court heard Betcke had marched at the head of recruiting processions in the Sydney suburb of Leichhardt. He wore the full uniform of a senior officer complete with 11 awards to which he was not entitled, including the DCM.[184]

In a number of cases, families in New Zealand were financially reliant on men serving in South Africa who were neither husbands nor fathers. John Graham, the MHR for Nelson City, noted during the First Contingent parliamentary debate that there were 'mothers, sisters, and other relatives who are dependent on many of the single men in the colony'.[185] Twenty-three-year-old Trooper John Aitken-Connell, who would be killed in action at Rensburg in 1900, sent a portion of his pay home to support his mother. John's brother Arnold requested Patriotic Fund assistance, telling Premier Seddon that he and his brother were their mother's sole source of financial support.[186] Mary Aitken-Connell subsequently received a yearly pension of £26, though she and Arnold, who served as a lieutenant in both the Seventh and Ninth contingents, later emigrated to Orange River Colony in South Africa.[187]

Other parents dependent on the earnings of their sons serving in South Africa were less fortunate. The father of Sergeant Frederick Wylie was critical of the government's actions after his son was killed in action at Klipfontein. In a letter written more than four months later, Galatea schoolmaster Joseph Wylie said he found it strange that the authorities took so long to pay money owed to men who had endured hardship and 'sold their lives so dearly fighting for their King & Country'.[188] Wylie said nothing about his son's actions at Vaal Bank earlier in the war, when Frederick was mentioned in General Babington's dispatches.[189] Despite Joseph Wylie's concerns, delays in processing pay claims were inevitable, with an understaffed Defence Department having to calculate pay entitlements from records held in multiple locations. Men who had served in more than one contingent only added to the complexity of the task. The many New Zealanders who fought in irregular corps raised in South Africa also faced difficulties, and they were often advised to address pay enquiries to their former officers in Africa.

Captain Barron of the Rotorua Rifles described the elderly Joseph Wylie as steady and industrious but in straitened circumstances. Noting that Wylie earned little as a teacher, Barron added that he believed Wylie relied on his son's wages to support his family, though confusion over Frederick Wylie's rank and service in the Seventh Contingent added to delays in Joseph receiving the money he was owed.[190] Communications between the War Office in London and the New Zealand Agent General regarding funds due to Joseph Wylie were still ongoing more than two years after Frederick's death in action.[191]

Among those gravely injured in the Machavie train accident was 30-year-old Trooper Patrick Lee, whose leg was so severely damaged that it required amputation.[192] After being invalided home, Lee received a £50 compassionate allowance from Lloyd's Patriotic Fund; £35 from the Otago Patriotic Fund; an imperial pension of 2s 6d per day; and 2s per day from the government for a fixed period, after which his case was to be reassessed.[193] In 1903, Lee sought further financial assistance. Noting the size of his imperial government pension, he complained to Joseph Ward that the New Zealand government had been less generous. Lee claimed his injury was classed as third-degree, which meant it only caused temporary disablement. The former trooper indignantly informed Ward that he considered the classification an injustice: 'the loss of my right leg, which was amputated five inches above the knee, [is] rather more than a slight or temporary disablement'.[194]

'The most deadly scourges'

In 1899 Taranaki MHR Henry Brown claimed in Parliament that there was no more unhappiness during war than in peacetime; there was more misery, Brown claimed, due to diseases and other causes.[195] While Brown was referring to the impact of disease in normal circumstances, Henry Cleary, the editor of the Catholic *New Zealand Tablet*, commented on what he considered the avoidable and unnecessary South African campaign. Noting the destructiveness of modern weapons, Cleary also drew attention to 'the world of untold suffering from mere disease'. The editor claimed these issues alone should give politicians and journalists reason to consider their actions before raising 'the howl for blood'.[196] Once hostilities commenced and military forces began operations in the field, Cleary's observations proved all too true as disease thinned contingent ranks.

For at least one New Zealand recruit the war was over almost before it began. Ravaged by severe dysentery contracted at Te Papapa Camp in Auckland,

Trooper Percy Leary died days before his contingent left New Zealand shores.[197] Though several men in the Ninth Contingent were afflicted with the illness, Leary, who had easily passed his medical on joining the contingent, proved the most serious case.[198] He was reportedly placed under the care of Surgeon-Captain King in a tent that served as the camp hospital.[199] After his condition deteriorated, King elected to move Leary to Auckland Hospital, where the recruit subsequently died. The 21-year-old West Waituna farmer's total period of service amounted to only 20 days, several of which were spent in hospital.[200]

On being notified of his death, Leary's sister travelled to Auckland by ship.[201] She claimed that Leary had been determined to serve in the war, adding that if he had been rejected by the Ninth he would have made his own way to South Africa and joined a local corps there.[202] On the day of Leary's death, the flag at Te Papapa Camp was flown at half mast for only the second time — the first being after the Seventh Contingent's mauling at Langverwacht.[203] After his funeral, Leary's coffin was conveyed from Onehunga Presbytery to the wharf in sombre procession aboard a gun carriage with an escort of mounted troopers, and then loaded aboard the *Takapuna* for shipment to Wellington.[204] From there, the trooper's body was transported on to Waituna West in Manawatū for burial.[205]

The Public Trustee sought details regarding any outstanding pay owing to Leary in order to expedite the distribution of the trooper's estate to his relatives, who were 'in poor circumstances'.[206] However, the trustee's initial requests went unanswered and officials claimed the funds could not be disbursed until the Ninth Contingent paymaster returned from South Africa. Five months after Leary's death the tangle of red tape was finally unravelled and a voucher for £1 1s 3d was forwarded to the Public Trustee — small recompense for the loss of a family member.[207] In the interim, the Defence Department had also questioned the £7 it was billed to transport Leary's body to his home.[208] As Leary never reached South Africa, his sister was informed her brother was not posthumously entitled to the Queen's South Africa Medal.

Newspapers gave conflicting accounts of the aftermath of Leary's death. The *Manawatu Evening Standard* suggested that Dr Murray, one of the contingent's medical staff, had suffered a 'grave injustice' when his services were dispensed with at the last moment.[209] Calling for an inquiry into the case, the newspaper further claimed that the general feeling within the Ninth Contingent was that

Leary's death was due to inadequate medical care and that Murray had been made a scapegoat for another's actions. Careful to avoid naming the individual in question, the newspaper stressed that the story was anecdotal. Nonetheless, it believed an inquiry would ensure that blame for Leary's death, if indeed there was any, would be attached to the person responsible. An earlier *New Zealand Herald* article gave a different version of events, reporting that Murray had retired and been replaced by the aged Dr Bakewell.[210]

The need for trained doctors in South Africa led to a number of expatriate New Zealand medical students in the United Kingdom volunteering their services. Among these was Angus McNab, who served in a 'line of communication' hospital in South Africa.[211] McNab, who hailed from Gore, was the brother of Mataura MHR Robert McNab.[212] He travelled to South Africa with the Edinburgh and East of Scotland Hospital accompanied by fellow New Zealand medical students and doctors Donald Murray, James Elliott, Alan Owen and Dr St Leger Gribben.[213]

A combination of professional interest and patriotic duty led the young medical professionals to temporarily suspend their medical studies. Murray welcomed the opportunity to gain first-hand experience of military surgery, while Elliott, the son of Reverend James Elliott, voiced support for the empire and praised the British soldiers he tended.[214] The younger Elliott expressed disdain for the Natal and Cape Colony rebels, and condemned the role which the 'disloyal colonist' had played in the conflict. He claimed to have seen the full horrors of war in his hospital at Norvalspont, adding that he had collected examples of the 'explosive' and soft-nosed bullets the Boers were frequently accused of using.[215] In September 1900, the *Otago Witness* reported that thousands of portable steel shields were being produced in Sheffield as protection from Boer bullets.[216] Accompanying illustrations showed a variety of potential designs and suggested uses for the shields including attaching them to soldiers' helmets as a type of sun shade. Only days before Christmas 1900, H. V. Drew from Timaru sent his own design for a portable bulletproof shield to the Defence Department.[217] While soldiers would no doubt have welcomed protection from Boer rifle fire, the idea of lugging hefty shields across the veldt in addition to their usual equipment was possibly less appealing.

McNab stated that he and his fellow New Zealand medical staff served in

South Africa as civilians.[218] While this was true in his case and that of Gribben, others such as Murray, Elliott and Owen enlisted in the Royal Army Medical Corps, with Murray holding the rank of corporal.[219] New Zealand surgeon Hugh Acland, who was based in England and was the son of former Legislative Council member John Acland, also volunteered. Acland worked in Woodstock Hospital near Cape Town alongside Nurse Emily Rowley from South Canterbury.[220] For some of the young New Zealand surgeons, doctors and nurses, the experience of battlefield injuries they garnered in Africa would prove invaluable during the First World War. In McNab's case, however, his involvement in the later conflict was short-lived. He was reportedly bayoneted to death in France by German soldiers while tending to wounded on the battlefield in 1914.[221]

After receiving a report from Major Pilcher in which the Agent General claimed 73 men had died of illness per week over the preceding five weeks, Seddon used similar language to Rose Shappere when he told Parliament 'our men were dying there like sheep'.[222] Though Pilcher was not referring solely to New Zealanders, the disease that killed Leslie Melville ravaged both imperial and New Zealand troops alike. James Elliott described conditions in the hospital where he worked: 'By far the largest number of cases are enteric, and the poor beggars rave away in fever, sometimes muttering incoherently of home.'[223] This grim assessment was repeated by Bessie Teape, a New Zealand nurse stationed in Bloemfontein Hospital:

> *We lose a lot of patients from the dreaded enteric, not less than thirteen to fifteen deaths every day. Fifty of our New Zealanders are in the hospital with it at present, and six of our nurses are also very bad, and we are always wondering whose turn will come next.*[224]

Teape later reported treating New Zealander Lieutenant Mellish prior to his death from typhoid in the hospital.[225] By the time her letter describing conditions in Bloemfontein Hospital appeared in New Zealand newspapers, the severity of the typhoid problem was evident. The premier said that the number of men contracting the disease indicated the seriousness of the problem and assured Parliament that if necessary the government would send nurses and pay for private hospital accommodation.[226]

Prior to sailing for England via South Africa in 1900, Dr William Hosking believed his age would prevent him obtaining a commission in the Fourth Contingent. Nonetheless, he offered to assist 'our lads at the front' in exchange for his passage.[227] In fact, although Hosking, who was born in 1841, was 58 at the time, he was still more than ten years younger than the 70-year-old Dr Bakewell. Hosking's earlier attempt to obtain a surgeon's commission in the First Contingent had been unsuccessful as the position had been filled, but he was accepted as consulting surgeon with the Fourth.[228] On Seddon's instructions, Hosking visited hospitals in Natal and Cape Colony to investigate the effectiveness of a typhoid 'antitoxin'. He carried with him 900 doses of the drug and syringes to administer it to New Zealand troops, though its impact is unknown.[229]

The day after Seddon discussed the typhoid problem in Parliament, the brother-in-law of New Zealand nurse Mary Warmington wrote to the premier. He said that he and his wife were anxious as they had received a letter informing them that Warmington had contracted typhoid in Bloemfontein Hospital, but they had heard nothing more.[230] Warmington was fortunate to survive the disease, but was nonetheless invalided home. *The Outlook* also printed a letter from Janet Williamson in Bloemfontein. Williamson wrote 'fancy having 40 enterics to look after, fearful cases usually!' and commented that she and her fellow nurses were working among the worst forms of typhoid.[231]

New Zealanders were understandably alarmed by such reports, and the letters they exchanged with family members, loved ones and friends fighting in South Africa reflected their concerns. After receiving information 'from outside sources' that Third Contingent trooper Simon Fraser had contracted typhoid, Marilen Hegglun of Manaia wrote directly to Seddon seeking confirmation. The anxious Hegglun, whose younger brother served in the Fifth Contingent, told the premier that she had scoured the newspapers fruitlessly for news.[232] Fraser was among the lucky; he survived the war and in 1906 married Hegglun. In a letter to her brother William in the Second Contingent, Penelope Farquharson mentioned reading Teape's letter and commented that typhoid seemed to be far more prevalent than combat injuries, while in another letter William's brother observed, 'It is a good job you are escaping the fever.'[233]

One of the youngest soldiers to serve, Trooper Douglas Corson, was less fortunate. The 17-year-old, who had falsified his age on his attestation form

Sergeant Rupert Hosking's father, Dr William Hosking. Dr Hosking travelled to South Africa aboard the S.S. *Gymeric* with the Fourth Contingent before accompanying his son on to the United Kingdom for medical treatment.
Otago Witness, supplement, 31 May 1900, p. 7. Alexander Turnbull Library, N-P-2176-7

in order to serve, died of typhoid at Krugersdorp early in 1901.[234] The Defence Department awarded his mother, 60-year-old Helen Corson, a compassionate allowance of 10s per week in September 1901.[235] At the time, the department was aware that Helen's husband, 'an aged Tramway driver', was earning 30s per week. As frequently occurred in such cases, the department asked police to conduct investigations into the circumstances of bereaved families without their knowledge, with the resulting police reports potentially having a long-term impact on those under investigation. In early 1904, the Defence Department instructed the commanding officer of the Dunedin Volunteer district to make enquiries through the local police into the occupation and earnings of Douglas Corson's father, William Corson.[236]

Detective John Cooney reported that 62-year-old William Corson was receiving £2 per week working as a groom at the City Corporation tram stables.[237] Cooney noted, however, that William would soon be made redundant as the introduction of electric trams would render horse-drawn trams obsolete. The detective found that of the Corson's four surviving children, one was working as a tailor in Invercargill but did not assist his family, the eldest daughter was in bad health, and the two remaining daughters worked as seamstresses in Christchurch on low wages. Cooney added that Helen Corson paid 12s 6d in rent and was dependent on her husband's wages. As a result of the police investigation, the Pension Board found that Helen Corson was ineligible for a pension under the Military Pension Amendment Act 1903.[238] The government cancelled her allowance and replaced it with a special one-off grant of £30 from the Patriotic Fund.[239]

In late July 1900, the *Nelson Evening Mail* reported Seddon's receipt of a telegram from Major Cradock in South Africa notifying the premier of the death from typhoid of Frederick Broome.[240] Yet it was almost exactly three months after Broome's death, and two months after the newspaper report, that Seddon sent his own telegram to the Broome family confirming that their son, who gave his occupation as 'gentleman' when he enrolled, was dead. Apologising for keeping the family 'in suspense so long', the premier told the family that confirmation of their son's death had only reached the High Commission in Cape Town on 21 September, the day the news was cabled to Seddon. For the Broome family, who had spent three months in emotional limbo, hoping for the best but fearing the worst, the news came as a particularly cruel blow. Despite the tragic circumstances, Seddon's telegram, in which he offered his 'heartfelt sympathy', was markedly more perfunctory than the condolence telegram he sent to Captain John Harvey's brother.[241]

Broome's death deeply affected the 22-year-old trooper's elderly mother, with her doctor reporting that her health had deteriorated rapidly after she learned of her only son's fate.[242] He added that she had been dangerously ill and was a complete invalid, requiring constant care, with little chance of recovery. Lieutenant-Colonel Penton informed the chairman of the Wellington mayor's Patriotic Fund Committee that while Mrs Broome's three daughters earned just enough to keep themselves, Trooper Broome had been his parent's principal means of support.[243]

The premier adopted a more sympathetic tone when extending commiserations to James Colvin in a telegram following the death of the latter's son from typhoid in Johannesburg. The premier and Colvin, the MHR for Buller, were personally acquainted and the telegram provides an insight into the Seddons' own concerns as parents. With their son serving in the Fourth Contingent, the premier and his wife felt the same trepidation as would any parents with a child in a war zone:

> *Mrs Seddon and our family join me in sympathising most heartily with you in this your irreparable loss. There is consolation however in looking forward to meeting again in a happier and brighter world and also in knowing that he died in his country's cause. You must be a sad home just now. We feel deeply for you all and have our own anxieties.*[244]

For some New Zealanders, the eagerly awaited return of contingent members from South Africa was not the happy homecoming they envisaged. On 6 July 1902, more than a month after hostilities ended, the troopship *Britannic* carrying men from the Eighth, Ninth and Tenth contingents weighed anchor in Durban, leaving the bustle and congestion of Port Natal in its wake.[245] In a last-minute change of plan, Captain Henry Heckler and the 86 Tenth Contingent men under his command were ordered aboard the ill-fated vessel at half an hour's notice under what were later described as 'unfavourable circumstances'.[246] Heckler was an experienced and apparently competent officer, having first served as a private and then as a sergeant in the Fourth Contingent, before rising to the rank of lieutenant in the Seventh and finally receiving a captain's commission in the Tenth. Heckler later testified that before leaving Fort Hay on the outskirts of the northern Natal town of Newcastle, 22 men in his own squadron were on the sick list suffering from measles and influenza. As if the presence of highly infectious diseases among men cramped in the close confines of a crowded troopship was not enough, Heckler's men reportedly entrained for Durban without their kits or adequate clothing for the voyage.[247]

The vessel had been at sea for five days when the first death occurred; Lance-Corporal Tyrell De Labrosse, a Whakapirau farmer in civilian life, succumbed to 'pneumonia' in the Indian Ocean while the *Britannic* was en route from Durban to the Western Australian town of Albany.[248] After the *Britannic* left Melbourne, Trooper William Lawrence died in the Tasman Sea of what was also described as pneumonia only days before the vessel reached Wellington early on the morning of 1 August.[249] Both De Labrosse and Lawrence were buried at sea.

When Health Department officials boarded the *Britannic* in Wellington Harbour, they reportedly found around 20 men suffering from 'pneumonia' and a further 40 cases of measles among those in the ship's hospital.[250] These cases were transferred to Matiu Somes Island, the city's quarantine station. Following the war, the chief health officer claimed that at the time members of the public were critical of the Health Department's decision not to quarantine the *Britannic* on account of the number of cases of measles on board, but he noted that measles had 'a long incubation period (from two to three weeks), and quarantine would have necessitated an irksome detention of the troopers for an indefinite period'. The official observed that in ordinary circumstances measles resulted in 'a comparatively trifling mortality'. While acknowledging that the 21 per cent mortality rate among those afflicted with the disease who had sailed aboard the *Britannic* had been bad, he believed that had they remained confined aboard the vessel the death toll would have been significantly worse.[251]

The outbreak of measles that occurred on the *Britannic* had particularly tragic consequences for James and Wallace Nicholson, two Christchurch brothers. James, who served in the Tenth Contingent, contracted the disease while on board the vessel.[252] He claimed that although he was given medicine, he had been turned away from the vessel's hospital by the surgeon-captain due to a lack of space and was then forced to sleep on the deck.[253] When his younger brother, Wallace, who was on the ship with the Eighth Contingent, also came down with the disease, James was convinced he had infected his sibling.[254] He stated that his brother 'caught the measles off me, and now he is dying at Somes Island'. Though James eventually recovered, Wallace would never again set foot on the New Zealand mainland. Already weakened by measles, he finally succumbed to pneumonia on the island and died within sight of Wellington on the same day his brother James gave evidence before the Contingents Transport Commission convened to enquire into conditions aboard the troopships.[255]

The government was criticised by newspapers for excluding the press from the commission hearings, with the *Star* suggesting witnesses who gave evidence behind closed doors would be influenced by their former officers, who had a vested interest in the commission's findings.[256] When it was tabled in Parliament, however, the commission's report was far from a rousing endorsement of military efficiency. Discussing conditions aboard a second troopship, the *Orient*, the commissioners claimed 'there was lax, if not an almost entire absence of, discipline on board'.[257] They also noted that whether the *Britannic* men had been medically examined prior to embarkation was 'unknown', despite the prevalence of measles at Fort Hay prior to their departure. The commissioners also reported a 'misunderstanding' between Health Department officers and the commanding officers on board the *Orient*, resulting in a 'lamentable display of incompetency, or unwillingness to maintain discipline, on the part of regimental officers'.[258]

Though acknowledging the presence of measles in his Newcastle camp, when questioned by Commissioner John Millar, Heckler appeared to vacillate, seemingly reluctant to concede that infected men may have been among those who boarded the *Britannic*. When the commissioner asked whether it was probable that many in the squadron might be infected if there were men among them who had earlier contracted measles and been hospitalised, Heckler claimed that the men were permitted to board the *Britannic* 'after getting well'. While conceding that 'no doubt' the diseases 'would be throughout the squadron', he added that this was 'not necessarily to any extent'. When Millar asked if, despite men in the squadron having had measles, a number of them had been ordered aboard the *Britannic*, Heckler replied 'Yes, sir.'²⁵⁹

Heckler testified that medical officers in Durban never enquired whether his men had been exposed to measles (information that Heckler himself apparently did not think necessary to volunteer). However, the infected men were not entirely blameless in the events that followed.²⁶⁰ Both before departure and on arrival in New Zealand, a number of them failed to report their symptoms in the hope of either not being left behind or avoiding quarantine on arrival.²⁶¹ Lieutenant-Colonel Edward Chaytor stated that men hoped to pass the medical officer and get on shore.²⁶² James Nicholson and three or four others with measles paraded before the Port Health Officer, Doctor Henry Pollen; they were not quarantined and were permitted to disembark.²⁶³ In his own evidence, Pollen claimed that both the principal medical officer, Surgeon-Major Walter Pearless, and Brevet-Colonel Richard Davies 'pooh-poohed' the idea of quarantining the vessel.²⁶⁴

The commission's findings highlighted the severity of the measles outbreak and the pneumonia cases that followed. Pearless noted that the *Britannic* had one case of measles before the vessel reached Albany and that three Australians infected with measles were dropped at Melbourne.²⁶⁵ In Tasmania, the Hobart *Mercury* reported that on arrival in Wellington the vessel had 55 cases of sickness on board and mentioned claims of overcrowding on the *Britannic*.²⁶⁶ The newspaper claimed that there had been three deaths en route from Durban and one since the ship docked.

Though the *Mercury*'s figures were at odds with those of the commission, the disease became especially virulent during the final leg to New Zealand. The commission heard that there were about 18 men in the ship's hospital on arrival at Melbourne, which was considered normal. However, the number of measles cases rose from around 16 when the vessel departed for Wellington to 28 the following day. The total then increased 'by seven and eight a day' until there were 51 patients when the ship reached New Zealand, nine of whom had pneumonia.²⁶⁷

The commission found that Heckler's Tenth Contingent men, rushed aboard the *Britannic* 'at the last moment', were the source of the measles infection.²⁶⁸ The Wellington district health officer noted that measles had been endemic in New Zealand since 1847, and acknowledged that cases of measles had existed in Auckland prior to the *Britannic*'s arrival. However, he said there was little doubt that the return of the troopers aggravated, if not caused, the outbreak of the disease.[269]

In Parliament, Acting Premier Joseph Ward defended the decision to quarantine the soldiers. He claimed it was essential that infected men were landed on Matiu Somes Island as soon as possible to prevent the spread of smallpox, or the 'epidemic pneumonia' that had caused deaths among the *Britannic* troopers.[270] A total of 13 soldiers died on Matiu Somes Island in August 1902 of measles, septic pneumonia, or a combination of both diseases.[271] Two men from the *Britannic* subsequently died in Wellington (one at Wellington Hospital and the other at a city boarding house), while a third died in Auckland.[272] Other health issues were also raised by troopers during the commission hearings. Michael Mulhern testified that he had seen dogs on board urinating on a pile of fish that had been left lying on the deck for two days prior to it being cooked for the men, and claimed that despite protests to their officers, soldiers were served rotten meat that was so bad he refused to eat it and threw his portion overboard.[273]

The *Orient* sailed from Durban five days after the *Britannic* and docked at Dunedin's Port Chalmers on 5 August 1902.[274] When health authorities once again found measles on board, 20 men were confined on Kamau Taurua Quarantine Island in Otago Harbour.[275] More than 200 *Orient* men permitted to disembark at Dunedin were housed at Tahuna Park, and when the vessel reached Lyttelton the following day a further 20 measles cases were identified on board.[276] Troopers intending to disembark were initially kept on the ship until a tent camp was established at Bottle Lake, near the Infectious Diseases Hospital outside Christchurch.[277]

Arguably of more immediate concern aboard the *Orient* was Lieutenant Walter Callaway exhibiting the symptoms of smallpox.[278] Fearing the deadly disease might spread to the general population, the authorities initially quarantined men aboard the vessel and on Matiu Somes Island on arrival in Wellington, with some troopers reportedly 'in a state of open mutiny', refusing to set up camp on the island and destroying clothing.[279] James Mason, the chief health officer, reported that he experienced difficulty getting the *Orient* men to parade for vaccination and struggled to convince them of its benefits.[280] Mason stated that he had to agree to a £300 bond and guarantee the vaccine was safe before one soldier would submit to the injection.

The following month Legislative Council member William Jennings raised the case of New Plymouth trooper Ernest Wray of the Ninth Contingent.[281] Wray, who was later diagnosed with acute bronchitis, claimed that he and other men were ordered to strip and bathe in the sea at Matiu Somes Island despite the season. The men had been landed from the *Orient* to prepare the tents, and Wray claimed that due to the cold he and others fell ill.[282] The chief officer of the Board of Health denied Wray's allegations, claiming some of the men had washed in the sea voluntarily prior to being issued with new clothes. Jennings, however, said it was common knowledge that the men had been ordered to bathe in the harbour.[283]

Callaway disembarked at Matiu Somes Island together with more than 400 other *Orient* men, who were housed in tents to separate them from the *Britannic* patients.[284] His stay on the island was brief. He was transferred to a nearby defence facility at Māhanga Bay on Miramar Peninsula, where he remained in isolation for two months under the care of Surgeon-Captain Frederick King and Nurse Winton, with a 'Permanent Militia' guard placed on the building in which he was housed.[285] Claiming that the lieutenant had never actually had smallpox, Auckland City MHR Joseph Witheford criticised the Defence Department in Parliament.[286] Although Callaway was a non-smoker, the department had attempted to deduct money from his wages for cigarettes he received while in quarantine.[287] Seddon indicated he was prepared to reimburse Callaway as he believed the lieutenant's friends (including the officer in charge) had helped him smoke the cigarettes. The premier claimed those involved had indicated they were scared of plague and smoked as a preventative measure.[288]

Prominent Australian temperance campaigner Bessie Harrison-Lee, who was visiting New Zealand at the time of the *Orient*'s arrival, opposed the establishment of a canteen selling alcoholic drinks to the men quarantined at Bottle Lake, and instead called for assistance in distributing cakes and refreshments.[289] In Parliament, Sir William Stewart asked Acting Minister of Defence William Hall-Jones whether he was aware that men being sent to Bottle Lake were arriving at the facility on public trains.[290] Stewart claimed that 50 to 60 troopers en route to the facility were transported on a crowded Race Day train travelling from Dunedin to Christchurch.[291] A Christchurch *Star* editorial described the situation at Bottle Lake as a 'farce', with both patients and visitors coming and going at will despite absconders being threatened with prosecution and fines of up to £300 for breaking quarantine under the Health Act.[292] Seventy-five men reportedly left camp in defiance of orders, and several troopers were detained at Riccarton Racecourse.[293]

Newspapers claimed police had received instructions to arrest all 'deserters', though most troopers grudgingly returned to Bottle Lake of their own volition.[294] An extract from a poem that appeared in the *Star*, 'The Shores of Bottle Lake', captured the lingering resentment of the men confined in the quarantine camp:

> Let others boast of Bothasberg or famed New Zealand Hill,
> Of Longbowspruit, or Liarskloof, or Ananiasville;
> We of the last Contingents proudly claim to take the cake,
> For we braved the cheerful troopships and the shores of Bottle Lake.[295]

Former Christ's College Rifles marksman Sergeant Michael Dixon of the Ninth Contingent was one of those hospitalised at Bottle Lake after contracting measles and pneumonia aboard the *Orient*.[296] After Dixon spent 18 days in quarantine, his doctor discharged him with the recommendation that he spend an additional month at Hanmer Springs Sanatorium.[297] Almost three months after Dixon's return to New Zealand, Dr W. H. Symes noted that the sergeant had only recently been well enough to resume work.[298]

Following the departure of the *Orient* from Durban, the balance of the Tenth Contingent sailed on the *Montrose* on 15 July (four days after the Newcastle brawl).[299] Scarlet fever appeared on board almost immediately, with measles close on its heels.[300] A Department of Public Health report claimed that 'a state of chaos reigned' on the vessel, with medical staff unable to cope with the diseases' rapid spread.[301] Farrier-Sergeant Sidney A'Court and Trooper Rudolph Manning died on board the *Montrose*. They were buried at sea following services conducted by Chaplain-Captain Luxford, who had spoken to A'Court the evening before his death, when A'Court spoke of looking forward to arriving in New Zealand.[302] Captain Charles Browne reported that Manning had been repeatedly hospitalised in South Africa before finally succumbing to pneumonia complicated by scarlet fever on the voyage home.[303]

As the *Montrose* neared Auckland on 19 August, the lighthouse keeper on Tiritiri Matangi Island noted the vessel was signalling the presence of sickness on the vessel.[304] When this news was made public, anxious crowds gathered on Queen Street Wharf eager for information.[305] Their concerns would hardly have been allayed by newspaper articles reporting the death of Trooper John Lunn two days earlier on Matiu Somes Island. Lunn was an Eighth Contingent member from Collingwood who had been diagnosed with measles aboard the *Britannic*.[306] In a letter to Joseph Ward, Margaret Lunn described herself as a 'Heart Broken Mother'.[307] She told Ward that her son was her 'main stay in life' and requested the return of his letters and possessions, with the Defence Department forwarding Lunn's bandolier and waterproof sheet to his bereaved mother the following month.[308]

Until the port health officer and the district health officer had conducted a thorough examination of the *Montrose*, the men remained under quarantine, though they were permitted to briefly leave the vessel to stretch their legs on the docks at Auckland.[309] Medical authorities found more than 80 patients in the ship's hospital, around 50 of whom had scarlet fever or measles, with the majority of the balance suffering from pneumonia.[310] A decision was made to transport most of the sick to Motuihe Island in the Hauraki Gulf with the exception of the most serious cases, who were taken to Auckland Hospital. One of these, Inglewood trooper Albert Blyde, died of pneumonia in hospital four days after arriving back in New Zealand.[311] On learning of their son's illness the day before he died, Blyde's parents immediately travelled to Auckland, arriving in time to see him in his final hours.[312] The trooper's body was conveyed to Inglewood, where he received a military funeral.[313]

Despite the authorities being forewarned, the landing at Motuihe taxed available resources. The troop surgeon and 10 orderlies, together with 38 men with scarlet fever, nine with measles, and 12 with pneumonia, were put ashore and housed at the Quarantine Station, which had been constructed in the early 1870s.[314] The *Montrose* remained in Auckland for nine days, during which time an additional nine new cases of measles were identified, with the invalids sent to join those already on Motuihe. The Department of Public Health reported a total of 60 cases quarantined on the island, and it was not until the end of September that Motuihe was finally evacuated.[315]

The month before the *Britannic*, *Orient* and *Montrose* reached New Zealand, the chief health officer attributed a substantial increase in measles cases to troopers who had arrived in Dunedin aboard the *Tagus*.[316] Trooper Arthur Kendall died of an embolism on board *Tagus* as the vessel neared Dunedin Heads in July 1901.[317] Two days later, Trooper Frederick Forbes, who had been hospitalised aboard the *Tagus* for five days, died of pneumonia on the ship.[318] In Parliament, Seddon made no mention of measles but said he had declined a request from 'Brigade-Surgeon' De Latour to land 11 men at Kamau Taurua Quarantine Island from the *Tagus* suffering from malaria, typhoid and pneumonia. Instead, he ordered that the sick soldiers be housed in Dunedin Hospital and in private hospitals in the city.[319] The Health Department's resources were further tested by the need to arrange medicine and supplies for sick contingent members, as well as send telegrams replying to queries about the condition of patients.[320]

Though disease presented significant challenges for New Zealand health officials, the spread of disease on troopships was no revelation to the imperial authorities. The 1902 measles outbreak aboard the *Britannic* largely mirrored another sailing of the same vessel almost exactly a year earlier. On that occasion, the *Britannic* left Albany for Adelaide transporting the Imperial Bushmen back from South Africa, with the Melbourne *Argus* reporting that 51 cases of measles had been detected on board.[321] During the war, vessels taking troops *to* South Africa were also responsible for the introduction of disease. In his May 1902 report on the health of the troops, the principal medical officer for Natal district observed that a mild epidemic of measles had been detected in Australian and New Zealand troops who had recently disembarked. The surgeon-general noted that such outbreaks invariably appeared among colonial forces newly arrived in Africa.[322]

Along with gunshot wounds, the New Zealand chief health officer noted the large number of malaria cases found among returning contingent members.[323] The Department of Public Health was sufficiently concerned by the incidence of the disease and its potential impact that it conducted experiments to ascertain whether it was possible for a New Zealand mosquito to spread malaria from an infected trooper to a healthy individual.[324]

The impact of diseases contracted either in Africa or aboard the transport vessels extended well beyond the contingents' ranks. As the Health Department noted, 'War exacts a penalty in some form or other from non-combatants.'[325] Just as it had been inevitable that, once introduced, measles would spread among men on the crowded troopships, so it was certain the highly contagious disease would spread within the New Zealand population.

The chief health officer reported that there was every reason to believe the return of the troops was responsible for a severe epidemic of measles in the Auckland district.[326] Schools were identified as hotbeds of the disease, with Wellesley Street and Newton East particularly affected. In 1902, a total of 92 deaths from measles were recorded in the four main centres of Auckland, Wellington, Christchurch and Dunedin, with the North Island cities bearing the brunt of the epidemic.[327]

Although the Department of Public Health described the case as a 'strange coincidence, if nothing more', it reported an incident where a child contracted typhoid in a town that had previously been free of the disease.[328] The child had been wearing underclothes formerly belonging to an uncle who had served in South Africa. The trooper brought the clothes he had worn while convalescing from typhoid back to New Zealand and gave them to his nephew.[329]

It remains unclear whether concerns Henry Feldwick raised in the Legislative Council in 1901 regarding the possibility of diseases introduced by men returning from South Africa spreading among the Māori population were warranted. Areas with significant Māori populations appear to have been spared the worst of the measles outbreak compared to major urban areas. This may have been due to the comparatively small number of Māori in the contingents. However, the evidence is inconclusive and population density may have been a more significant factor. While Waikato County had only four reported cases, there were 43 in Rotorua and 158 in Hawke's Bay District.[330] In Auckland District, however, a total of 3544 cases were reported, with 1392 in the city, 1526 in the suburbs and 626 in surrounding rural areas.[331]

A 1903 Department of Public Health graph showed the negligible number of notified measles cases in the Wellington region prior to the arrival of the troopships, and the explosion of reported infections that followed in their wake. A total of 32 cases were notified a little over three weeks after the *Britannic* arrived. This fell to seven by the fourth week, but roughly three weeks after the *Orient* men were released from quarantine on 13 August, the number rose

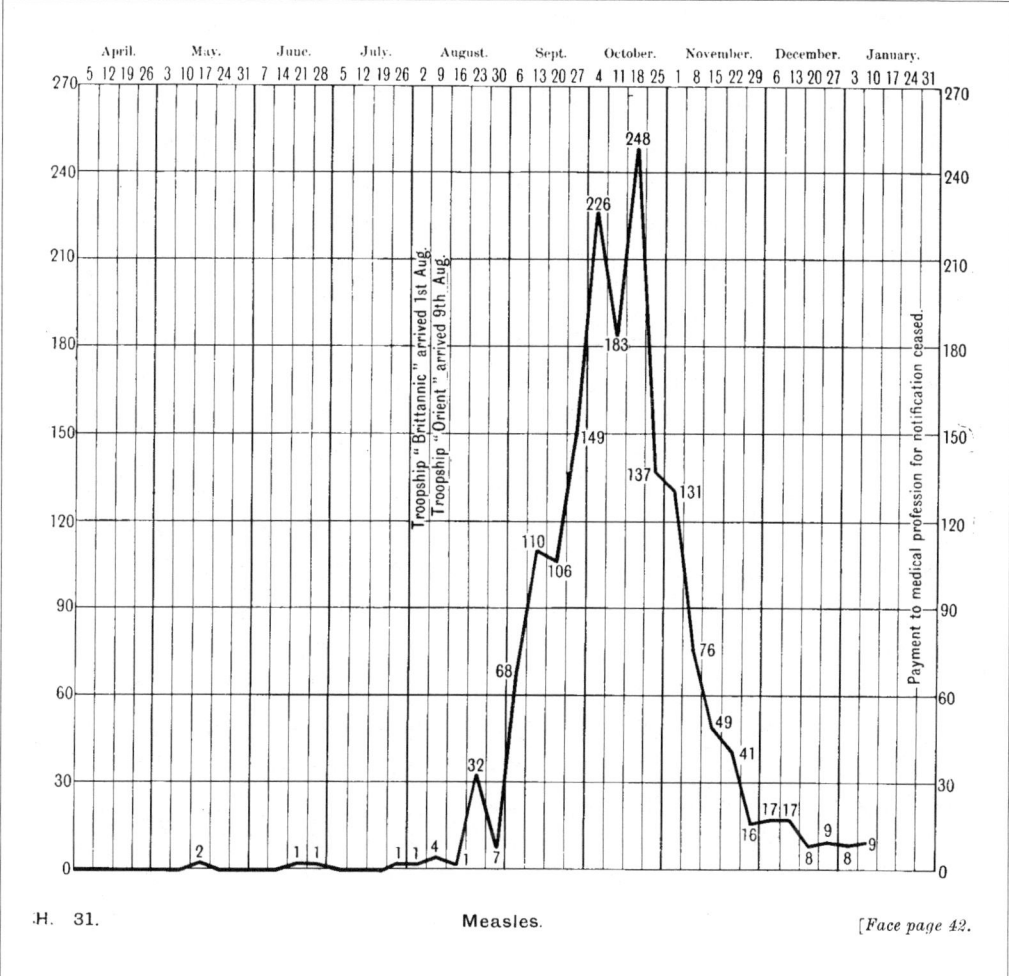

A Department of Health graph showing the increase in measles cases following the arrival from South Africa of the troopships *Britannic* and *Orient* in 1902. *AJHR*, 1903 H-31, FACING PAGE 43, NATIONAL LIBRARY OF NEW ZEALAND TE PUNA MĀTAURANGA O AOTEAROA

significantly to 68, and then continued to climb (with two brief pauses) before peaking at 248 cases in the city in mid-October.[332]

The section of the report that discussed the Wellington region noted that four cases of measles had been identified in the second week of August.[333] These could not be attributed to the disbanded troopers due to the disease's incubation period. Although there were concerns that a considerable loss of life would result from a malignant strain of measles, in fact the epidemic was considered 'extremely mild' compared to the disease's mortality rate in the United Kingdom.[334] A total of 134 deaths attributed to measles occurred throughout New Zealand during the 1902 epidemic, compared to 525 deaths during a more virulent 1893 outbreak.[335] Nonetheless, the mortality rate among the *Britannic* men who contracted measles was still alarming.[336]

Early in the war, Private James Elliott, the New Zealander serving in the Royal Army Medical Corps, summed up the situation in a letter to his father: 'War is a terrible thing. So is disease.'[337] As Corporal Robert Drinnan discovered, other ailments also had the potential to profoundly affect soldiers' lives. The Fourth Contingent soldier's commendation from Lord Kitchener and mention in the dispatches of Major-General Babington counted for little in the post-war years. Drinnan was hospitalised three times in South Africa with dysentery and malaria, and also lost all his molars, but his health deteriorated still further on his return to New Zealand.[338]

The year after the war ended, William Russell asked the minister of defence if he would arrange for men like Drinnan, who had fought in South Africa and were permanently debilitated through wounds or disease, to be employed as parliamentary doormen. At the time Russell raised the issue, members of the Permanent Artillery were used in the role and Russell was critical of able-bodied members of the corps 'lounging about the entrance-door to Parliament'.[339] Despite his support for the war, it exacted a heavy toll on Russell's family; his own son, Lieutenant Lionel Russell, died in 1901 of wounds received while leading his men of the West Yorkshire Regiment in eastern Transvaal.[340]

By 1915, Drinnan was suffering from spinal paralysis, which the doctor said was probably attributable to the effects of campaigning in South Africa.[341] In March 1916 Drinnan wrote to James Allen, the minister of defence, stating that his health had seriously deteriorated and enquiring whether he was eligible for a pension.³⁴² His letter followed an earlier letter from M. W. Fleming to Allen drawing the minister's attention to the former corporal's plight.³⁴³ Fleming claimed that Drinnan was crippled, unable to earn a living and costing his family £200 per year in medical fees. In 1916 Alfred Robin, by then a brigadier-general, asked whether funds were available for such cases.³⁴⁴ Captain Rockstrow of the General Staff replied that a little over £5 was left for disbursement to individuals like Drinnan, adding that the money was all that remained of the thousands of pounds raised to support New Zealand soldiers during the South African War.³⁴⁵

Unfortunately for Drinnan, a combination of government red tape and departmental ineptitude resulted in his letter being misplaced and the slow-moving wheels of bureaucracy grinding to a halt.³⁴⁶ No action was taken for a further two years until Fleming once again wrote to Allen in 1918. Noting that Drinnan had been hospitalised for the previous six years, Fleming questioned the Defence Department's inertia.³⁴⁷ For Drinnan, however, it was too little and too late. After enquiries were once again resumed, Major Daniel Hickey, the decorated survivor of Langverwacht, informed headquarters that Drinnan had died on 22 April.³⁴⁸

Possibly influenced by his brother, who had earlier served in the South African Light Horse, 20-year-old John Lund enlisted in the Tenth Contingent in 1902.³⁴⁹ One of 11 children, Lund saw no action, and after spending two months in Africa he contracted bronchitis aboard the *Montrose* as the vessel neared Auckland.³⁵⁰ His health did not improve following his return, and he died on 26 October 1904.³⁵¹ After his death in Oamaru Hospital, his father sought government assistance to pay the £3 the railways charged to return his son's body to Pleasant Point in Canterbury.³⁵² Despite John Lund's wartime service, the Defence Department once again requested that police make enquiries into the 'circumstances and character' of Walter Lund without his knowledge after he also sought government help with the cost of his son's funeral.³⁵³

Though the contingents contained significant numbers of men who had previous experience of warfare, the majority of New Zealanders in South Africa did not. For those who survived engagements like Slingersfontein, Rhenosterkop, Ottoshoop and Langverwacht, or witnessed the carnage at Machavie, the experience was not easily forgotten. Although it is unclear whether his military service was a factor, former trooper Robert Moore viciously assaulted his father less than three months after returning from South Africa, while Trooper John Salter reportedly committed suicide during the *Orient's* voyage to New Zealand.[354] Although some newspapers reported that Salter had suffered a brain injury after falling from his horse in South Africa and had jumped overboard, others portrayed his death as an unfortunate accident.[355] In his draft history of the war, James Shand also recorded the trooper's death as accidental.[356]

A military inquiry convened to investigate the incident heard that the deceased had been classified a 'lunatic', while his medical record identified his condition as 'melancholia'.[357] Government reports provided conflicting accounts of Salter's death. A 1901 report claimed he 'fell overboard', though eyewitness accounts said he climbed the vessel's railing and jumped.[358] The *Evening Post* reported the incident in an article titled 'Suicide of a New Zealand Soldier', and in 1903, a second government report tabled in Parliament stated that the trooper 'jumped overboard' from the *Orient*.[359] At a time when mental health issues and the effects of post-traumatic stress disorder were not well understood, the role the war played in these events is difficult to gauge. However, soldiers' diaries and letters occasionally reveal the psychological impact of the disturbing events they witnessed.

In a letter to his sister, Trooper Frederick Shaw described the aftermath of battle: 'While the fight is on we feel allright [sic], but as soon as it is over and we see the dead lying about it gives a different feeling.'[360] Another trooper told his parents that his sleep was affected because 'one is unable to deface the awful sights from one's mind'.[361] Trooper Wrigley, who witnessed the accidental shooting of Trooper Byrne, told his brother that after the incident 'it would have been hard to find a more wretched and unhappy lot of men'.[362] Whether Patrick Fitzherbert's harrowing account of the aftermath of Paardeberg gave Wanganui Collegiate School students borne along by the tide of patriotism pause for thought is unknown. Although Fitzherbert also acknowledged the interesting sights he had seen, and getting 'his first taste of plunder', after Paardeberg he clearly did not view war as a glorious enterprise:

> Yes, I will be a strong supporter of the disarmament scheme after this. And I say this, that never mind how ardent a military spirit a man may have, let him see a sickening sight like that and his ardour will be quenched forever.[363]

Soldiers exhibited varying degrees of willingness to share their wartime experiences on their return to New Zealand. While Trooper David Waldie, a Dunedin blacksmith, gave a speech to a large audience in the Garrison Hall, and Sergeant Matson addressed North East Valley School, others were reluctant to publicly relive their time in South Africa.[364] In a voice audible throughout the hall, Waldie, dressed in his uniform, described his experiences during the conflict, though the *Otago Daily Times* was quick to point out his shortcomings as a public speaker. The paper reported that Waldie 'spoke of a number of British killed and wounded being picked up and made as comfortable as possible'. While relating another incident where troops had taken a farmer's cattle, Waldie reportedly claimed they 'killed them, and then roasted them alive'.[365]

Rupert Hosking could not be induced to speak to the boys at Wanganui Collegiate School about his experiences at Mafeking during a visit to his old school.[366] Hosking had travelled to England in 1897, and from there made his way to South Africa, where he joined the Cape Mounted Rifles.[367] Reportedly tired of waiting for the war to begin, Hosking considered returning to New Zealand, but instead enlisted in the Protectorate Regiment.[368] He served as a sergeant in D Squadron and was besieged in Mafeking along with fellow Old Boy Edward 'Teddie' Jollie.[369] Although Jollie survived the siege unscathed, in a letter published in the *Wanganui Collegian* Hosking described having his toe 'nearly blown off'.[370]

Hosking's Protectorate Regiment bore the brunt of an unsuccessful sortie on the Boer forces occupying Game Tree Fort near Mafeking when his force pushed forward only to discover the position was strongly defended. Jollie, who described the Boers shelling Mafeking from afar as 'nothing less than murder', reported that the Boxing Day attack left 23 men of the attacking force dead, including several officers.[371] Shortly before the siege was lifted, the Boers launched a final attempt to take Mafeking in the early hours of the morning with a concerted attack led by Field Cornet Eloff. On this occasion Hosking sustained a serious injury requiring the extraction of several bullet fragments and pieces of clothing from a wound to his hip.[372] The *Evening Post* reported that a bullet had splintered Hosking's femur.[373]

After reaching South Africa in 1900 and testing the typhoid antitoxin, William Hosking removed Rupert from hospital and took his son to England, where he slowly recovered before returning to New Zealand.[374] The *Wanganui Collegian* noted that when Rupert visited the school after his return to New Zealand he 'proved obdurate to all requests that he would write some description of the Siege' and strongly objected to being termed a 'hero'.[375] Hosking did not explain his reticence, but it is possible his injury changed his view of the war. In 1900 Lieutenant Michael Lindsay was equally reluctant to discuss his wartime experiences and, like Hosking, declined requests to address his former school.[376]

Injuries that a Tenth Contingent private sustained in Africa appear to have adversely affected his later life. The private, who received an 'indifferent' character assessment rating on his 1902 discharge certificate, had earlier served in Brabant's Light Horse.[377] He was repeatedly arrested for alcohol-related offences after returning from South Africa, where he sustained a serious head injury. Giving evidence at one of the veteran's trials, Constable Shaw noted that since the private had been shot in the head, 'drink made him lose control of himself'.[378] Imprisoned in Mount Eden for a month in 1910 after facing charges of drunkenness and being a vagabond, he briefly served overseas during the First World War, though by 1937 his address was Ward B, Tokanui Mental Hospital.[379] Other soldiers also struggled with alcohol issues following the war. After the Defence Department asked police to conduct a confidential investigation into Patrick Lee's 'mode of life', Constable Emerson noted that Lee had purchased a 200-acre farm and built a small cottage, though he claimed Lee was not cultivating the land and was living on his government pension. Emerson stated that Lee was of good character but drank heavily at times.[380]

Hugh McDonagh also turned to drink after suffering a serious injury in South Africa. John Adams of Lloyd's Patriotic Fund claimed McDonagh had 'rather suffered in the head' following his spinal injury and was prone to drunkenness. This was supported in a letter from G. E. Way of the Patriotic Committee. Way expressed concern about McDonagh being made aware of the funds held on his account as the trooper had been drinking heavily and it was feared he would fritter his remaining money away on alcohol.[381]

Return to Maoriland

Though the press and politicians like Kaiapoi MHR David Buddo expressed concern about the potential economic impact of New Zealanders staying in

South Africa after the war, less prominence was given to the social disruption this would cause. Buddo drew attention to the Rhodesian Charter Company offering farms to Australian men as an inducement to settle in the country.[382] Frank Swanwick of the Fourth Contingent claimed colonials were being offered cash and 3000 acres if they agreed to remain.[383] While Joseph Ward felt the government should give men their discharges in Africa if they so desired, he also believed they should be encouraged to return, as New Zealand was 'undoubtedly an infinitely better country to live in'.[384]

A 1902 newspaper editorial dismissed as a 'fallacy' suggestions that many soldiers would not return. It claimed if a soldier survived both the war and disease he would almost certainly 'find his way back to his Maoriland again'.[385] Nonetheless, the New Zealand under-secretary of defence said he had heard that many men were indeed returning to Africa.[386] Despite warnings that wages and conditions were often poor, men such as Lieutenant Henry Heywood took their discharges in South Africa and indicated they intended remaining there.[387] Noting that he could get a 'fair screw' in the South African police, George Leece told his mother that although he wanted to return to New Zealand, he dreaded the prospect of unemployment.[388]

Many of those who did return to New Zealand appeared reluctant to return to the mundanity of their former occupations. The *Ashburton Guardian* observed that even if men did come back, it would take a long time for them to settle down to civilian occupations after experiencing the excitement of war.[389] This difficulty reintegrating into civilian society extended beyond soldiers who resumed their former trades in New Zealand. Sergeant-Major Gillespie, who had been besieged in Kimberley, was invalided back to New Zealand. After a period of recuperation, he sailed back to Africa and his old job at the *Kimberley Advertiser*. However, his tenure at the newspaper proved brief: in a letter home, Gillespie said 'soldiering upsets a man, and office work was simply unbearable to me'.[390] After reportedly declining a commission in Denniston's Scouts because the corps was mainly 'Cape Boys', Gillespie instead secured a captaincy in Scott's Guards, a corps formed from employees of De Beers, the Kimberley diamond mining company.[391]

A number of soldiers worried about their prospects in New Zealand, with some like Gillespie either returning to South Africa or seriously considering doing so. Private Denis Hickey, formerly of the Fifth Contingent, wrote to Lieutenant-Colonel Stuart Newall requesting his pay as he intended to try his luck in South Africa.[392] Despite being invalided home after being crushed by a horse and receiving a modest pension, former trooper Joseph Culling also indicated he wanted to seek employment in South Africa.[393] Arthur Stuckey of the Eighth Contingent not only returned to Africa, but also returned to soldiering. Stuckey fought during the 'Boer Rebellion' of 1914 as well as serving in the German South West Africa (present-day Namibia) and German East Africa (present-day Burundi, Rwanda and mainland Tanzania) campaigns.[394]

Shortly after the war ended, Herbert Hart noted that 'Landgrabbers' in his camp from nearly every colonial corps were awaiting land grants.[395] Although he returned to New Zealand via the United Kingdom, Hart himself considered volunteering for service in Somaliland.[396] Private Cooper, invalided home with malaria, intended settling in the Transvaal after securing employment at Pretoria Prison; Ronald Saxby considered returning to the work in the South African mines; and John Duigan, after recovering from injuries sustained at Wepener, went back to South Africa to join another contingent.[397]

To soldiers like Duigan, Hart and Walter Borlase, the military life clearly appealed. Duigan, as noted earlier, would become Chief of the General Staff in the New Zealand Army, and Hart became a temporary brigadier-general during the First World War. Borlase, who served in the Tenth Contingent but did not see active service, said he was sorry to leave Africa.[398] He took his discharge in England and became an officer in the 7th Royal Sussex Regiment.[399] Percy Cohen, an Eighth Contingent trooper who before the war had been a steward with the Federal Steam Navigation Company, remained in South Africa and in 1903 gave a Johannesburg address, while Prosper Rain Berland, the First Contingent quartermaster sergeant captured at Sanna's Post, later settled in Africa.[400] Berland took part in the occupation of German South West Africa and died at Windhoek in 1930.[401]

Like Cohen, William Dunnet, an Otago shepherd who enrolled in the Tenth Contingent in Durban in the final week of the war, remained in Johannesburg when it ended.[402] Together with former New Zealand Railways shunter John Sanderson, who served in both the Fifth and Tenth contingents, Dunnet found work in the railways.[403] A thorn in the premier's side during the war, William

Hutchison junior resigned his position at the Christchurch *Press* after the conflict and emigrated to the Transvaal. By 1904 Hutchison had joined forces with Ashburton farmer and former Fourth Contingent trooper Robert Simmers in a Johannesburg news and real estate business.[404]

Despite Africa's attraction, former soldiers from New Zealand who chose to establish themselves there after the war did not always prosper in their new home. Piers Tudor took up farming at Nylstroom but contracted typhoid and died on 12 December 1902.[405] By 1904 William Dunnet had left his job with the African railways and was serving in the field artillery in Brisbane.[406] Other former troopers apparently experienced difficulties. The public accounts for 1905–6 show the government paid £15 6s 6d for passages from Cape Town to New Zealand for 'three distressed ex-troopers of New Zealand Contingents'.[407] Nonetheless, for at least some New Zealanders South Africa repaid the faith they had placed in it. A New Zealand nurse and several teachers settled there permanently; Nurse Gertrude Littlecott stayed on in Durban, while Dora Webb taught at a school at Gezina in Pretoria.[408]

Graves of the fallen

The bodies of men who died in South Africa remained there, sometimes in poorly marked battlefield graves. When the family of Sixth Contingent trooper Irving Hurrey was informed he had died of wounds received in action, Hurrey's father requested that his son's body be exhumed and returned to New Zealand.[409] John Hurrey informed the premier that it was his wife and daughters' wish that the body be repatriated, and that he would cover any associated costs. Seddon was sympathetic but exhibited a poor understanding of conditions in the field. He instructed the under-secretary of defence to consult either the New Zealand agent in South Africa or the proper authorities about returning Hurrey's body to New Zealand, noting that 'formal quarantine' would be required.[410]

John Hurrey also requested that a war correspondent photograph his son's grave, but once again inaccurate press reports caused unnecessary consternation in both New Zealand and Australia.[411] Newspapers published conflicting accounts claiming that either 'J. C. Hurrey' or 'J. S. Hurrey' of the New Zealand Mounted Rifles had died of wounds at Bethel.[412] Adding to the confusion, Irving's younger brother, Ernest Charles Hurrey, was a provost-sergeant also serving in the Sixth Contingent, while his older brother, John Alexander Hurrey, was a private in the Third.[413] After reading in a Melbourne newspaper that 'J. S.' Hurrey had died at Bethel, J. W. Hazeldine (the concerned father of Irving's fiancée) wrote to the Defence Department from his home in Victoria seeking clarification.[414] A Defence Department staff officer informed Hazeldine that regrettably it was indeed an error and it was his daughter's fiancé who had been killed.[415]

Despite the family's request and Seddon's willingness to assist, Hurrey's body does not appear to have been returned to New Zealand during the conflict. Several months after John Hurrey wrote to the premier, the commanding officer of the Sixth Contingent, Major Andrew, informed the Defence Department that Hurrey had been buried at Trichardtsfontein, west of Bethel.[416] Andrew said that Ernest Hurrey had attended his brother's funeral and, as was customary on active service, Irving's remains had been buried without a coffin. Ernest Hurrey, who was with General Plumer's column, had ridden across to check on his brother's condition after hearing he had been shot. In a scene that the *Evening Post* special correspondent described as 'pathetic', Ernest arrived just in time to see his brother buried. Andrew added that he believed recovering Hurrey's body, which had been buried under a grove of blue gum trees alongside that of Royal Horse Artillery gunner Wye, would be impossible until after peace was declared as Boer forces remained active in the area.[417]

For at least one family, the fate of their son remained a mystery. Trooper William Mathews of the Sixth Contingent disappeared without trace in September 1901 while on patrol. Although Mathews' body was never recovered, the Timaru trooper was believed to have drowned in the Caledon River and was officially listed as 'missing'.[418] Even when the precise location of graves was known, few bereaved families had the time or money to travel to South Africa. Captain John Harvey and Trooper Septimus McDougall, who was also killed at Ottoshoop, were initially buried at the base of a kopje near a road in western Transvaal.[419] In 1904, however, the New Zealand government reportedly paid for a marble memorial commemorating the pair to be erected in South Africa.[420]

Even the comparatively few New Zealanders able to visit the graves of their loved ones in South Africa could still face difficulties. Despite physician and renowned author Arthur Conan Doyle doing his best to save the young trooper's life, Lionel Smith of the Second Contingent died of wounds he received at Vet River in May 1900 and was buried in the field.[421] However, when a party from Smith's contingent later returned to erect a cross they were unable to locate the grave.[422] Smith's father wrote to Seddon thanking him for supplying details of his son's death, but asked for information about where he was buried so his family might have an opportunity of visiting the grave.[423] Though Seddon instructed the agent general in Cape Town to make enquiries, Lieutenant Todd reported in 1901 that he had been unable to find its location.[424]

Shortly before the war ended, Otago nurse Bessie Hay sent photographs to the *Otago Daily Times* (which were then distributed by Major Robin to the dead men's families) of some of the New Zealand soldiers' graves in the cemeteries of Pretoria and Johannesburg.[425] Hay said that all the graves she encountered were well cared for, had flowers and were marked by an iron cross and shield. In a 1904 letter, Frances Davis, one of the New Zealand teachers who had accepted a position in South Africa, assured the premier that the teachers had not forgotten their 'fallen brothers'. Davis said the teachers tended New Zealanders' graves in the vicinity of their schools and at other locations they visited.[426]

The Soldiers' Graves Guild was established by New Zealand women in 1902 with Louisa Seddon as its acting president.[427] The organisation was inspired by the Guild of Loyal Women of South Africa, whose members tended the graves of imperial troops and sent photographs of them to relatives. Louisa Seddon solicited funds from New Zealand towns as well as county and borough councils, which were then passed on to the South African Guild and the New Zealand Soldiers' Graves Upkeep Fund.[428]

In 1908, the Soldiers' Graves Guild still had £660 to its credit in the National Bank, and although the accuracy of the claim must be questioned, given deaths often occurred in remote areas where bodies were hastily interred before forces moved on, in 1910 its members were assured that all New Zealanders' graves in South Africa had been located and marked.[429]

Until it was dissolved and its functions transferred to the Victoria League of Great Britain in 1912, the Guild of Loyal Women of South Africa (or 'Daughters of the Empire') did a significant amount of work marking and tending the graves of imperial soldiers.[430] Eight years after the war ended, the New Zealand Soldiers' Grave Guild received permission from Wellington City Council to erect headstones on the graves of all South African War veterans buried in Karori Cemetery.[431]

The adverse impact of the war in human terms was manifold. Injury and illness could have lasting implications for both soldiers who served in South Africa and their family members. Adding to the emotional burden of dealing with grief, those who were financially dependent on soldiers who lost their lives also faced economic uncertainty, while other soldiers experienced psychological and alcohol-related issues. As well as the toll it exacted on New Zealand soldiers, war-related disease also affected a number of New Zealanders who had never set foot in Africa.

The conflict dislocated families; and in cases where men remained in South Africa, returned there or travelled further afield, these separations were prolonged or even permanent. As a newspaper editorial noted shortly after the war, 'Many of the troopers who set out from our shores during the last few years in full flush of young manhood's strength have returned to us maimed, crippled and completely broken in health.'[432]

CHAPTER SEVEN

'THESE WARS WILL ALWAYS BE POPULAR'

THE ECONOMIC IMPACT OF THE SOUTH AFRICAN WAR

Two months prior to the outbreak of war in 1899 Dunedin's future mayor, George Denniston, addressed the city's Chamber of Commerce as political tensions in South Africa mounted. On the same day, the *Otago Daily Times* captured the mood of the British population in Pretoria when it reported that 'a great scare' existed following the government of the South African Republic distributing Mauser rifles to Boer burghers.[1] Yet the focus of Denniston's speech was commercial opportunities rather than political intrigue. He supported the establishment of a 'lucrative permanent trade' between Dunedin and Britain's South African colonies and expressed satisfaction with steps taken by the Chamber of Commerce to increase awareness in Cape Colony of the quality and suitability of Otago goods.[2]

Denniston was not alone in recognising the economic opportunities of trade with South Africa. Only a matter of weeks later and with war looking increasingly likely, New Zealand parliamentarians discussed the country getting its fair share of valuable War Office contracts. William Massey asked Seddon whether he was aware of sizeable orders for fodder reportedly placed by the British War Office with Australian firms. Eager to capitalise on the situation, Massey observed that supplies of forage could probably be sourced more cheaply in New Zealand than in Victoria.[3] The premier's reaction to Massey's suggestion that the British War Office be informed of New Zealand's ability to fill forage orders was characteristically dismissive. He saw no need to draw the War Office's attention to New Zealand's existence, or to remind it that New Zealand could supply oats and chaff; 'They knew that we had chaff in abundance,' and during the previous year the United Kingdom had imported over 21,000 bushels of New Zealand oats.[4]

Three days after Massey addressed Parliament, Seddon announced the imperial authorities had placed an order for 2800 tons of oats, and claimed that the bulk of a similar order placed in Australia would in all likelihood be filled by oats sourced in New Zealand.[5] Barely a week later, the *Feilding Star* reported that a Christchurch company had secured an imperial government order for 30,000 sacks of oats to be loaded aboard the steamer *Vinebranch* at Lyttelton, Tīmaru and Bluff.[6] With the approaching war presenting the tantalising prospect of an economic boon for the agricultural sector, the more sanguinary aspects of warfare were temporarily set aside. A *New Zealand Tablet* article titled 'Good News for Farmers' predicted that the situation in South Africa was likely to benefit New Zealand's farming community.[7] Legislative Council member Henry Scotland, however, remained suspicious of the reasons for the conflict, claiming, 'These wars will always be popular in the colonies, as they are profitable.'[8]

'An excited market for oats'

There was a desire on the part of politicians to ensure that, where possible, their constituencies directly benefited from any increase in South African trade. Following Seddon's announcement on 29 September that tenders had been called for the shipment of New Zealand produce to South Africa, Ashley MHR Richard Meredith claimed that farmers in Canterbury could provide 200,000 tons of Derwent potatoes.[9] When Meredith asked the premier if orders had also been placed for other foodstuffs, Seddon replied that encouraging information had been received from South Africa indicating that flour, wheat and tinned meat were in demand.[10] The *Evening Post* considered the market for forage even more promising: 'In the quality of [New Zealand] oats, however, it is conceded that she is above competition.'[11]

Napier MHR Alfred Fraser urged that all New Zealanders should be informed of imperial oats orders so that farmers rather than speculators could benefit. Pātea MHR George Hutchison agreed, though Seddon was reluctant to comply out of concern that full disclosure might inflate prices. New Zealand farmers and businesses were quick to capitalise on the situation, with the *Rangatira*,

'Tons upon tons of oats for Tommy's faithful friends', at De Aar, Cape Colony, South Africa. RICHARD STOWERS COLLECTION, AUCKLAND WAR MEMORIAL MUSEUM TĀMAKI PAENGA HIRA, PH2002/1

a cargo steamer bound for South Africa, having to decline further produce for want of space in her holds, and with sufficient goods remaining to load a second vessel.[12] The government, however, was initially reluctant to assume any form of financial liability in developing the South African trade, with Seddon stressing that 'the Government had not incurred the slightest responsibility in regard to [the *Rangatira*]'.[13]

While it was prepared to advertise the availability of vessels to carry produce to South Africa, Seddon considered shipping arrangements should remain the domain of the companies involved.[14] This unwillingness to expose the government to potential financial losses placed the burden of liability squarely on the shoulders of shipping companies and New Zealand exporters. Seddon's concerns were not entirely unwarranted, as the produce the *Rangatira* carried reportedly received a mixed reception. While the vessel's general cargo was satisfactory, the meat it carried was not competitively priced, raising the possibility that some would need to be shipped to England.

The day after war broke out, the *Ashburton Guardian* reported that the situation in Transvaal was already having a positive impact on the New Zealand grain market.[15] Exports of oats soon reflected this, with a government report detailing shipments to South African ports recording 12,000 tons loaded aboard three ships at Bluff in June 1900.[16] Subsequent orders totalling 15,000 tons were shipped aboard an additional five vessels in March, April and May 1901 after loading at Lyttelton, Tīmaru and Ōamaru.[17] Although New Zealand was not the only source, demand for New Zealand oats remained high as long as mounted forces' horses required feed. A Tenth Contingent trooper claimed New Zealand oats were better than those sourced in Canada, which he described as dirty and inferior to the worst New Zealand sample.[18]

In response to a July 1901 request by Waitematā MHR Richard Monk, a parliamentary report detailing exports of oats, flour and wheat indicated that a total of 7,491,407 bushels of oats had been exported to South Africa.[19] By 1902 demand showed little sign of abating, with the *Otago Witness* claiming that buyers were 'chasing growers all over the paddocks'. Noting that South Canterbury farmers had produced a bumper crop, the newspaper claimed growers were holding out for higher prices.[20]

Almost all vessels carrying New Zealand produce to South Africa did so via

Australian ports to re-coal and tranship cargo, thereby extending delivery times. In 1900 the *Kilburn* loaded 45,000 sacks of oats at Ōamaru, Tīmaru and Lyttelton for Friedlander Bros Ltd, before sailing for South Africa via Western Australia, while the *Longships* carried 96,000 sacks of Southland grain to the Cape via Melbourne.[21] The *Marlborough Express* noted this brisk trade in New Zealand oats, chiefly for transhipment at Melbourne.[22] Despite heightened demand, exporters faced challenges in the booming oats trade. Having an abundance of produce was one thing, but transporting it to Africa in compliance with War Office timetables was another matter entirely.

Demand for oats in South Africa played directly into the hands of South Island farmers in a position to grow the crop. In 1898 no oats, wheat, butter, mutton, frozen beef or horses were exported to either Cape Colony or Natal.[23] Even New Zealand's tiny trade with Norfolk Island represented a bigger market for her oats than the South African colonies, with 24 bushels going to the remote Pacific island.[24] All that changed during the war years. While wheat, horses, butter, mutton and frozen beef periodically featured among the top five New Zealand export products by value to Natal and Cape Colony, oats retained the top position for the duration of the war.

Given the central role horses played in military operations, demand for horse feed naturally surged. While a percentage of New Zealand oats sales to Australia was destined for the domestic market, it appears a sizeable portion was re-exported to South Africa. New Zealand exports of oats to the Australian colony of Victoria increased in volume by a massive 985 per cent between 1898 and 1900, while financial returns from oats sales to Victoria rose by 870 per cent for the same period.[25] This phenomenal rise largely mirrored the growth in New Zealand exports of oats directly to the South African colonies.[26] The *Star* confirmed the role of some Australian buyers as middlemen in a March 1900 report, noting that in less than three months South Africa had taken 140,000 bags of Victorian oats and 160,000 bags from New Zealand and Tasmania through Melbourne shippers.[27] Though in 1902 Seddon bemoaned the conditions of an imperial oats contract, which he termed 'an insult', New Zealand continued to ship oats to South Africa in large quantities.[28]

Heightened export demand did not result in the inflated prices that Seddon initially appeared anxious to avoid. In 1900, the *Evening Post* noted subdued business in the oats market but predicted that an imperial government order for 5000 tons might provide a fillip to values.[29] Prices of New Zealand oats exported to Cape Colony and Natal experienced comparatively minor fluctuations during the first three years of the war, with a high of 2s per bushel in 1900 and a low of 1s 8d in 1901.[30] Only in 1902 did the price exceed 2s, reaching an average of 2s 6d per bushel in Cape Colony and 2s 7d in Natal.[31] According to a Department of Labour report by William Boase, the war did affect prices Greymouth farmers received for their produce. Boase claimed that West Coast farmers faced more difficulties than their East Coast counterparts, but added that the war had allowed local farmers to raise their prices and they were consequently 'cheerful'.[32]

The importance of the oats trade was clear in the primary growing areas. In 1900 a Southland correspondent reported that oats prices had eased off since harvesting commenced, adding that further reductions were likely with the end of the war in sight.[33] In fact, in March 1900 the end of the war was far from imminent and War Office orders for oats increased rather than declined. Combined oats sales from 1901 to the end of the 1903 statistical period totalled £1,345,234, compared to combined sales of £369,221 for the 1899 and 1900 periods.[34]

Just as oats dominated exports to South Africa, so the South Island ports from where they were shipped cornered the export trade to Natal and Cape Colony for the duration of the war. Of the southern ports, Invercargill and Bluff experienced the largest percentage increase in exports to South Africa in relation to total foreign exports during the war. In 1898, £359 worth of goods left these ports for South Africa, which represented a tiny portion of the ports' total foreign exports.[35] In 1901, this had soared to £320,448 or 32 per cent (before falling slightly to 27 per cent in 1902). From 1898 to 1901 inclusive, total foreign exports from Invercargill and Bluff increased by 59.65 per cent, and exports destined for South Africa increased by a staggering 89,161 per cent. Ōamaru experienced a similar trend. In 1901, exports shipped from Ōamaru to South Africa represented 23 per cent of the port's total foreign export trade.[36]

Given the importance of the South African market to Invercargill MHR Josiah Hanan's constituents, it is hardly surprising that he 'warmly' supported Seddon's proposal to dispatch the Second Contingent in 1900.[37] Either directly or indirectly and to a greater or lesser degree, many in Southland benefited from this surge in both demand for oats and increased shipping to South Africa. However, the wartime windfall of oats sales to South Africa inevitably declined, and as the ports of Invercargill and Bluff had benefited the most from war contracts they also had the most to lose.

The local confidence in the superiority of New Zealand oats was not always shared by the imperial authorities purchasing them. Noting the magnitude of the graders' task, the *Otago Witness* quoted a report from the secretary of the Department of Industries and Commerce. He proudly claimed that although 30 steamers had loaded a total of 115,907 tons of oats between January 1901 and September 1902, not a single complaint had been received regarding the quality or condition of the grain.[38] However, in 1907 the War Office somewhat belatedly criticised the quality of New Zealand oats supplied in the last months of the war.

The imperial authorities claimed that shipments of oats sent from South Island ports to Durban early in 1902 were found, upon examination in February 1903, to be in such poor condition that they were unsuitable for military use and had to be sold by public tender at a considerable loss.[39] The Royal Commission on War Stores in South Africa said that although the cost of the oats to the imperial government, inclusive of freight charges, amounted to £45,000, the War Office had only been able to recover £5000 through their sale.[40] On learning of this, the New Zealand government assured the imperial authorities that the oats were passed by the official graders prior to loading. The British *Standard* newspaper reported that the Army Council was not prepared to accept this statement in its entirety.[41] The implication that New Zealand government inspectors had passed inferior quality oats irritated Otago politician James Allen, who said that if New Zealand was responsible then it should pay for the oats, but if it was shown to be blameless, the War Office statement should be withdrawn.[42]

Joseph Ward raised the matter again in Parliament when he quoted directly from the commission's report presented in the British House of Commons. The suspect oats were loaded aboard three vessels in January and April 1902 at Bluff, Port Chalmers and Lyttelton.[43] The New Zealand Agent General in London stated, 'The Army Council feel bound to come to the conclusion that the loss of these cargoes must be attributed primarily to the indifferent quality of the oats passed by the graders of the New Zealand Government.'[44]

Approached for comment, the graders who had examined the oats in all three ports strongly refuted this claim and stated that the oats were accurately graded and shipped in good condition. Ward further stressed that the New Zealand government was not advised of any problems in relation to the oats until three years after their shipment. The Royal Commission implied that New Zealand may have passed inferior-quality oats in order to overcome difficulties in filling imperial contracts. It noted that the annual report of the New Zealand Agriculture Department had drawn attention to the ongoing problems it faced sourcing the enormous volume of grain required by the War Office.[45]

Temporarily setting aside their political rivalry, Allen and Ward made common cause regarding the issue. Both parliamentarians were clearly irked at the War Office's apparent suggestion that the New Zealand government had knowingly supplied a substandard product. Ward pointed out that by its own admission the War Office had discovered that large stockpiles of its emergency military rations stored both in Britain and overseas had deteriorated beyond use. He claimed it was unreasonable to blame the New Zealand government and said he was not prepared to accept the commission's findings.[46]

In the post-war years, trade to South Africa from the ports of Invercargill and Bluff plummeted, with no exports at all to South Africa in 1905, and South African trade once again representing a tiny percentage of total exports from these ports in 1910.[47] In 1907, when overall trade volumes to South Africa had fallen to a fraction of their wartime levels, the South African trade imbalance finally swung in the North Island's favour, with northern ports exporting more to the African colonies than their southern counterparts. An article titled 'Big Drop in Oats' predicted that 1908 would be 'the most disastrous year in the oats trade for a good many years'.[48] As far as the oats trade to South Africa was concerned, the halcyon days of the war years were well and truly over.

'An unlimited field for New Zealand enterprise'

The importance of South Africa as a potential export market for New Zealand produce was heightened in October 1901 when the Australian Commonwealth government introduced a federal tariff on a wide range of imported goods. Alarmed by its potential impact, the New Zealand government applied the new tariff to its 1900 exports to Australia and found that its duty liability would have amounted to £277,822 — more than twice that of the former tariff regime.[49]

While acknowledging that the increased tariff had created dismay in many quarters, a Christchurch *Star* editorial expressed some sympathy for Australia's position, seeing the tax as the penalty New Zealand paid for declining federation with the Australian colonies. The newspaper claimed duties imposed on potatoes, onions, bacon and other exports produced by small farmers would, except in lean years, effectively shut the Australian market to New Zealand producers.[50] In fact, while exports to Australia failed to increase at the rate that New Zealand producers hoped, 1902 statistics showed that the total value of New Zealand exports to Australia rose by 34.63 per cent compared to 1901, and did not fall below the 1900 level until 1904.[51]

In October 1901, a representative of a prominent Canterbury merchant claimed that the proposed Australian duty on grain was crippling unless unforeseen circumstances created a special demand. The Cantabrian believed the answer lay in developing South African trade.[52] Even though dairy sales to some Australian states increased, a deputation of major North Island dairy exporters met Seddon in mid-January 1902 and reiterated the claim that the Australian market was largely closed to New Zealand farm exports.[53] They informed the premier that 'exporters looked to the Cape for an outlet'.[54]

Seddon did not need to be reminded of the potential for trade with South Africa. Not only was New Zealand already exporting sizeable orders of oats but also, in the year prior to the introduction of the Australian federal tariff, the premier was considering trade opportunities beyond New Zealand's traditional export markets. In late 1900 in his capacity as colonial treasurer, Seddon identified South Africa as a market with 'vast possibilities' for New Zealand produce.[55] He also saw potential trade opportunities arising from the situation in China. Seddon believed European forces remaining in China following the Boxer Uprising would require supplies of foodstuffs unobtainable in the Asian nation.

In 1901 the government dispatched Trade Commissioner J. Graham Gow abroad with a wide-ranging brief to explore new export markets for New Zealand produce. Prior to his departure, Gow toured New Zealand, gathering information from potential exporters and collecting product samples. He first investigated the South African market, where he sought to obtain lower freight costs that would permit New Zealand products to compete on equal terms with other colonies. His interim report was promising, with Gow claiming that South Africa offered 'an unlimited field for New Zealand enterprise'.[56] Like many before him, Gow stressed the trade required the establishment of a direct, reliable shipping connection between New Zealand and the South African colonies.[57]

Exporters' pressing need for cold storage facilities in South Africa was another factor affecting export opportunities for perishable New Zealand goods. Shortly after the outbreak of war, an *Evening Post* article outlined the difficulties New Zealand exporters faced. It detailed the pitfalls of the South African market identified by James George, who had resided in South Africa for 10 years and in 1899 was acting as an agent for three Christchurch produce exporters. While George claimed there was huge potential for South African trade, he also advised caution: 'The difficulty in the way at present is the want of cold storage accommodation.'[58]

George, who was visiting New Zealand on business, said that the Australians had rushed into the South African market and formed business relationships with unscrupulous agents whose actions had undermined local confidence in Australian products. He also claimed that South African butchers had existing contracts with Combrinks, a wealthy proprietary company that he alleged held a near-monopoly. George noted that until these contracts expired, potential buyers of New Zealand mutton and dairy products were contractually obliged to deal solely with the company. He added that although a large South African firm had been impressed by the Kaiapoi woollen blankets he had shown them, the blankets' quality was too good and the market wanted cheaper products.[59] New Zealand did export frozen meat to South Africa, though during the war years the scale of these exports never came close to rivalling the dominance of oats.

Other producers also benefited from the South African market, though pricing vagaries, competition and availability issues meant that trade could be fickle. Wheat sales rose from a respectable £20,855 in 1899 to £23,940 in 1900, but then declined in 1901, before reaching a 1902 low point of £5997.[60] Butter and cheese represented the only New Zealand exports to South Africa that enjoyed yearly growth for the entire war. Export sales of butter to South Africa soared from £335 in 1899 to almost £28,000 in 1902, then jumped still higher to nearly £78,000 in 1903.[61]

While cheese sales to the South African colonies rose, the increases remained relatively modest. In June 1901, newspapers reported that the War Office had placed an order with the Agriculture Department for five tons of cheese for the troops in South Africa sourced from the Kaūpokonui factory in Taranaki.[62] If the quality and price proved satisfactory, a much larger order was anticipated.[63] Though exports did increase by approximately 132 per cent between 1899 and 1902, the trade's value was only £2096 at its peak in the final year of the war, and then fell to £1714 in 1903.[64]

In 1902 the New Zealand Agent General in the United Kingdom, William Pember Reeves, reported his efforts to promote New Zealand's trade interests during the war and obtain a 'fair share of Imperial Government orders'. Some in New Zealand were critical of his performance in the United Kingdom, and, as noted earlier, he was accused of being a Boer sympathiser in 1900.[65] While Reeves defended his work, he also drew attention to constraints that limited his effectiveness, stressing that he could not act as a commercial enterprise and approach imperial departments directly with prices and shipping dates.

His intended audience seems to have been disgruntled New Zealand producers, as he noted that the government was already familiar with his report's contents from his regular dispatches. Reeves then explained that after approaching the War Office he had succeeded in securing the trial consignment of New Zealand cheese that was shipped aboard the *Indramayo*. This led to a further two orders of 10 and 15 tons respectively, with Reeves predicting these shipments would result in additional orders. He went on to outline sizeable War Office orders he had secured for oats and an order for 700 horses shipped aboard the *Cornwall*.[66]

Before the First Contingent departed in 1899, Seddon noted a shortage of horses for the men and urged parliamentarians who intended showing 'their patriotism' by giving horses to contact the commander of the forces.[67] Once the war began, the steady demand for horses in South Africa continued to tax New Zealand's ability to supply them, with the *Evening Post* noting in 1901 that 'the drain on our horse supplies for South Africa is beginning to tell its tale'.[68] In June that year, the *Cornwall* arrived in Auckland from Wellington carrying a cargo of 560 remounts.[69] Arthur Nathan, the shipping agent, arranged for the vessel to load a further 150 horses at Railway Wharf before it sailed for South Africa.[70]

New Zealand contingent members on the road from Bulawayo to Fort Tuli on the Shashi River in Matabeleland, Rhodesia. ALEXANDER TURNBULL LIBRARY, 1/2-025283-F

Following discussions with two imperial remount commissioners who were visiting the colony, suitable horses for the *Cornwall* were selected by the New Zealand Agricultural Department and purchased on behalf of the War Office.[71] No grey horses were considered due to concerns regarding their visibility in the field.[72] While Reeves successfully facilitated the purchase of horses for the *Cornwall*, he conceded there were difficulties obtaining regular orders for other products, including tinned meats, butter, hay and preserved milk. He secured an order for 100,000 pounds of corned mutton, but the War Office rejected New Zealand preserved milk as it was sweetened.[73]

Exports of horses varied considerably during the war, both in number and price. Excluding horses that contingent members took with them, only two horses were exported to South Africa in 1899 for a combined price of £100.[74] These were presumably not intended for military use, as the maximum Defence Department price for remounts was £25.[75] In 1900 600 horses were sent to Cape Colony with a combined value of £12,900, or an average price of approximately £21 per animal.[76] The 1901 statistics recorded only two animals worth a total of £30 being sent to Cape Colony.[77] In 1902 exports once again surged to 1031 horses, the highest number dispatched in a single year during the war. The animals, which were valued at nearly £22,000, were shipped to Natal.[78]

The government's 1900 statistics indicated that the total value of exports from the colony excluded contingent horses and supplies shipped to South Africa for New Zealand forces.[79] While there were numerous cases of private individuals donating horses for the war effort, many horses sent to South Africa were either bought directly by the government or purchased using donations collected by patriotic groups like the Otago 'More Horses' fund.[80] In 1900 the Otago fund and the government each bought 10 remounts for New Zealand contingents.[81] When the *Undaunted* sailed for South Africa in the same year she carried 'free of freight' 30 horses gifted by the people of Otago and 66 purchased by the New Zealand government.[82]

A parliamentary return indicated that the *Tropea* also carried 180 horses free of freight, while the *Ormazan* carried 502 horses purchased for the imperial government.[83] In total, the *Undaunted* and the *Tropea* also carried 39 officers and men, including grooms to look after the animals at a cost of £15 per man.[84] Though 1900 government statistics recorded a lower number, a government report indicated that up to the 16 June 1900 sailing of the *Ormazan*, 1701 horses had been purchased by the government and shipped to imperial military forces in South Africa.[85] Perhaps reflecting Otago, Southland and Canterbury reaping the majority of the benefits during the early stages of the conflict, in March 1900 the *Otago Witness* considered New Zealand's expenditure on items including horses to be of little consequence compared to the promising overall economic situation:

> *It is not surprising that there are signs of unwonted prosperity all around us when our exportable commodities give us a million and three-quarters more to spend within nine months of the financial year, and we can well afford a few horses and a few thousand pounds for the Imperial exigencies of the war. A footnote to the statistics informs us, with a touch of unconscious humour, that the export returns 'do not include the value of horses, fodder, etc., sent per Waiwera.' These are, in fact, unconsidered trifles.*[86]

Six months later, Seddon criticised the short notice New Zealand suppliers received of a War Office order for 4000 tons of meadow hay. Although Christchurch farmers informed Riccarton MHR George Russell that they had stocks of hay available in various parts of Canterbury, Seddon claimed it was impossible for small New Zealand farmers to get the hay on board vessels in the time allowed, and noted that the Agriculture Department doubted their ability to supply so large a quantity. Clearly annoyed by the situation, Seddon instructed Reeves to seek an open order for similar contracts, or at least to secure a deadline extension. Seddon favoured the establishment of a mutually beneficial relationship with the War Office where the New Zealand government would do its best to meet imperial needs but expected a quid pro quo arrangement.[87]

He felt that as New Zealand had already been forced to decline an earlier contract, it was belittling that the colony was unable to fill the current order. Despite the considerable increases in exports to South Africa that New Zealand enjoyed during the first months of the war, Seddon nonetheless expressed dissatisfaction with the division of imperial war contracts. Revisiting his earlier concerns, Seddon considered New Zealand was being denied its fair share of direct orders. He claimed such orders were instead placed with other colonies that often used middlemen who sourced quantities of the produce from New Zealand.[88]

The 1899 Victorian oats contracts were not the only examples of this. In November 1901 the *Wanganui Herald* reported that an enterprising Sydney firm sought to purchase all available poultry in New Zealand for shipment to South Africa.[89] Possibly as a result of this, the following month the *Thames Star* reported that the Agriculture Department, acting on behalf of unnamed firms, intended shipping around 3000 frozen birds and several thousand eggs to South Africa aboard the *Otarama*.[90] The vessel's manifest highlighted the diverse mixture of New Zealand produce it carried to South Africa. Stowed in its holds were 1046 boxes of butter, one case of cheese, 321 crates of poultry, 648 sides of pork, three cases of eggs, six cases of turkeys, 20 sacks of oatmeal, 90 crates of rabbits, 40 casks of cider, 25 sacks of malt, 2500 sacks of oats, 492 cases of preserved meats, and nine packages of sundries.[91]

Though it failed to feature among the top five New Zealand export products to South Africa during the war, poultry was clearly seen as an area of potential growth. D. D. Hyde, the 'Government Poultry Expert', visited poultry farms in Burnham and Milton where fowls and ducks were selected for breeding purposes.[92] The *Evening Post* claimed that '[t]he demand for New Zealand poultry for South Africa is greater than the supply'.[93] Another newspaper accused Seddon and the government of 'palpable negligence' for their inertia regarding the South African shipping link, claiming farmers were being robbed of the opportunities Hyde had identified for poultry sales.[94]

As well as New Zealand produce, the country's farming expertise was also in demand. In the final months of the war, Lord Ranfurly was approached by the Colonial Office with a proposal for Boer farmers to travel to New Zealand and observe pastoral methods. Carefully selected burghers would be chosen from the internment camps and permitted to visit New Zealand, Australia and Canada. The colonial secretary asked Ranfurly to consult the New Zealand government to ascertain the feasibility and desirability of the idea.[95] The government agreed, and in early 1903 a group of Boer farmers accompanied by an imperial officer toured New Zealand, where they attended agricultural and pastoral shows.[96]

Rivalry and recriminations

During the First World War, individuals seen as profiting from the war were criticised by both New Zealand newspapers and the country's trade unionists, but during the South African conflict such criticism was muted.[97] Henry Scotland's remarks in the Legislative Council linking the war's popularity to the prospect of financial gain remained the exception, with few newspapers commenting on war profiteering.

A 1900 *Pearson's Weekly* article that appeared in the Christchurch *Star* did, however, examine the issue. *Pearson's Weekly*, published in the United Kingdom, limited its discussion to the British experience, but many of its comments also rang true for New Zealand. The article identified a raft of commercial enterprises and trades that benefited financially from the South African situation, and it claimed that both capital and labour employed in certain industries reaped 'a magnificent harvest'.[98] In one respect, the article failed to accurately reflect New Zealand's position. It claimed that farmers did not find the war particularly profitable compared to other industries. For South Island farmers and agents filling sizeable War Office oats contracts, this was clearly not the case.

Given the opportunities on offer, competition and regionalism inevitably surfaced in New Zealand in relation to the war contracts, with some politicians actively championing their electorate's cause. Reeves' secretary, Walter Kennaway, wrote to the under-secretary of state at the Colonial Office in April 1900 recommending the imperial government seek tenders for oats and forage contracts. Kennaway noted that New Zealand farmers were unhappy that fodder was being sourced by the imperial authorities almost exclusively through a single firm. He proposed allowing the New Zealand Agricultural Department to arrange the tender process and inspect the oats.[99] Manawatū MHR John Stevens asked the minister of agriculture to consider sourcing portions of oats contracts from farmers in Rangitīkei and Manawatū. Up to that point, Stevens claimed, orders had been filled solely in the South Island.[100]

A 1901 letter reportedly written by a representative of a northern grain-exporting firm highlighted competition for the oats trade. It revealed the suspicion with which some North Island agents viewed both their Southland and Australian rivals. The letter claimed that an oat ring existed in Southland, but added that it would not succeed in cornering the market on account of the tremendous quantities of oats available. The writer accused the 'Southland ring' of naivety in thinking that, because of reports of substantial orders being placed in Melbourne, their own oats were wanted. According to the writer, reports of orders were wired to New Zealand to inflate prices, while Australian farmers won War Office oats contracts at prices New Zealand could not match.[101]

Concerns about the inequitable division of the oats trade had existed since the war began. In an October 1899 letter to the *North Otago Times*, 'Diligent Tom' queried whether North Otago was getting its share of the oats contracts then being filled in Tīmaru and Lyttelton.[102] Conversely, a March 1900 article attributed to the *Lyttelton Times* claimed there was little chance of ships loading at Lyttelton for South Africa at that time as the vessels which had already been engaged were to load at southern ports.[103] The following month the *Oamaru Mail* reported that a 500-ton order for oats had caused 'a considerable flutter in the market'.[104]

But as the North Island grain company noted, New Zealand faced direct competition for war contracts from Australia, which in the early years of the conflict dominated the South African meat trade. Canada was another competitor as it also exported quantities of oats, with the *Hawera & Normanby Star* reporting a 500,000-bushel shipment of Canadian oats to South Africa in October 1901.[105] Even Russia was reportedly competing strongly for South African contracts.[106] The *Evening Post* advised caution, observing that though New Zealand products were known and appreciated in South Africa, New Zealand exporters still had numerous competitors, and it was 'in waging war with these' that New Zealand needed to exercise care.[107]

A report by a correspondent in Cape Town claimed South African cold storage facilities and distribution networks were dominated by three major players: the Imperial Cold Storage and Supply Company; De Beers Cold Storage; and Frank Hudson acting as an agent for Bergl and Co., which had branches in London, Argentina and Australia.[108] Assessing South African trade opportunities, Gow cautioned that it would be folly for New Zealand to challenge the combined supremacy of the formidable South African cold storage and meat ring.[109] He warned that of the three companies, the 'gigantic monopoly' of the Imperial Cold Storage and Supply Company alone had £1 million at its disposal and was capable of either crushing or absorbing any competitor.[110]

Gow added that new companies in South Africa might seek New Zealand government assistance in challenging the three corporations' dominance, but advised the government to avoid involvement as the venture involved risk and would be better left to private enterprise in New Zealand. A 1900 *Evening Post* article claimed that New Zealand producers had been offered the opportunity to invest in cold storage facilities in South Africa in the past, but no action had been taken at the time, as African trade was not seen as particularly important.[111]

As it transpired, Gow's advice was sound. Although meat exports to South Africa remained at higher levels after the war than they had been prior to it, New Zealand meat was in direct competition with products from Argentina and Australia. South America's comparative proximity to Africa gave it an edge over its more distant South Pacific rivals. In 1902 Ranfurly informed Chamberlain that Argentine meat contracts were the source of 'widespread irritation' in New Zealand.[112] Though sympathetic to New Zealand's complaints, Chamberlain informed Ranfurly that the situation involved a number of unspecified large issues over which no general consensus had been reached.[113]

Seddon remained upbeat during a 1902 visit to South Africa, informing Ward there was no need to look elsewhere for markets for New Zealand produce and claiming there was a 'famine in N.Z. produce' in South Africa.[114] The premier discussed the steamship service with representatives of the Bucknall Line, and met representatives of Bergl and Co. regarding meat exports with 'very satisfactory results'. Seddon's optimism regarding the African shipping connection was misplaced: in September 1902 Bucknall Brothers informed the New Zealand Agent General that the South African steamship service terms and conditions made tendering impossible.[115]

Four months after the war ended, W. White, the general manager of the Hawke's Bay Freezing Company, identified the Weddell Company and another 'very powerful' meat exporter as businesses in the Argentine that were rapidly increasing their trade with South Africa.[116] Similar concerns were expressed by Andrew Cleland of the Timaru Agricultural and Pastoral Society. While not opposed to entering the South African meat market, Cleland suspected the trade would be short-lived as Argentine competitors had superior distribution systems.[117] As with the hay orders, Reeves claimed that short time frames given to prospective suppliers to tender for lucrative War Office contracts made it almost impossible for New Zealand meat exporters to compete.[118]

During a 1902 joint deputation to the under-secretary of state for war in London, the Australian and New Zealand Agents General were assured that as much time as possible would be afforded Australasian businesses eager to tender for meat contracts.[119] However, a statement made in the British House of Commons regarding War Office meat contracts stopped short of providing an ironclad guarantee. It merely stated that contractors had given an undertaking in writing to source meat 'so far as possible' from New Zealand and Australia rather than from Argentina.[120] This loose commitment was confirmed by Chamberlain in a dispatch to Lord Ranfurly, with Chamberlain indicating the agreement had been signed by the directors of the Imperial Cold Storage and Supply Company.[121]

Despite the threat that Argentine beef represented, New Zealand beef sales to South Africa did experience respectable growth after the first wartime exports occurred in 1901.[122] Sales of lamb and mutton fluctuated, with no exports at all occurring in 1900, but both then increased for the remainder of the conflict.[123] Nonetheless, White and Cleland's fears regarding South American meat proved well-founded, with the War Office placing orders destined for South Africa with Argentine suppliers in January 1902.

Seddon now railed against the injustice of what he considered the imperial authorities looking beyond the empire's borders to fill orders 'to the detriment of their own flesh and blood'.[124] While diplomatically avoiding names, he accused anonymous individuals in authority of 'aiding and abetting' the colony's foreign rivals to the prejudice of New Zealand.

The premier stated that if the imperial authorities believed Australia and New Zealand were unable to furnish the required meat supplies then they 'needed to be woke up, and the colonies would have to shake them up to let them know with no uncertain voice'.[125] On the same day that the government finally advertised for tenders for a direct shipping connection to Cape Colony and Natal, the *Evening Post* reported that Seddon had cabled the imperial authorities complaining that the Argentine contracts could jeopardise the proposed New Zealand–South Africa steamship trade.[126]

Despite Seddon's bombastic statements, he remained a strong supporter of the empire and both before and after the Argentine beef issue arose he championed the imperial cause ahead of New Zealand interests. His reluctance to divulge the full details of imperial contracts in 1899 out of concern that doing so might inflate prices, and his somewhat dubious claim that if meat for the troops in Africa was sourced entirely from Australia 'there would be no jealousy or heartburning on the part of the people of New Zealand', could be viewed in this context.[127] Nonetheless, his indignation regarding Argentine beef proved popular in New Zealand.

In a February 1902 dispatch to the Secretary of State for the Colonies, Lord Ranfurly included resolutions from the Canterbury, Auckland, Hāwera, Wanganui, Pātea, Westport, Feilding and Eltham chambers of commerce strongly supporting Seddon's stance. The Auckland Chamber of Commerce hoped that Seddon would be successful in securing more favourable conditions allowing New Zealand to benefit from war contracts.[128] The Chamber of Commerce heartily supported Seddon and trusted he would be successful in inducing the War Office 'to treat New Zealand in a fair spirit, which is due to us as members of the Empire'.[129]

Yet the New Zealand government was not entirely immune to accusations of sourcing war-related materiel beyond the empire. In 1902 Dunedin City MHR John Millar brandished an 'American denim' clothing tag in Parliament that he claimed had been removed from a pair of dungarees issued to a trooper on his return to New Zealand. The parliamentarian questioned Ward about the origin of clothing supplied to returned soldiers quarantined on Matiu Somes Island, and claimed that where clothing of the same quality could be manufactured in New Zealand preference ought to be given to locally made articles.[130]

In response, the acting minister of defence stated that the Defence Department had been obliged to source the clothing at short notice from Sargood and Sons and the Wellington Woollen Company. He added that the companies' managers had opened their businesses on Sunday and provided clothing for 650 men in the space of six and a half hours, only completing the order in the early hours. While agreeing that where possible government contracts should be filled using locally manufactured articles, Ward stated that given the urgency of the situation, department staff could be forgiven for overlooking the origin of the clothing.

'These wars will always be popular'

While the rural sector largely mirrored its urban counterpart in the provision of troops for the war, this division did not extend to the economic benefits of the conflict. As we have seen, the greater part of New Zealand profits from South African exports accrued to South Island farmers, shipping companies and their agents. A 1902 government report detailing trade with South Africa for the previous 12 months indicated that the port of Invercargill was the largest beneficiary. Though the figures excluded horses and forage shipped with the contingents, Invercargill exported products to Cape Colony and Natal valued at £240,290, followed by Lyttelton (£217,034), Tīmaru (£101,802), Dunedin (£68,295) and Ōamaru (£50,128).[131]

Wellington's total of £41,333 represented the largest amount by value exported from a North Island port, with Waitara a distant second at £4722. Notwithstanding Waitara's comparatively small contribution, it still overshadowed Auckland — the main port that benefited the least during the same period. Despite its size, Auckland shipped only approximately £2000 worth of beer, whisky, tobacco, cigarettes, stationery, drugs, furniture, hardware, paints, pumice and a piano.[132] The report showed that a quantity of frozen beef valued at £1323 represented the most valuable product exported from the port of Auckland to South Africa during the final months of the war.[133] The cargo of the *Pakeha* illustrated the glaring disparity that existed between exports to South Africa from the North and South Island ports. The vessel, which sailed from Lyttelton for South Africa in September 1901, bulged with the usual selection of oats, meat, butter and assorted other produce from southern ports, while Wellington added 70 packages and Auckland contributed three cases of chairs.[134]

If Southland benefited most from the oats trade, Wellington and Christchurch seem to have initially enjoyed the lion's share of contingent contracts for uniforms and equipment. However, the need to supply comparatively large numbers of contingent items in a short time affected workmanship and materials. In New Zealand, complaints surfaced regarding the quality of the locally manufactured uniforms supplied to the Sixth Contingent. An *Evening Post* article that criticised the uniforms' poor fit claimed they were made to a standard size with little consideration for variations in the men's physiques.

The criticism touched a raw nerve with at least one supplier who considered that the article insulted the colony's manufacturers. While conceding that the fit of some uniforms was less than ideal, 'Manufacturer' claimed that approximately 1200 uniforms had been made in seven days by some of New Zealand's best outfitters to standard measurements that would fit an average man.[135] Other manufacturers were proud that their products had been selected for the contingents. Whanganui newspapers repeatedly featured advertisements that lauded the supremacy of R. Hannah and Company boots after the manufacturer received an order to supply 'the Transvaal contingent'. Suggesting that even in October 1899 the government envisaged sending further contingents, the order was reportedly for 1000 pairs — four times the number of men in the First Contingent. The advertisements boasted that the substantial order had been filled in a week.[136]

However, in the case of the First Contingent's boots, complaints quickly surfaced. Major Robin, the contingent's commanding officer, was critical of the footwear supplied to his men, with the soles of many pairs wearing out after a single month in South Africa.[137] The issue of poor-quality boots lingered; Dunedin parliamentarians John Millar and James Arnold referred to the boots supplied to the Ninth Contingent by a Christchurch supplier as 'a disgrace to the colony'. The pair hoped that Dunedin manufacturers would receive their fair share of contingent orders. After warning the premier to 'beware of dummies', they then supplied him with a list of Dunedin bootmakers they claimed were capable of completing the work. They also lobbied Seddon on behalf of Dunedin manufacturers for Tenth Contingent saddlery and clothing.[138]

Regional advocacy appears to have been effective, with Seddon expressing concern regarding the equitable distribution of contingent clothing and equipment contracts. Noting that Wellington and Christchurch had 'done very well' from wartime contracts, the premier indicated that as far as possible he would like to see Dunedin and Auckland receive their fair share.[139] Seddon's recognition of war contract inequities went some way towards appeasing dissatisfied businessmen who felt they were being denied a slice of the trade. However, his recommendation that a sizeable portion of future contracts be awarded to businesses in Dunedin and Auckland was unlikely to have gone down well in other parts of the country.

The South African steam service

An awareness that South Africa might offer a market for New Zealand produce pre-dated the war, but the accompanying challenges were also well understood. In 1896 the *Otago Witness* observed that if a reliable shipping connection could be established, there was considerable scope for New Zealand produce in South African markets.[140] A Cape Colony resident claimed the problem lay in transporting New Zealand goods when there was no direct shipping connection to Africa, whereas Aberdeen Line vessels carrying Australian produce to African ports were fully laden every voyage.[141] For most of the war, these two issues — shipping and competition — would remain significant hurdles.

In 1899 Seddon, too, noted the obstacle presented by the lack of a regular shipping connection.[142] Early the following year the government was urged by produce shippers to establish a direct steamer connection with South Africa that bypassed Australian ports.[143] Though the *Otago Daily Times* reported in January that Minister of Commerce and Industries Joseph Ward was attempting to establish such a link, progress proved desultory.[144]

In November 1900 Ward announced that the government proposed calling for tenders from shipping companies interested in providing a regular South African service.[145] The contract conditions required vessels to call at five New Zealand ports and at least three in South Africa. Scenting commercial opportunities, the Wellington Chamber of Commerce applauded the proposal. Not surprisingly, it also stated that Wellington should be one of the five ports stipulated in the agreement.[146] Yet despite the air of optimism that accompanied the government's announcement, there was no long-term, direct shipping service as a result.

In an apparent change of heart, the government decided to delay the tendering process.[147] Instead, Shaw, Savill and Albion; New Zealand Shipping; and the Tyser Line (which in 1900 subscribed £200 to the 'More Men Fund') reached an ad hoc agreement to service South Africa jointly on a six-weekly basis.[148] The shipping companies used vessels bound for the United Kingdom, though this too was considered less than satisfactory by shippers.[149] The government and shipping companies failed to reach agreement on both government subsidies and the shipping companies' stipulation that vessels be permitted to continue on to London with produce they could not discharge in South Africa.[150] During the period in which the six-weekly service operated, the government held back to see if private enterprise could profitably operate the connection.[151]

In 1901 Waihemo MHR Thomas Mackenzie expressed concern that New Zealand lagged behind Australia in the establishment of reliable shipping connections with South Africa, and advocated a government subsidy to address irregular shipping schedules. Mackenzie gave the Tyser Line vessel *Indradevi* as an example of the problems New Zealand exporters faced, and claimed the vessel's agents had encouraged New Zealand shippers to find cargo.[152] However, Mackenzie claimed, shippers who entered into agreements with factories to deliver produce to the South African colonies then found the *Indradevi*'s sailing schedule changed.[153]

A cargo manifest for the June 1901 sailing of the *Indradevi*'s sister ship, the *Indramayo*, is indicative of the types of goods traded with South Africa. The *Indramayo* loaded sheep and lamb carcasses at Lyttelton, along with quarters of beef, sides of pork, cases of poultry, tongues, dripping, butter, whisky, potatoes, sacks of wheat and oatmeal, 5988 sacks of oats, and assorted packages. After loading cases of cheese at Dunedin, boxes of butter at New Plymouth, additional quarters of beef at Waitara, and more butter and cheese in Wellington, together with condensed milk and sundries, the *Indramayo* sailed for South Africa, unloading at Cape Town, East London and Durban.[154]

Seddon told Parliament that the irregular South African service combined with higher freight costs paid by New Zealand exporters compared to their Australian counterparts left small producers and traders with 'no chance whatsoever'.[155] In an attempt to reassure ordinary New Zealanders, Ward stressed that no agreement of any kind existed between the government and the three shipping companies involved, which, unable to profitably continue the six-weekly service, eventually withdrew their vessels.[156]

The issues highlighted by Mackenzie were compounded by wartime shipping demands. The need to transport both imperial and colonial forces, as well as the vast quantities of supplies needed to sustain them, adversely affected some New Zealand industries. Vessels formerly available to transport products from the colonies to Britain were often diverted to the South African route. Businesses unable to ship produce originally destined for Britain were instead obliged to try to sell their wares on the domestic market, resulting in oversupply. A *Manawatu*

Farmer article highlighted the impact of the situation. It claimed that roughly 250 workers employed in connection with Foxton flax mills had lost their jobs and four mills had closed. The newspaper attributed this to a glut in the Wellington flax market reportedly caused by insufficient shipping capacity to the United Kingdom after vessels were rerouted to South Africa.[157]

Shipping shortages also affected the availability of transport ships for New Zealand contingents. In an attempt to circumvent the problem, Seddon sought Joseph Chamberlain's permission to commandeer 'by legislation if necessary' suitable British Merchant Navy vessels, an idea he had earlier mooted to former premier Sir John Hall.[158] However, neither the Colonial Office nor the Admiralty were willing to consent to this. Accepting colonial offers of military assistance to bolster the appearance of imperial solidarity was one thing, but giving colonial governments legal authority over British commercial shipping, albeit in specific circumstances, was a completely different matter.

In reply to a dispatch that included Seddon's proposal, the under-secretary of state at the Colonial Office made clear the stance taken by the Lords Commissioners of the Admiralty. In their opinion, seizing vessels and compensating the owners when circumstances required was preferable to legalising 'such high-handed action as "commandeering" in anticipation of an emergency'.[159] Perhaps still smarting from this rebuke, Seddon again complained to Ranfurly in October 1900, highlighting the difficulties involved in obtaining transports and complaining of the expense of securing these vessels on the owners' terms.[160] Seddon was possibly also irked that a comparatively small amount of the profits from these sailings accrued to New Zealand shipping companies.

A report tabled in Parliament listed vessels chartered by the New Zealand government to carry troops, horses or cargo from New Zealand to South Africa for the two years ending June 1901.[161] Of the five vessels listed, four were registered to companies in Sydney, with the *Monowai* alone registered to New Zealand's Union Steam Ship Company.[162] The report indicated the government had chartered the *Undaunted,* the *Ormazan* and the *Monowai* for £8500, £12,500 and £12,000 respectively, while the *Gymeric* and the *Tropea* were chartered for £12,375 each.

Early in 1902 Seddon announced to dairy exporters that he expected to establish the sought-after direct steamer link with South Africa within a month, though this was to prove optimistic.[163] In late January the government finally called for tenders in Australian, New Zealand and United Kingdom newspapers. The conditions stipulated a three-year monthly service direct from four New Zealand ports to Durban, Port Elizabeth and Table Bay. The ships could stop at one Australian port to top up, but only if insufficient cargo was available in New Zealand. They had to be suitable for carrying refrigerated produce as well as passengers, livestock and mail; had to weigh between 2000 and 4000 tons; and had to be capable of steaming at a minimum speed of 10 knots.[164]

Abandoning its earlier hands-off approach, the government offered a maximum yearly subsidy of £30,000 and fixed a scale of freight charges. In a move that furthered the service's economic benefit to the colony, the contract stipulated that, where possible, steamers would be repaired, refitted and supplied in New Zealand.[165] In a letter to the Agent General in London, Joseph Ward stressed the need for a rapid service and advocated vessels coaling at Westport to reduce delays caused by bunkering in Australia.[166] Tenders were also called for an alternative service to South Africa via Western Australia, allowing vessels to discharge a fixed percentage of their cargo at Fremantle.[167]

None of the established shipping companies that had traditionally carried New Zealand produce to the United Kingdom expressed interest, but two parties did submit tenders: the New Zealand and African Steamship Company, and Harold C. Sleigh, operating under the title of the Blue Star Line.[168] Although the former was prepared to accept a lower subsidy of £27,000, it could only provide vessels capable of 9–10 knots with third-class accommodation, while Sleigh offered vessels capable of 11½ knots with saloon accommodation for 30 passengers.[169]

However, the Blue Star Line existed only on paper, as Sleigh did not actually have the necessary vessels. He was attempting to use the £30,000 subsidy as an inducement to attract investors and float a company in the United Kingdom.[170] Several months of prevarication followed as Sleigh sought more favourable terms, until negotiations finally lapsed in early August 1902, two months after the war had ended.[171] In the same month, Bay of Plenty MHR William Herries claimed that trade with South Africa had been 'a burning point ever since the war began'.[172]

Sleigh's failure to secure the contract did not, however, spell the end of the South African trade. The government instead established a direct shipping connection in December 1902 with the New Zealand and African Steamship Company. Though this was greeted with an understandable degree of scepticism following the Blue Star Line debacle, it at least temporarily addressed the transport issues that had plagued the export trade during the war years.[173]

The subsidised New Zealand and African Steamship Company service drew criticism in Parliament. Two years into the contract's three-year term, Wairarapa MHR Walter Buchanan asked the government how much notice was required before the 'unfortunate' service could be terminated.[174] Buchanan maintained that reduced freight volumes meant taxpayers were subsidising the service to the tune of £1 10s per ton. Disputing Buchanan's figures, Ward defended the service, claiming that since its inception its vessels had carried £1,195,547 worth of New Zealand produce, and pointing out that South Africa was the only viable market for New Zealand poultry.[175]

Massey countered by tabling damning figures in Parliament showing that the subsidised service's freight rates on wheat, flour, barley, oats, bran, potatoes and grass seed were all higher than those of Australian charter vessels.[176] Buchanan also questioned the worth of the New Zealand Commercial Agency established in Durban by former trade commissioner Gow. In 1903, Gow had been appointed commercial agent in South Africa by the Department of Industry and Commerce.[177] Citing the £1000 he claimed it cost the New Zealand taxpayer to have Gow stationed in South Africa, Buchanan expressed scepticism about whether New Zealand was getting value for money.[178]

A Department of Industries and Commerce report two years earlier had painted a rosier picture of continued trade prospects, referring to a 'very appreciable increase' in South African trade that included a portent of things to come. The report noted that 3094 live fat sheep were exported to South Africa aboard the New Zealand and African Steamship Company's vessel the *Essex*, with this nascent trade pre-dating by more than 70 years New Zealand's foray into live sheep exports to the Middle East.[179] The livestock, loaded at Wellington, were destined for slaughter on arrival, though there was also speculation that New Zealand sheep would be used to restock farms following the war.[180]

In 1906 the *Evening Post* reported that the government was once again considering calling for tenders for a steamer service to South Africa.[181] Though the 1902 contract had been discontinued after its three-year term expired — and after the government had spent £90,000 subsidising it — Seddon was urged by 'an influential deputation' in Christchurch not to allow the maritime connection with South Africa to permanently lapse.

Labour wanted?

The war had a significant effect on labour availability and demographics, though the extent varied from region to region. Discussing the situation in Eketāhuna, Charles Grey, a Department of Labour inspector of factories, noted that the war in South Africa had withdrawn a number of young men from the labour market whose places were being filled by older workers. Grey claimed that a considerable amount of money had been removed from circulation due to the shortage of manpower, adding that this would be felt for a time.[182] Similar concerns were expressed elsewhere, with a Wellington businessman claiming a farmer acquaintance had lost £500 of cocksfoot grass due to a lack of labour to harvest the crop.[183] The shortage was attributed to the absence of 'some 2500 of [New Zealand's] best working men'. John Bayne, the director of Lincoln Agricultural College in Canterbury, claimed in 1901 that many young men who would have attended the college had instead gone to war.[184] After resigning from Lincoln, Bayne himself travelled to South Africa and Europe shortly before the war ended to examine agricultural techniques.[185] Another letter from a farmer claimed that the difficulty in finding labour was seriously affecting the dairy industry and forcing farmers to dry off cows that could not be milked. He blamed Seddon, accusing the premier of having no idea of the hardships he was inflicting on the farming community.[186]

In Canterbury, unemployed labourers reportedly took exception to remarks made by Hugo Friedlander, the mayor of Ashburton and one of the most prominent Jewish businessmen in New Zealand, to Acting Premier Joseph Ward. Friedlander claimed there was a shortage of men to harvest crops in the region. A labourers' deputation that met Friedlander maintained there were 200 unemployed men in Ashburton and that sufficient workers could be found in the town without employing men from other regions. Friedlander claimed his concerns lay with farmers' ability to harvest their crops quickly, adding that six weeks earlier he had advertised for workers without receiving a single reply. He assured the labourers that during the harvest they would be much in demand and that every idle man in Ashburton who wished to work would get full employment.[187]

Friedlander added, however, that if workers attempted to adopt collective action to demand 'unreasonable wages' he would assist the farmers. His motivations were not entirely altruistic. He, too, had a vested interest in bringing in the harvest as quickly as possible. Together with his male siblings, Hugo Friedlander was a co-founder of Friedlander Brothers, one of the largest New Zealand agents for oats, which made thousands of pounds from exports to South Africa to fill imperial contracts. Another wealthy Cantabrian whose business benefited from the demand for oats was grain exporter and racehorse owner George Stead. The *Bushmills*, which loaded 49,663 bags of oats at Timaru in late 1900, was just one of a number of vessels carrying oats to South Africa on account of Friedlander and Stead.[188] Stead actively supported the formation of a Canterbury contingent and personally subscribed £500 to the Canterbury contingent fund.[189] Like Robert Heaton Rhodes, he also contributed 10 horses to the Second Contingent.[190]

In 1901 Dunedin City MHR John Millar asked Seddon what the government was doing to assist troopers invalided home and unable to pursue their pre-war occupations. Seddon assured Millar that the government would do all it could to help men find work and noted that 22 troopers had already received jobs in government departments, with others returning to their former government jobs. When Auckland politician Joseph Witheford asked if the government would obtain a list of contingent members seeking employment, Seddon assured him that the government had already compiled a list and claimed contingent members seeking government work would be preferred ahead of other applicants.[191] This arrangement was formalised in 1906 when Internal Affairs under-secretary Hugh Pollen instructed department heads to give preference when filling public service positions to suitably qualified former contingent members.[192]

For some, however, Seddon's promises had a decidedly hollow ring. In 1903 former trooper Charles Baré expressed frustration at being unable to obtain work in a government department. Baré indicated that he was still unemployed, with 'not a single penny' of his contingent pay and gratuity left. He said he had complied with all regulations but his credentials had been returned from Wellington. Surely, Baré claimed, between the railways and the government workshops something could be found for a Seventh Contingent trooper. He indicated that during the premier's visit to Natal, Seddon had assured the Seventh Contingent men that they would find work on their return. The disgruntled trooper said he would have remained in South Africa had he known he would instead face unemployment.[193]

DCM recipient James Langham also struggled following his return to New Zealand. He suffered a relapse of the typhoid he had contracted in South Africa and, despite seeking work for almost two months, he remained unemployed.[194] For one former Second Contingent trooper, financial difficulties had more tragic consequences. Despite suffering from malaria in South Africa, the trooper enrolled again in the Ninth Contingent when he returned to New Zealand. A recurrence of his illness prevented him returning to South Africa, and in an October 1902 letter to Seddon the trooper described himself as 'hard up'.[195] His situation did not improve and, unable to settle his debts, he took his own life in a Kaikōura hotel early in 1904.[196]

One solution put forward to the problem of employment for returned soldiers was either giving them land grants or preferential treatment in land ballots. Seddon was dismissive of both suggestions, claiming that a large percentage of returned soldiers were single young men who 'would not care' to become farmers.[197] While this may have been true of many returned soldiers, in 1902 Witheford introduced a deputation of returned soldiers to Thomas Duncan, the minister of lands.

The deputation was led by two former officers who claimed to represent 50 to 60 men who had been encouraged by the imperial authorities to settle in South Africa. Lieutenant William Lorigan outlined the imperial government's proposal, which he claimed allowed men to select their colony of choice and obtain land wherever they wished. Lorigan added that a man with £500 would be offered a pound-for-pound subsidy. He also said they had been offered free seed for two years, free use of ploughs, veterinary services, arms, rations, harness and tents, imported farm implements at cost, and a guaranteed market for their produce for five years.

When Witheford asked what the officers wanted from the New Zealand government, Lorigan explained that while they didn't expect the same terms they had been offered in Transvaal and Orange River Colony, they wanted 'first-class' land in New Zealand without delay, as the imperial offer of cheap passages for the men and their families stood for only a month. Another officer mooted the idea of establishing special settlements on 57,000 acres in the Whanganui district and south of Kāwhia. Duncan remained reticent; pointing out that a decision of that magnitude could not be made at short notice without consultation, he declined to give a firm commitment, but said he thought the government would do what it could to assist them.[198]

Three months later, the government set aside Crown land at Ōtanake in the King Country as a soldier settlement.[199] The 7253-acre block comprised 36 smaller blocks ranging in size from roughly 100 to 300 acres, and in January 1903 Lord Ranfurly formalised the establishment of the Otanake Special Settlement Association.[200] The blocks were allocated by ballot to those of the 295 applicants who were successful. Many former contingent members who applied had been farmers or had experience working on the land, though one applicant was a cycle mechanic.[201] Witheford continued to advocate putting returned soldiers on the land and again raised the topic during the unveiling of Trooper Bradford's memorial in Paeroa in 1903.[202]

In one of the more unusual events of the war, a group of Chatham Islanders encouraged the settlement of soldiers on their isolated archipelago, though unlike Witheford's plan the men they proposed relocating were not New Zealanders. As noted earlier, the imperial authorities had approached the New Zealand government in 1901 with a proposal to intern Boer prisoners on Rakiura Stewart Island. While the imperial government confined thousands of Boer prisoners on the remote South Atlantic island of Saint Helena, many were also sent to other locations, including Barbados and Ceylon.[203] Seddon, however, questioned the logic of New Zealand sending contingents to fight the Boers if the enemy was then going to be placed on New Zealand soil, and opposed adding Rakiura to the imperial list.[204]

Seddon's rejection of the Rakiura proposal did not discourage a deputation of Chatham Islanders, who were introduced to the premier at Warner's Hotel in Christchurch by former Speaker of the House Major William Stewart. Stewart believed the majority of Chatham Islanders would welcome the establishment of an internment camp for Boer prisoners as it would create an internal market for the islands' sheep. Noting that the government owned only 18 acres on the islands, he said it could buy thousands more at comparatively low prices from Pākehā and Māori, while the Boers would provide a boost for dairy farming.[205]

Chatham Islander Edward Chudleigh, whose employee, Trooper Edgar Clough, served in South Africa, claimed that although he had not consulted Māori, all Europeans on the islands supported the idea. He agreed with Stewart's proposal but suggested the islands could accommodate 3000 to 4000 Boer prisoners rather than the 1000 Stewart envisaged.[206] Delegation member William Hoban, who had a 4000-acre estate on the Chathams, believed the introduction of Boer prisoners would be a boon that could lead to the establishment of bacon and dairy factories.[207]

Seddon, however, remained unconvinced; while indicating he would raise the proposal with Cabinet, he informed the delegation that he did not support the idea. The premier added that he thought the Boers were close enough to New Zealand where they were, and recommended instead locating the prisoners in New Guinea. Following their discussions, Chudleigh stressed to a *Press* reporter that the Chatham Islanders were not pursuing the proposal out of self-interest; it would, he claimed, benefit New Zealand as a whole.

The Chatham Islands idea went no further, but concerns about unemployed New Zealand troopers remained. A deputation of 28 returned troopers from the Eighth, Ninth and Tenth contingents sought assistance from the mayor of Christchurch following fruitless attempts to find work.[208] The issue was raised again in Parliament by Liberal MP William Field. Field felt that returned troopers should not be required to go cap in hand to their parliamentary representatives seeking assistance. He suggested the government make 'strenuous and systematic' attempts to find work for veterans, through, for instance, the establishment of an employment bureau. In reply, Seddon noted that the Christchurch *Press* had published advertisements free of charge for troopers seeking work, and urged other newspapers to follow suit, but added that some of the men were 'a little fastidious' and desired better jobs than those they had left when they joined the contingents.[209]

Seddon's claim that some former troopers were too particular about employment attracted both support and indignation. The *Press* refuted the claim; the men who had advertised seeking work were, the newspaper reported, anxious to obtain steady employment. Most were ready 'to begin at a step or two lower on the ladder than they occupied when they went to South Africa at duty's call'.[210] Nonetheless, in 1902 the newspaper reported that of 50 unemployed returned troopers in Dunedin, only six had found work. James Park, who by this time had replaced Denniston as mayor of Dunedin, established an employment bureau and arranged a room where unemployed returned troopers could spend time while seeking work. Twenty were reportedly offered work on the Catlins, Central and Heriot railways, but, the *Press* claimed, none would go to the country as they objected to leaving their homes.[211] Five months after the war ended, Park asked that if any members of the public could alleviate the 'pressing needs' of returned troopers by offering them garden work, they should enter their names in a town hall register.[212]

A returned soldier using the pseudonym '15-Pounder' claimed he had been forced to seek assistance from the employment bureau, despite Seddon allegedly assuring veterans that they would receive government work.[213] The author of a letter to a Dunedin newspaper gave a different account of the situation. He claimed that former contingent members who had previously worked as shepherds and labourers in rural areas were disinclined to return to the 'monotony' of farm work after the excitement of South Africa.[214]

'Anti-Humbug' was also unsympathetic regarding the plight of unemployed veterans: 'it is about time a halt was cried to all the jingoism that has been going on about what these troopers have been doing for the Mother Country'.[215] While conceding that some men had left good jobs to serve in South Africa, the writer claimed many troopers earned more in the contingents than in the jobs they had left, and that their motivation in going was 'more from love of adventure than from patriotism'.[216] Testifying before the 1903 Royal Commission on the War in South Africa, the under-secretary for defence, Arthur Douglas, said it was a desire to serve in South Africa rather than the prospect of monetary gain that led men to enlist. He claimed labourers working in the Defence Department stores in Wellington were earning 8s per day — double the base rate for a contingent trooper — and added that 'everybody was very much excited about the Empire's troubles at the time, and the men were only too anxious to go'.[217] In the case

The luncheon for returned First and Second contingent members held at the Dunedin Agricultural Hall on 23 January 1901. Acting premier, Joseph Ward, Dunedin's mayor, Robert Chisholm, and the speaker of the Legislative Council, Sir Henry Miller, were among the dignitaries who attended. The celebrations were comparatively subdued, however, as the arrival of the soldiers at Port Chalmers aboard the *Tutanekai* coincided with receipt of the news of Queen Victoria's death. ALEXANDER TURNBULL LIBRARY, PAColl-6075-02

of Charles Hagensen of the Dannevirke Volunteer Rifles, this appears to have been true. In a letter recommending Hagensen for service in South Africa, his commanding officer noted he had a good job at a sawmill and wealthy parents but wanted to go 'purely for a liking for the undertaking'.[218]

Several unemployed former soldiers who placed advertisements in the *Press* seeking work were looking for jobs involving horses, such as drivers, carters or ploughmen. The majority did so anonymously, many using pseudonyms that highlighted either their military service or their employment concerns. 'Mafeking', who claimed to have been severely wounded, had wholesale and retail experience; 'Pom-Pom' sought work as a groom or dray driver; and 'Waterval' was looking for employment as a 'horse-shoer'. 'Beira' would accept any employment, while two different troopers used the pseudonym 'Anxious'. Another 21-year-old seeking work as a farmhand ambiguously informed prospective employers that he had 'two years among sheep, and can kill'.[219]

In service of the Crown: The war and government departments

When *Pearson's Weekly* listed the organisations and trades that benefited from the war, it included the railways. In Britain, war-related railway profits largely stemmed from the inevitable surge in the movement of troops and supplies to ports. New Zealand's comparatively small response meant that the railways reaped correspondingly smaller returns in these areas. Nonetheless, the New Zealand Railways Department benefited from the war, given that rail was the main method of moving the large quantities of grain bound for South Africa to the various South Island ports where it was loaded onto vessels. While this kept the railways busy, problems arose due to congestion and logistical issues. This was raised in Parliament by Kaiapoi MHR David Buddo, who noted that delays in the arrival of vessels for South Africa had caused time-consuming bottlenecks. According to Buddo, these delays adversely affected farmers, who were obliged to await the return of empty wagons.[220]

The conflict also affected passenger movements. A railways official pointed out that an apparent decrease in holiday excursion traffic could be explained by the abnormally large numbers of passengers who travelled to Port Chalmers to witness the departure of the Fourth Contingent the previous year.[221] Increased sales of holiday excursion tickets in 1902 were at least in part the result of large numbers of people moving by train to see off the Eighth and Ninth contingents.[222] The chief traffic manager also noted the 'abnormal and exceptional' passenger traffic associated with the royal visit and the dispatch of contingents to South Africa.[223]

Though Seddon was quick to scotch the suggestion, Waikato MHR Frederick Lang suggested that as a mark of the nation's gratitude soldiers who had served in South Africa receive a medal entitling them to free travel on the railways. While supporting the issuing of medals and claiming that the government would always acknowledge New Zealand contingent members' services to colony and empire, Seddon baulked at issuing free rail passes. Giving the example of former soldiers who might be employed as commercial travellers, he claimed the distribution of free rail tickets could give their employers an unfair business advantage.[224]

Other government departments also saw their workloads increased or their operations affected due to the war. In 1901 the Public Works Department was tasked with constructing the loading stages necessary to ship horses aboard the *Cornwall*.[225] The following year, the department was again involved in shipping horses, as well as carrying out troopship repairs and work on a contingent drill-shed.[226] The Department of Lands and Survey lost 11 employees, including clerks, draughtsmen, chainmen and cadets, when they enlisted for service in South Africa.[227] This included men like John MacDonald, who served in the First Contingent, remained in South Africa and subsequently joined the Sixth Contingent.[228] Other Lands and Survey workers who enlisted were draughting cadets Cyril McGowan of the Eighth Contingent and Robert Mitchell of the Tenth. Francis Fisher worked as a clerk in the department's Christchurch office prior to receiving his captain's commission.[229]

Despite the inconvenience caused by their departure, the department was clearly proud of the contribution its staff made to the war effort. An entire page of its 1901 annual report was devoted to a photo of Lieutenant Robert Collins, who had entered the department as a cadet in 1894.[230] The report, which listed the contingents of another six former employees and expressed regret that the service details of former department chainmen were unavailable, gave an account of Collins' wartime service and the circumstances surrounding his injury at Ottoshoop.[231] It added that the wound would probably prevent Collins accepting the lieutenant's commission he had been offered in the Oxfordshire Light Infantry.[232] After four additional officers volunteered for service, the department emphasised 'the loyalty and ardour' of its staff, and when David Watt, an assistant draughtsman in the department's Dunedin office, enlisted in the Otago Contingent his friends and fellow workers purchased his equipment and horse.[233]

For some government departments, the impact of the conflict on staff appears to have been negligible. In 1900 Lieutenant Henry Tuckey of the Native Land Court, and Sergeant William Parsons of the Government Printing Office resigned to join the Fifth Contingent.[234] Ninth Contingent sergeant Henry Power had been employed by the Customs Department prior to enlisting.[235] During a ceremony at the Auckland Customs House, the collector of customs presented Power with a watch and chain, noting that to the best of his knowledge Power was the only employee 'to fight for King and Empire' in South Africa.[236]

Despite the demand for police in South Africa, only two police officers were recorded as resigning to serve in New Zealand contingents.[237] The Wellington Police Force lost an experienced officer when Constable James Poland travelled to South Africa, where he served in Brabant's Horse and the Johannesburg Mounted Rifles.[238] He had joined the Permanent Artillery in 1893 and transferred to the police force in 1894.[239] Poland never returned from South Africa, dying of pneumonia in Johannesburg in 1903.[240]

Many civil servants who did not enlist were nevertheless eager to display their patriotism. In 1900 contributions from members of the public service paid for the manufacture of an 18-carat gold salver for Baden-Powell.[241] After considerable delays, the salver was completed and reportedly engraved with scenes of New Zealand and an inscription in Māori.[242] At Baden-Powell's request, the gift was presented by Seddon to the British officer's mother when the premier visited the United Kingdom in 1902.

Early in 1900 the *New Zealand Herald* reported that Ashburton Railway Department employees had mooted the idea of dispatching skilled railway workers to South Africa.[243] Joseph Ward reportedly contacted the Cape Colony authorities regarding the proposal and was informed that the services of such men would probably be required. A Railways Department circular seeking the names of men interested in going attracted considerable interest, though it appears a railway 'contingent' as such was never dispatched. Despite this, a number of New Zealanders did serve in the South African railways and at least one NZR employee left to try his luck in the African colonies.

George Sim had been employed as a porter for several years prior to emigrating and joining the South African Light Horse.[244] It is possible the prospect of long service for low wages in New Zealand drove Sim to seek excitement and opportunity abroad. The *Otago Witness* reported that after applications were sought from railwaymen prepared to work in South Africa, Sim decided to 'try his fortune' there and left in 1900.[245] The decision would prove fateful, as Sim was killed in action at Laing's Nek shortly after his arrival.[246]

His demise did not deter others and only days before the war ended Norman Keane of New Plymouth applied for a permit to enter South Africa in order to obtain employment with the South African railways.[247] In January 1901 Frederick Harcourt, the Wellington share broker and Wanganui Collegiate School Old Boy who served in the Second Contingent and lobbied Seddon regarding receiving his discharge in South Africa, gave his postal address as the Imperial Military Railways, Johannesburg.[248]

As the main conduit of communication between New Zealand and other nations, the New Zealand Post and Telegraph Department was also influenced by the South African conflict. Departmental resources were stretched due to increased telegraph and postal traffic, as well as staff members enlisting in the South African contingents. Ward noted the 'severe strain' placed on the General Post Office's operations caused in part by employees like Charles Tasker, Manson Chant, William Warner and Edgar Spiers leaving for South Africa. All gave their occupation as 'letter carrier' on joining their contingents.[249]

A 1900 report noted that 11 department employees had been accepted for service overseas, adding that the men's jobs would be kept open and their seniority preserved until their return.[250] At least 19 telegraph operators enlisted, including George Mann of the First Contingent, Private Peter Dewar of the Sixth and Tenth contingents and Corporal Francis Hodge of the Ninth.[251] Added to this exodus was Charles Coutier, who worked as a telegraph linesman.[252]

By the time hostilities ended, a total of 51 Post and Telegraph employees had served in the contingents, while other department employees were considering seeking employment in Transvaal.[253] The department even had its own

Volunteer corps in Wellington — the Post and Telegraph Rifles.[254] Of the 71 men on the corps' roll in 1902, six cadets and three letter carriers, including Tasker and Chant, had either fought in South Africa or were currently serving there in New Zealand contingents.[255] Nineteen-year-old Clyde McGilp, a telegraph operator in the Wellington Telegraph Office, had spent a year in the corps prior to his departure for Africa.[256]

The Gisborne office of the Post and Telegraph Department was particularly affected by staff defections. Determined to 'try his fortune' in Africa, 20-year-old telegraph operator Harry Holford quit his job in Gisborne and sailed for Portuguese East Africa as a deck-hand aboard the *Waimate*. In a letter to his older brother, who also worked for the department, Harry wrote that on arrival in Beira he had found employment at Pauling's Railway Telegraph Company on a salary of £20 per month. Holford, who enlisted in the Ninth Contingent in the Natal town of Newcastle a month before the war ended, was dismissive of what he claimed was Beira's reputation as the 'White Man's Grave' and considered that although it had been built on swampy land it was possible to stay healthy.[257]

Possibly swayed by Harry's tales of adventure and prosperity in an exotic location (which included hunting 'tigers'), Fred Holford and a co-worker also decided to try their luck in South Africa. A *Manawatu Evening Standard* article titled 'Exodus' reported that Fred and Alfred Cox, both young Post and Telegraph employees in Gisborne, had resigned and intended leaving for South Africa.[258] Fred had worked as a telephonist, while Alfred had been employed as a messenger.[259] Harry was later employed by the Mashonaland and Beira Railway and served for a time as stationmaster at Inchope on the Umtali Line.[260] On a return visit to New Zealand in 1902 he claimed there were 'splendid prospects' for New Zealanders who had chosen to remain in South Africa.[261]

The burden placed on the Post and Telegraph Department by war-related telegraphic communications was to a degree alleviated in November 1901 with the completion of a new undersea cable between Durban in Natal and Fremantle in Western Australia. The cable, which followed a route from Natal, via the Indian Ocean islands of Mauritius, Rodrigues and Cocos to Fremantle, more than halved the cost of transmitting telegrams from New Zealand to South Africa. From Fremantle, messages from South Africa were transmitted to Glenelg in South Australia, and then on to Sydney for final transmission to the Wakapuaka Cable Station.[262] Even with the new link, and the accompanying reduction in cable charges, communication expenses remained high; a 250-word cable sent from the Agent General in South Africa to Seddon in March 1902 cost the government £12.[263]

The war also resulted in the establishment in early 1900 of a parcel post with the South African colonies that was reportedly 'much used and appreciated'.[264] Yet even with the new service, the successful delivery of parcels to individual New Zealand soldiers remained problematic. In July 1900, David Buddo claimed in Parliament that parcels for the troops sent by members of the public in Timaru had failed to reach their intended recipients as there was no way of ensuring delivery in South Africa.

According to the parliamentarian, the Union Steam Ship Company had refused to carry the parcels, while the New Zealand Shipping Company would do so only if they were sent via London. Seddon's response was unlikely to have satisfied the public; he assured Buddo that the government would endeavour to dispatch parcels to the troops on condition that they would be distributed unconditionally, and not delivered to individuals or smaller units, as doing so could cause resentment among soldiers who received nothing.[265]

Nonetheless, the service initially proved popular. In 1900 a combined total of 1052 parcels was sent to Cape Colony and Transvaal, though the yearly total steadily decreased, with 941 parcels in 1901, 622 in 1902 and 485 in 1903.[266] From 1901 onwards, these figures included parcels sent to Natal.

Though parcels sent from South Africa to New Zealand never matched the volume sent in the opposite direction, a steady flow nonetheless continued.[267] They contained an assortment of souvenirs and other articles posted by soldiers to friends and family. Ostrich feathers, which were in vogue at the time, also appeared among the trickle of South African goods imported by New Zealand milliners.[268] A trooper's sister, who asked her brother to catch an ostrich and keep its tail, told him the feathers were selling for £1 each in Dunedin.[269] New Zealand soldiers were aware of the demand, and when Trooper Burnett's contingent camped beside a Dutch farm that kept ostriches the men wasted no time in collecting the birds' plumage.[270] One of the more unusual parcels sent from Africa came from Lieutenant Arthur Bauchop of the Fourth Contingent.[271] It contained a pickled chameleon Bauchop had come across in his contingent's camp and posted to Otago High School. The unfortunate lizard was eventually added to the collection of the Otago Museum.

In addition to affecting the parcel service, the conflict influenced New Zealand postage stamp design. In 1900 the Post and Telegraph Department introduced the first of many New Zealand postage stamps commemorating foreign wars in which New Zealand took part. The khaki-coloured stamp featured the words 'The Empire's Call' and a picture of contingent members framed by the New Zealand flag and a fern.[272] A selection of Post and Telegraph Department postcards depicting the departure of New Zealand contingents was reportedly also popular.[273]

As well as the women teachers already discussed, several contingent members worked as Education Department teachers in civilian life.[274] These included Sixth Contingent captain David Cossgrove, who taught at Tuahiwi Native School at Kaiapoi and would later become the head of the New Zealand Boy Scouts movement, and Trooper John O'Reilly, who was employed by the Wanganui Education Board.[275] Although O'Reilly exceeded the leave of absence he was granted to serve in South Africa, the Education Board reportedly put his name forward to fill a headmaster's vacancy at Hurleyville School so he could return to teaching.[276]

When war comes to town

Provincial imbalances in the war's financial benefits were not lost on the New Zealand public, with some alleging disparities between advantages accruing to urban and rural areas. Wellington resident George Dutton painted a very different picture to the *Otago Witness*'s claims of 'unwonted prosperity' in a letter to his niece in Australia. Dutton claimed that New Zealand farmers selling horses and cereals had received good prices, whereas in towns the war had adversely affected business except in the case of those companies fortunate enough to win contingent contracts.[277]

Whether Dutton's claims were accurate or not, companies like the Ashburton Woollen Factory, which received an order for 5000 yards of fabric for Seventh Contingent uniforms, did benefit from war-related orders.[278] Opportunities were in fact presented to a diverse range of businesses in New Zealand; along with the larger and more profitable orders of oats and foodstuffs, there were small-scale exports of items as diverse as decks of playing cards, confectionery and cigars.[279]

Financial benefits could be short-lived though as companies that received contingent contracts seldom had the opportunity of resupplying the troops once they left New Zealand shores. On one of the few occasions when orders for war-related materiel were received from Africa, wartime communications presented problems. Wellington jewellers Rash and Gooder reportedly produced approximately 12,000 of the brass fern-leaf badges contingent members wore on their uniform collars and hats.[280] In August 1901, Lieutenant-Colonel Porter cabled the Defence Department from the South African town of Adelaide.[281] The garbled telegram read 'Commandant Send Six hundred mans hab bouges Sixty officers'. As Porter did not send messages in code, department officials assumed the message had become 'mutilated' during transmission.[282] Interpreting the cable as a request for 600 small badges for troopers and 60 large badges for officers, the order, worth £18, was sent by the defence storekeeper in October.[283] When the badges finally arrived in Cape Town it was nearly five months after the order had been placed and only four months prior to the end of hostilities.[284]

The day after war was declared, the government was criticised in the press by businessman Vincenzo Almao, who may not have been a mad hatter but was certainly a disgruntled one. The Wellington hat-maker indignantly observed that the government had awarded the contract for First Contingent slouch hats to his competitor, Hill and Son, without calling for tenders — a process he described as neither fair nor just.[285] Almao's concerns were unlikely to have been assuaged when subsequent orders for 500 felt hats for the Volunteers and an additional 600 for the Sixth Contingent were placed with Stafford and Collins, another of the hat-maker's business rivals.[286]

Almao, an Italian who claimed to have been an officer in the forces of Francesco II, king of the Two Sicilies, had served in the Dunedin Cavalry Volunteers in the 1880s.[287] In 1885, he was dismissed from the South Island corps for insubordination following an altercation with Sergeant-Major Alfred Robin — later Major Robin, the commanding officer of the First Contingent.[288] By the time of the South African War, Almao had relocated to Wellington, where he traded from The Hat Shop, his Cuba Street premises.[289]

Although he alleged favouritism in the acceptance of Defence Department contracts, Almao was not entirely shut out of them, as he won the contract to supply the Rough Riders.[290] He also supplied 150 hats and two plumes (all he had in stock) for the contingent sent to Melbourne for the Federal Parliament opening in 1901.[291] Despite the sense of injustice he felt regarding the First Contingent hats, Almao contributed £2 2s to the More Men Fund early in 1900.[292]

Though unlikely to retire on the proceeds, in urban centres across the country shop owners included war-related products among their stock, with the imminent departure of contingents presenting financial opportunities. On Wellington's Lambton Quay, H. W. Lloyd's offered Eighth Contingent watches, while four shops away the Goldsmiths' and Silversmiths' Depot advertised an assortment of presents suitable for contingent members.[293] Gifts included 'military hairbrushes', Kruger coins, wristlet compasses, field glasses, telescopes, patriotic brooches and dram flasks.

In Masterton, shoppers could buy contingent medals and badges from Vibert's Book and Stationery Depot, or a wide range of contingent mementoes from J. P. Elliot, Book, Stationery, Music and Fancy Goods Importer.[294] In Christchurch, a store on Colombo Street sold contingent souvenir badges for 6d, while a block away at Schlesinger's tobacconist customers could purchase contingent badges or get one free if they spent sixpence in the store.[295] Wallace's Golden Hat Shop in Dunedin sold war maps, George and Kersley in Whanganui sold Fifth Contingent badges, and Wesney Brothers Leviathan Gift Shop in Invercargill offered 'Mafeking Relief Flags' for sixpence, as well as Fourth Contingent badges and medals.[296]

'Ladies' Kharki Hats', 'Baden-Powell Hats', 'Kharki Cigarettes' featuring a New Zealand contingent picture in every packet, patriotic ribbons, contingent ties and hatbands, boys' khaki 'contingent suits' advertised as exact reproductions of those 'our soldiers are wearing in South Africa', and 'Contingent quality' men's underclothing were all offered to New Zealanders eager to display their patriotism.[297] In one of the more unusual attempts to capitalise on the war, an advertisement for Thorp's in Hawke's Bay made repeated references to the conflict and described its range of footwear as 'Our Lyddite shells', a reference to the high explosive widely used by British forces in South Africa.[298]

When the Fourth Contingent left Dunedin, an 'enterprising syndicate' reportedly made £15 by leasing a rooftop vantage point for £5 and charging spectators half a crown to view the parade below.[299] The departure of contingents could also be a boon for businesses providing food and accommodation for visitors from outlying areas. In the small Otago town of Tapanui, a hotel reportedly made £400 in a week from people visiting for the Fourth and Fifth contingents' send-offs.[300]

Like Thorp's, other businesses lacking any discernible connection to the war nonetheless used references to it in their advertising.[301] In Auckland, a plumber marketed the 'Mafeking Ventilator', boldly claiming it was the best thing of its kind ever invented.[302] An advertisement placed in the *Southland Times* by George Findlay assured his customers that as he was not a pro-Boer he could not supply 'De Wet' but could supply a range of waterproof clothing.[303] Vanity Fair cigarettes provided portraits of Lords Roberts or Kitchener to customers who sent in labels from its cigarette tins, while Smith's in Ashburton assured its customers it did 'not wish to BOER' them with its price list.[304]

The conflict in South Africa also provided a rich vein of material for public entertainment for New Zealand audiences eager to engage with war. Waxworks and the comparatively new kinematograph (an early form of film projector) were among the biggest attractions as they offered the public visual images of key figures and events. When it was displayed at the 1896 Palmerston North Industrial Exhibition, the kinematograph was described as 'the most marvellous instrument of the age', and by 1899 its popularity had not diminished.[305] Within weeks of the outbreak of war Montgomery's Kinematograph Company showed scenes of the First Contingent send-off at Whanganui's Oldfellows' Hall, and in December 1899 Northcote's Kinematograph and Unique Specialty Company staged a show in the Wellington Exchange Hall.[306] Advertisements for the event promised audiences footage of 'actual battle, showing the glories as well as the horrors of war' for the first time in New Zealand.[307]

An exhibition by Fullers Royal Waxworks and Big Vaudeville Company featured 'Transvaal War Celebrities', including the president of the South African Republic, Paul Kruger; the leader of the failed Jameson Raid, Dr Leander Starr Jameson; and Secretary of State for the Colonies Joseph Chamberlain.[308] The show had a three-month season in Wellington at the Choral Hall in Courtenay Place and proved equally popular in Dunedin the following year.[309] In 1902, with the end of the war in sight, a Wairarapa exhibition by Rowley's

Waxworks and London Company featured likenesses of 'heroes' Baden-Powell, Kitchener and Roberts, as well as the Boer leaders Kruger, Cronjé and De Wet.[310]

The Wellington Opera House was the venue for a lecture on explosives that raised funds for the More Men Fund where Professor Thomas Easterfield of Victoria College demonstrated the properties of lyddite, nitroglycerine and guncotton by exploding all three before an enthralled audience.[311] A Gisborne myriorama and kinematograph show at the Theatre Royal also drew a crowded house. Less sophisticated than the kinematograph, the myriorama involved a moving painted screen that gave the audience the impression of a constantly changing panorama. During the show, images of Lord Roberts, Baden-Powell and Major Robin were applauded, while a picture of Kruger received 'a hostile reception'.[312]

In Northland towns more remote from the main recruiting areas and contingents' ports of departure, the commercial response to the war appears to have been comparatively subdued. Ten per cent of the proceeds of 'Broncho George's Wild West Show' in Whangārei went to the Third Contingent Marsden Rough Riders' Fund.[313] The organisers of the Brass Band Patriotic Concert at the town's Theatre Royal also contributed the money they raised to the fund, and further north in Kamō a patriotic concert at the town hall was advertised under the heading 'War, War, War'.[314]

Enterprising children could also profit from the South African hostilities. Boys willing to sell contingent badges were offered a 'liberal commission' by Whitcombe & Tombs, even though the company did not profit from the transactions.[315] The company sold badges commemorating the departure of the Second Contingent for 6d each, with one newspaper assuring its readers that all proceeds went to the Patriotic Fund.[316] In Wellington a Lambton Quay business also sought 'respectable Boys' to sell contingent badges.[317] Whether these youthful entrepreneurs contributed their earnings to the empire's cause was not recorded, but plans were in place for children to receive their own commemorative souvenirs. Reflecting the widely held belief in 1900 that the capture of Pretoria and Bloemfontein would signal the end of hostilities, the government prematurely budgeted £1250 to produce medals commemorating peace for distribution to schoolchildren.[318]

The conflict also spurred interest in the acquisition of war trophies and mementoes. A newspaper noted that some contingent members had become

Crowds gather on the Wellington waterfront for the departure of the Second Contingent on 20 January 1900. ALEXANDER TURNBULL LIBRARY, 1/2-062513-F

great collectors of curios, coins, stamps, 'pom-pom' shells, Boer bibles, books, horns, and African weapons such as knobkerries and assegais. The article correctly predicted that after the war New Zealand would have many relics of the campaign.[319] Trooper Frank Perham claimed that troopers wanting to retain their saddles as mementoes of the war were disappointed when they were ordered to hand them in.[320] Though denied his saddle, Perham sent photos, postage stamps, ostrich feathers and sjamboks back to New Zealand.[321]

Trooper William Farquharson's sister Penelope asked him to get 'Kaffir' bangles and spent bullet shells so she could make a brooch or buckle; 'Try and save one or two for me, especially ones you know have hit their mark.'[322] In a letter home, Trooper George Leece said he would send Boer cartridges that could be fitted to the top of a stick or used as a parasol handle as they were 'all the rage'.[323] Farquharson sent his father shell cases from Rhenosterkop and leaves from a tree under which the body of William Earle of Southland was found following his death in action at Bronkhorstspruit.[324]

Farquharson's father encouraged William to 'score some relics to bring back', enviously noting that Lieutenant George Crawshaw had acquired a Boer field cornet's bandolier.[325] Herbert Hart, whom some credit with the adoption of the lemon squeezer hat by the New Zealand Army, 'speculated' £3 on ostrich feathers and African curios, while Fourth Contingent trooper Richard Burnett acquired a revolver and Mauser rifle from a Boer prisoner to add to his sizeable collection.[326] Burnett described his collection as 'already rather a big item to cart about' and claimed the revolver, which he took from 'Captain Müeller' of the Staats Artillery, was loaded with dum-dum bullets. Fellow Wanganui Collegiate School Old Boy Trooper Edgar Galpin donated the collection of Boer coins he had amassed in South Africa to the school's museum.[327]

Soldiers like Hart were well aware that collectables from the war represented an investment. Edward Jollie wrote that empty examples of the chocolate tins that Queen Victoria had distributed to the troops were selling for £5 in Kimberley, while a set of 19 Mafeking siege stamps with a face value of 8s 6d was selling for £25. He added that he had already acquired one set that he thought he would sell and advised keeping any of his letters with siege stamps. Jollie had also retained two complete sets of Mafeking siege currency with a face value of £1 16s which he claimed were fetching £20.[328]

Some New Zealanders endeavoured to capitalise on South African opportunities to improve their situation once they returned to New Zealand. Norman Palmerston, who served in the Kaffrarian Mounted Rifles and the Eighth Contingent, saw employment in Portuguese East Africa as a way of advancing his position in New Zealand and claimed he had been offered £30 a month as a platelayer on the railway linking Beira and Rhodesia. While acknowledging that the combination of fever and heat drove wages up in the Portuguese territory, Palmerston maintained he could live relatively well on £10 a month and believed if he could endure the Beira climate for a year, he would have almost enough to buy land in New Zealand.[329]

The war also influenced book sales in New Zealand, with the editor of the *Tuapeka Times* drawing attention to the mass of literature available associated with the conflict.[330] More than a year before the war ended, a Whanganui bookstore sold *The Great Boer War* by Arthur Conan Doyle, the creator of Sherlock Holmes.[331] Doyle served as a doctor in South Africa and, as noted, tended to Second Contingent member Lionel Smith of Stratford after he was fatally wounded at Vet River.[332] In Nelson, bookstores sold copies of *The South African War*, while British war correspondent Bennet Burleigh's *Natal Campaign* and *Towards Pretoria* by Julian Ralph were available at Whitcombe & Tombs in Wellington.[333] Meeting demand in Wairarapa, Masterton bookseller J. P. Elliott listed *The South African Frontier* and *British Africa* among its new arrivals.[334]

Even newspapers diversified to profit from the war, with the *Tuapeka Times* advertising maps of South Africa and advising interested parties to be quick as numbers were limited.[335] James Birch's *History of the War in South Africa* features sections titled 'New Zealand Shows Her Patriotism' and 'Intense Enthusiasm in Wellington'. The book, first published in 1899, ended with the occupation of Pretoria, which the author optimistically claimed would be the final act of the conflict.[336]

J. P. Fitzpatrick's 1899 British best-seller, *The Transvaal from Within*, was an attempt to vindicate the role of the Johannesburg Reform Committee (of which Fitzpatrick had been a member) in the events leading to the Jameson Raid. The book was widely available in New Zealand book shops and received free publicity in Parliament and the press.[337] Wellington City MHR John Hutcheson alluded to it in a parliamentary address, and it was praised in at least one newspaper editorial.[338] Fitzpatrick's book was also mentioned in *The British Case against the Boer Republics*, the Imperial South African Association pamphlet circulated in New Zealand. With a price of 3d, the pamphlet was distributed to New Zealand newspapers and was also mentioned in Parliament, where George Fisher recommended the government reprint and distribute additional copies.[339]

Counting the cost

When Seddon proposed the dispatch of a contingent to South Africa in 1899, he acknowledged the financial burden that New Zealand's involvement would entail. As the offer of military support included not only providing the force but also equipping and transporting it, Seddon predicted participation would inevitably involve 'a heavy expenditure'. Exclusive of transportation charges, the premier estimated that to send a force that included a commanding officer, two captains, six lieutenants, four sergeant-majors, 12 sergeants, 12 corporals, four drummers and 168 privates would cost nearly £50 per day, though where possible troopers would provide their own mounts. The total cost to the taxpayer of maintaining the contingent for six months, according to Seddon, would be in the region of £20,000. He added that the personal expenses for the men of the First Contingent would amount to roughly £10 per head.[340]

Although New Zealand footed the bill for the earlier contingents, once they reached South Africa the imperial government met the majority of their financial costs. Replying to a query from the colonial secretary, in August 1900 Lord Ranfurly supplied pay rates for the New Zealand contingents in South Africa. Ranfurly informed Chamberlain that imperial pay was deducted and the difference alone paid by New Zealand. An addendum noted that the pay of Fifth Contingent privates serving beyond the borders of Natal and Cape Colony rose to 5s per day, with other ranks receiving a proportional increase.[341]

Giving evidence before the 1903 Royal Commission on the War in South Africa, Lieutenant-Colonel Penton noted that so much money remained following the departure of the Third Contingent from funds raised through public subscription that it was decided to send the Fourth. According to Penton, the Fifth, Sixth and Seventh contingents were paid on what he described as 'the Rhodesian scale of pay'.[342]

During his presentation of the 1901 Budget, Seddon stated that apart from the imperial portion of the soldiers' pay they received while serving in South Africa, the New Zealand government had borne the cost of the First, Second and Third contingents.[343] This assertion was presumably only in reference to the men's wages because, as Penton noted, the Third Contingent was almost entirely publicly funded, with Lord Ranfurly informing Chamberlain that the cost of outfitting and transporting the contingent had been met by voluntary contributions.[344] Seddon stated, however, that the Fourth and Fifth contingents were 'practically Imperial soldiers' and were entirely paid by the imperial government.

Three months earlier, a newspaper reported that the government had set the pay of the First and Second contingents at 4s 6d per day, adding that men who received an additional 1s 2d per day imperial pay while in the field had this sum deducted. The article, titled 'A Question of Pay', claimed the deduction was unpopular with the troops and noted that the pay rate had long been in dispute, with the men claiming they were entitled to 4s 6d in addition to their imperial wage.[345] These figures were contradicted by Sir Arthur Douglas in 1903 when he informed the Royal Commission that the pay rate for privates in the first four contingents was 4s per day, with all subsequent contingents receiving 5s.[346] Wages for the Fourth Contingent may have been met by the imperial authorities, but funds to equip the contingent came largely from public contributions in the Otago and Southland regions.[347]

Public contributions in the form of direct payments or donations to the assorted patriotic funds further reduced the government's financial burden. Wellington businessman John Plimmer gave £300 to help fund the Third Contingent; a meeting in Gisborne raised a further £150; and in Napier, Nelson Brothers, Cottrell and Humphreys and the Gas Company each contributed £150 towards expenses associated with Hawke's Bay men in the contingent.[348] In August 1900 a parliamentary return detailed the sources of £4350 held in the Consolidated Fund to help defray contingent costs.[349] It showed that the money was part of £25,720 earmarked for the contingents, and that the funds represented contributions to the More Men Fund from 21 districts and the 'Ngaiapa Tribe of Natives', with the largest contribution being a £4000 donation from Christchurch.[350]

For the duration of the war (and for several years after) imperial funds were lodged with the New Zealand government, which drew on this money as required to meet imperial war-related expenditure. When Seddon was criticised for delays in paying returned soldiers, he stressed that the government was only a trustee acting on behalf of the imperial authorities and could not pay out the imperial funds without authority.[351] In a 1902 cable from London, William Pember Reeves emphasised the difficulties involved in monitoring the usage of imperial money from England and indicated that the British secretary of state for war placed considerable importance on New Zealand authorities auditing expenditure.[352] However, unravelling the complexities of imperial pay taxed the Defence Department's limited resources.

In 1910 the imperial government was still making deposits to cover South African War expenses, and as late as 1911 an invoice for £7 was paid from these funds.[353] In total, between 1899 and 1910 £1,143,810 was deposited by the British government for New Zealand's 'South African Contingents', and by 1911 £1,114,299 of these funds had been spent.[354] Although the New Zealand government declined a War Office offer to pay the cost of shipping the Third Contingent, it accepted the same offer for the Fourth, and was fully reimbursed the £27,000 it cost to transport the Sixth to South Africa.[355]

These imperial contributions spared the New Zealand government considerable expense, but the nation's comparatively small outlay on the war was not exempt from Audit Office scrutiny. Determined to fulfil its own legal obligations, the Audit Office refused to be borne along by the wave of patriotic fervour sweeping the country.

In 1900 the controller and auditor-general informed Parliament that his office had declined a Treasury paymaster-general requisition involving war-related funds on the grounds that the unauthorised expenditure was illegal. The funds were released only after Lord Ranfurly exercised his powers under the Public Revenue Act and resolved the dispute by issuing an Order in Council recognising the legality of the requisition. Although the auditor-general initially expressed doubts concerning the governor's intervention, he finally relented and consented to the requisition's passage.[356]

The army council in England also questioned expenditure anomalies involving the use of imperial funds without supporting receipts. When

Major Henry Jackson of the Ninth Contingent claimed £15 for a luncheon for residents of Albany in Australia, refreshments for bandsmen, and dinner for the officers of the *Kent*, payment was disallowed.[357] The imperial government also refused to fund the funeral of Lieutenant Robert McKeich on the grounds that the expenses should have been recovered from McKeich's estate.[358] Jackson unsuccessfully claimed £6 for digging and filling in McKeich's grave, erecting a marker with McKeich's name and regiment on it, and employing the Fusiliers' band to play during the burial.[359]

In October 1901 the government was criticised in Parliament by Frederick Pirani for awarding bonuses to officers without including these expenses in the estimates. Pirani said he understood that the commander of forces had received a large bonus, while men who had done considerable work in relation to the dispatch of the contingents had been overlooked.[360] The politician claimed that one man who had worked 500 hours' overtime had received a paltry £2. In reply, Seddon conceded that both the commandant and the under-secretary for defence had received £50 each, with a large number of other individuals receiving smaller sums. However, he pointed out that the total amount paid in bonuses amounted to £667 and claimed a large portion of this sum was paid by the imperial government.

Contingent costs borne by the New Zealand government fluctuated, but every year during the war the sums voted from the New Zealand Consolidated Fund proved insufficient and required augmentation. The largest additional allocation occurred during the 1900 financial year when approximately £40,000 was required in addition to the £35,000 voted by the government.[361] The most expensive financial year was 1901, when the government allocated £96,000 for the South African contingents, but then added nearly £18,000 for a total of almost £114,000.[362]

In comparison, government expenditure on police and Armed Constabulary forces in New Zealand for 1900–01 was £117,744.[363] Coincidentally, this sum was only £29 less than the imperial authorities spent purchasing New Zealand oats in 1901.[364] The 1902 *New Zealand Official Year-Book* also indicated that spending on militia and Volunteers rose from £91,388 during the 1897–98 period to £229,704 in the 1900–01 period. Only during the 1904 financial year did government funds earmarked for contingent-related expenditure exceed the amount actually required, when £1500 was voted for the South African contingents but only £888 of this amount was drawn.[365] The 1905 financial year was the final time funds were voted for contingent costs, though once again the £500 allocated proved insufficient and an additional £817 was required to meet the £1317 expenditure.[366]

The government's share of contingent costs was also reduced by a number of businesses and individuals either donating services and items of equipment or supplying them at reduced rates. The Kaiapoi Woollen Manufacturing Company offered to donate material for 100 uniforms for the Second Contingent, while the New Zealand Clothing Factory offered to make 200 uniforms free of charge.[367] Bing, Harris, and Company tendered to make boots for the Fourth Contingent for 5s a pair, which the company claimed was less than half the true cost, with the balance being a patriotic contribution.[368]

In Dunedin, the Dresden Company donated a piano for the use of the Fourth Contingent during their voyage, though the government purchased another for £25 for the Fifth Contingent.[369] A Wellington 'Assyrian' rather ambiguously donated 360 white handkerchiefs for distribution to the contingent men.[370] Even horse racing was not immune to the patriotic fervour sweeping the country. The Manawatu Racing Club added the Transvaal Relief Open Welter Handicap to its summer race meeting to aid the Transvaal Relief Fund.[371] In some cases, the soldiers themselves contributed to costs associated with their service in South Africa. Two Wellington troopers offered to pay for their horses, equipment and passage to South Africa, while a further 12 indicated they would equip themselves and provide their own mounts.[372]

There were limits, however, to war-related largesse. Although the imperial government contributed £140 towards the bill, the Wellington City Corporation charged the government £390 for the use of Newtown Park and damage the park sustained while serving as a camp for the contingents and Volunteers.[373] In September 1901, George Fisher, who had twice been mayor of Wellington and claimed to take a 'fatherly interest' in the issue, asked in Parliament why the government had not honoured its promise to leave Newtown Park in good condition.

Given his role as minister of defence, Seddon's curious response was unlikely to engender confidence in his own department. He told Fisher that if corporation officials 'wanted anything done in the way of business they should have nothing to do with the military, otherwise there would be a mess made of it'. Noting that the bill for repairs had been £140 after seven contingents had camped there, but soared to £600 after Volunteers had occupied the park for a single week, the premier indicated the government would pay £250 towards repairs, but beyond that 'would not give a single shilling'.[374]

The government was equally inflexible when it was the one owed money. In 1904, when the Defence Department found that Captain John Pringle had been paid the £60 war gratuity twice, it sought to recover the funds.[375] After Pringle indicated he was unable to repay the outstanding sum, the Defence Department wrote directly to the Bank of New Zealand seeking details of Pringle's finances.[376] Rather than protect his client's confidentiality, the manager of the bank's Wellington branch advised the government to take action, as Pringle would probably pay 'if pressed'.[377]

The taxpayer met a variety of other ancillary, war-related expenses. The sister of former New Zealand Railways worker George Sim received a £7 government grant to pay for the return of the trooper's personal effects after he was killed in action at Laing's Nek while serving in the South African Light Horse.[378] Whether adverse publicity surrounding Sim's case influenced the government's decision to assist his bereaved sister is unclear, but a letter that appeared in several South Island newspapers decried the government's actions. In the letter, A. D. Bell of Shag Valley Station alleged that the box supposedly containing George Sim's possessions, including a watch and chain, arrived in New Zealand only after a circuitous and costly journey from South Africa. He indignantly accused the government of refusing to assist Sim's sister with the freight charges and claimed that to make matters worse, when opened the box was found to have been rifled and contained nothing more than some old clothes and a horse's feed bag full of oats.[379] The mother and sisters of Private Frederick Broome, the trooper who died of typhoid in Johannesburg in 1900, were given £50 by the government to pay for their passage to England, and also received an additional grant of £60 from the Wellington Mayor's Patriotic Fund Committee to assist with their relocation.[380]

A 1901 parliamentary report discussed the origin of two Boer guns that had been brought to New Zealand by members of the Fourth and Fifth contingents aboard the *Tagus*. The report indicated that Lieutenant-Colonel Davies, the men's commanding officer on the voyage home, believed the Vickers and Krupp weapons that had been given to the two contingents 'in recognition of the good service rendered by them' were weapons the New Zealanders had helped secure after the guns had been buried by the Boers.[381] Among smaller war-related government expenses was £3 spent on brass plates for the guns, which were temporarily installed at the entrance to Parliament House, and £2 the New Zealand Agent General in Cape Town spent on a case for Boer Mauser rifles — presumably also captured trophies destined for New Zealand.[382] Several years after the conflict ended, expenditure continued, with the government paying £17 for packaging and shipping war trophies to South Island locations.[383]

Other miscellaneous war-related expenses also continued after the termination of hostilities. Injured soldiers were required to attend medical boards, often in cities distant from their homes, and the government met costs associated with these examinations. Trooper Joseph Culling of Otago, who suffered internal injuries when a horse killed by an exploding shell fell on him, was paid 12s 6d for meals while travelling to and from Wellington, and in 1904 Sergeant-Major Lockett was reimbursed £2 9s for 10 days' board and lodgings at the City Buffet Hotel in Lower Hutt while awaiting his pensions board examination.[384] Medical board members who conducted these examinations were also paid for their services.[385]

On his return to New Zealand, Trooper Patrick Lee received £22 from the government for an artificial limb to replace the leg he lost in the Machavie train accident, with 'A. A. Marks: Inventors and Manufacturers of Artificial Limbs with Rubber Hands and Feet of Broadway, New York' urging Lee to buy two so he would be 'ready for any emergency'.[386]

During the 1904/05 financial period, the government earmarked £300 for the production of a history of the New Zealand contingents in South Africa (with an additional £83 spent during the 1905/06 financial period). In the same year, the government allocated £250 for the compilation of a Māori history. Though several potential authors were approached about writing the history of the New Zealand contingents, following lengthy delays Captain Francis Beamish finally produced a manuscript that then remained unpublished, with Major-General Edward Chaytor dismissing it in 1920 as 'of little use as an official

History'.[387] The task then passed to former war correspondent James Shand, who completed 'O'er Veldt and Kopje' in 1931, but once again the project was shelved after Shand's death, with the manuscript permanently lodged in the Alexander Turnbull Library.[388]

In 1903 the former under-secretary for defence, Sir Arthur Douglas, told the Royal Commission on the South African War that the net cost of the conflict to New Zealand was £195,000.[389] Douglas added that New Zealand still had contingent liabilities totalling £27,000 and an estimated pension bill of £3000 per year. The under-secretary agreed, however, that pension costs were a decreasing cost as veterans would 'die off'.[390] In 1905 newspapers reported that a total of £113,256 had been raised through public contributions to help finance the contingents during the war.[391] In his 1949 book *The New Zealanders in South Africa 1899–1902*, David Hall stated that exclusive of public contributions the war cost the New Zealand government £334,000, equivalent to 8s 8d per head of population.[392]

However, the total sum from the Consolidated Fund expended on South Africa contingents during the financial years from 1899/1900 to 1904/05 was £282,681.[393] Hall's claim that the £334,000 did not include public subscriptions is debatable. Although he did not cite his source, the figure Hall quoted appeared in a 1903 government report recording Canadian, Australian and New Zealand expenditure on the war.[394] A footnote in the report stated that where possible public subscriptions were included in the colonial figures. Whether this was possible in New Zealand's case is not recorded, although another footnote indicated that the £334,000 excluded pension obligations.

What is abundantly clear is that while the New Zealand government's war-related expenditure was not insubstantial, it never came close to the combined export revenue from war-related trade. Sales of oats to Natal and Cape Colony in 1900 alone amounted to £317,475, which easily eclipsed Consolidated Fund contingent expenditure for the entire war.[395]

From 1899 to the end of the 1902 financial year, oats sales amounted to £1,597,546 — more than five times the costs met by New Zealand taxpayers from the Consolidated Fund for the South Africa contingents.[396] Exclusive of parcel post, total exports to the South African colonies during the years 1899–1902 amounted to £2,073,188.[397] Obviously, export sales do not equate to net profit;

farmers had to subtract expenditure for seed, labour, machinery, transportation and storage from their earnings.

Although in 1899 the bulk of New Zealand exports to South Africa went to Cape Colony, following the occupation of Bloemfontein and Pretoria, Natal dominated the export trade. The British colony received 81.71 per cent of exports by value in the period from April 1899 to March 1903.[398] Even though exports to the South African colonies fell dramatically after the war, they nonetheless continued at considerably higher levels during the eight years after hostilities ended than at any time prior to the war. In addition, the trade connections New Zealand established in South Africa as a result of the war remained largely intact in the decade after the conflict ended.

Sections of the population undoubtedly profited from the conflict. In 1902 the *Sydney Morning Herald* quoted a speech Seddon gave on the troopship *Surrey* in which he expressed his indignation at the Argentine meat contracts.[399] Calling them a 'grave injustice', Seddon made a clear connection between participation in the war and the expectation that New Zealand would be able to financially benefit from it. He noted that the imperial government's action was enough to dampen the patriotic ardour of colonies sending men to fight in South Africa.

Perhaps regretting his words, the premier appeared to retract them less than two weeks later when he claimed that while sourcing produce from other countries would not affect the patriotism of Britain's colonies, the colonies had a right to complain when they did not receive 'fair play'.[400] For the entire war Seddon remained an ardent supporter of the imperial position in South Africa, but he also clearly considered New Zealand capitalising on wartime opportunities an attractive and legitimate adjunct to the nation supplying manpower for the imperial cause.

EPILOGUE

In 1899, as Parliament debated New Zealand offering a contingent for service in South Africa, Australian-born Walter Carncross, the MHR for the South Island electorate of Taieri, rose to speak. The 44-year-old Liberal politician claimed that if the offer was accepted, New Zealanders would feel 'a thrill of excitement'. Carncross believed similar emotions would be evoked in the United Kingdom, where both politicians and the general public would say, 'Well done, little New Zealand!' He further predicted the war would 'have a great affect [sic] for good':

> *More than that, it will create a feeling of patriotism — a much keener feeling of patriotism than at present prevails. Not only will it create feelings of patriotism and love for the Old Country, but many of the contingent will come back here proud of what they have done as representatives of this country. Their patriotism will be increased, and their love will be greater for the little country they represented.*[1]

In fact, as New Zealand's response to the relief of Mafeking showed, the unparalleled surge of patriotism that swept the country during the early stages of the war was most pronounced among its civilian population. Given the number of New Zealand residents with direct connections to the 'Mother Country', such patriotic displays were hardly surprising. The conflict's influence, however, extended well beyond fundraising, patriotic songs and flag-waving. For the duration of the war, South Africa was an integral part of New Zealand's social, economic and political fabric. From the largest urban centre to the smallest rural town, few, if any, New Zealanders could claim to have been completely unaware of, or untouched by, the conflict. When, in May 1902, the signing of the Treaty of Vereeniging finally brought hostilities to an end, it by no means signalled an end to the South African War's influence on New Zealand society.

Despite the intense focus on the First World War during the 2018 centenary, many New Zealanders remain largely unaware of their country's role in the war that preceded it. The grandiose South African War memorial in Dunedin reflects the importance of the conflict to the region, and the recent cleaning of South African War graves in Dunedin's Northern Cemetery suggests renewed interest in the conflict. But other memorials commemorating those who lost their lives, like the Johnsonville lamp honouring Leonard Retter, are often either ignored or mistakenly associated with later conflicts.

Retter's brother Hector, a South African Veterans' Association member and former Fourth Contingent trooper, returned to South Africa in the late 1930s. Though the monument to Langverwacht Hill erected in the vicinity of the engagement has since been restored, when Hector visited he found the site neglected and the names engraved on the stone slab illegible.[2] As time passed and memories faded, a similar fate befell other sites. Despite the earlier efforts of the Guild of Loyal Women, in 1938 a district commissioner of the South African Police claimed that cemeteries containing war dead were poorly maintained, graves were unmarked, and in some cases skeletal remains were exposed on the veldt.[3]

For at least the decade afterwards (albeit to a diminishing degree as the war lost its immediacy and other international crises vied for press attention), the South African War remained a factor in the lives of many New Zealanders. For some, Carncross's prediction was incorrect. With the lure of Africa stronger than any connection they felt to New Zealand, many New Zealanders either remained in South Africa after the war or chose to relocate there.

These individuals found employment in a diverse range of occupations including commerce, mining, the railways, the police, and even the theatre. Like the New Zealand teachers working among the Boers in the concentration camps, and the nurses caring for sick and wounded soldiers during the war, independent-minded women pursued career opportunities in South Africa either in the final stages of the conflict or soon after it ended. Numerous other

South African War veterans on Anzac Day 1930. MAJOR (RETD) NOEL W. TAYLOR ED** RNZIR

women travelled to South Africa to either marry or reunite with their partners.[4]

The war's impact on New Zealand was far greater, however, than this comparatively minor post-war emigration trend among its citizenry. It reinforced in the minds of New Zealanders both young and old their nation's place as an integral part of the British Empire. New Zealand's system of government largely mirrored that of the United Kingdom; its governor was British, as was the legal system on which New Zealand's laws were based, and on arrival in South Africa New Zealand contingent members operated as imperial troops.

Yet despite widespread support within New Zealand for British actions in South Africa, the war also caused divisions in New Zealand society, within the labour movement, the church and women's political movements, and among Māori. The suspicion and intolerance of those who either publicly opposed the war, or questioned the way it was being conducted, such as James Grattan Grey, Wilhelmina Bain and Rutherford Waddell, was an ominous portent of what was to come in later conflicts. In one of the war's contradictions, Charlotte Bewicke, who not only sympathised with Boer victims of the war but also actively and publicly raised money to aid them, largely escaped public vilification despite the prevailing climate of hostility towards anybody perceived as a 'pro-Boer'. In sticking to her convictions, Bewicke displayed considerable personal courage, yet she has remained an almost completely anonymous and unrecognised figure in New Zealand history.

The South African War was, in many ways, a coming of age for New Zealand on the international stage. Grey's articles in the *New York Times* drew Americans' attention to the level of jingoism in New Zealand's response to the conflict, while the *Times* in London praised New Zealand's loyalty to the empire. From a military perspective, perhaps the most significant outcome of the conflict was that it showcased New Zealand's martial capabilities in a foreign conflict to both an expectant New Zealand public and the imperial authorities. It also gave the Defence Department and New Zealand troops experience of being part of a wider imperial military response to a foreign war.

The conflict provided a core of experienced soldiers, such as Edward Chaytor and Lieutenant-Colonel Stewart Newall — respectively commanding officers of the South Island regiment of the Eighth Contingent and the Fifth Contingent. Following the war, Newall commanded the Volunteers in Wellington District.[5]

During the 1913 general strike, Newall, who as an Armed Constabulary subinspector had, like William Messenger, taken part in the 1881 invasion of Parihaka, led a force of mounted special constables that violently suppressed protests in Wellington.[6] The 'specials' were technically civilian volunteer police, but with support and training from the military. During the strike, several South African War veterans serving as special constables reportedly wielded Zulu knobkerries they had brought back from South Africa.[7] Former contingent members were also among those supporting the strike. James Thorn, who as a 17-year-old had enlisted in the Third Contingent, became a pacifist after witnessing civilian deaths in South Africa.[8] After becoming involved in the trade union movement, Thorn served on the Christchurch strike committee in 1913.[9]

Chaytor went on to command the New Zealand Mounted Rifles Brigade in Sinai and Palestine as a brigadier-general. Herbert Hart, the Ninth Contingent trooper who described contingent members looting, and taking part in the Newcastle brawl, also became a competent and respected senior officer in the First World War. Many of the lessons of mounted warfare learnt on the veldt by Chaytor and his fellow contingent members proved of value in the Middle East.

The same was true of the doctors, nurses and veterinary surgeons who served in South Africa. In several cases, the expertise these same individuals garnered in South Africa, aboard the vessels that transported men and horses, and in the hospitals and quarantine facilities in New Zealand, allowed them to make significant contributions to their country in the years that followed.

It would be crass to suggest New Zealand entered the war for pecuniary gain, and the financial returns that accrued from New Zealand's wartime exports benefited a comparatively small section of the population in specific regions. Nonetheless, revenue generated by exports to South Africa exceeded the New Zealand government's war-related expenditure. By maintaining a steady flow of men and essential supplies to South Africa, New Zealand displayed its ability to contribute meaningfully to the empire in a distant war.

Seddon's vocal and at times strident support of the empire was not, however, unconditional. His refusal to accept unquestioningly imperial policies he considered harmful to New Zealand interests, such as the War Office's purchase of Argentine beef at the expense of New Zealand producers, the imprisonment in England of Charles Tasker, and the imperial authorities' repeated rejection of

Māori military service, was a manifestation of nascent political autonomy.

Despite Māori support for the war being far from universal, and the repeated official rejections of Māori offers of military service, those of Māori heritage who did serve in South Africa, such as Henry Vercoe and Edward Broughton, contributed to the erosion of racial barriers within New Zealand society. By the First World War, the value of Māori soldiers was increasingly recognised, and both Vercoe and Broughton received commissions as second lieutenants on the same day in 1916, with Vercoe earlier mentioned in the dispatches of Sir Ian Hamilton for meritorious service.[10]

The challenges New Zealand faced in the provision of uniforms for the contingents were highlighted by New Zealand under-secretary of defence Sir Arthur Douglas during his testimony before the 1903 Royal Commission appointed in England to inquire into the prosecution of the South African War.[11] Even so, New Zealand manufacturers' production of uniforms and equipment during the war in short time frames provided Defence Department officials and civilian contractors with experience for the significantly larger manufacturing demands of the First World War.

The war also left its mark on social welfare. The contingent military pension legislation enacted in 1900 provided the legal precedent for the awarding of pensions to New Zealand military personnel and their dependants for injuries sustained and illnesses contracted while serving overseas. However, these pensions came at a cost to civil liberties. The government's clandestine investigations of the personal circumstances of soldiers and their families when assessing pension eligibility brought police and Defence Department officials into the lives of law-abiding New Zealanders without their knowledge.

The South African War gave New Zealand soldiers their first experience of looting in foreign lands, a practice that continued unabated in the larger conflicts that followed. During his own evidence before the 1903 Royal Commission, Lieutenant-Colonel Penton claimed he had received nothing but flattering reports and not a single complaint about New Zealand soldiers' behaviour while he was in Africa.[12] While Penton's claim may have been true, it overlooks government compensation payments for looting by contingent members, the role New Zealanders played in cases of theft, the Newcastle brawl, and the burning of the Malay mosque in Worcester.

Even though the Worcester incident did not involve violence on a comparable scale, the racial intolerance and unsanctioned use of force against Worcester's Malay population has parallels with the 1918 killing of Arabs by New Zealand troopers in the Palestinian village of Surafend. Similarly, the rioting and looting that took place in Newcastle while contingent members awaited transportation back to New Zealand could be seen as a forerunner of the rioting by disgruntled New Zealand troops awaiting repatriation at Sling Camp on Salisbury Plain in 1919. Yet despite the events at Worcester and Newcastle, and the well-documented instances of looting and insubordination, overall the majority of New Zealand soldiers acquitted themselves well in what was for many their first experience of participating in a distant and protracted conflict.

Though valuable experience was gained from the war, other opportunities were missed. Little appears to have been learnt from the accidental wartime deaths of William Byrne in South Africa and Percy Crawford in Feilding, with cavalier attitudes towards firearm safety again resulting in death and injury following the war.[13] After a Volunteer was accidentally shot in 1904, the *Observer* claimed such incidents were 'woefully frequent'. The newspaper reported other cases of both new recruits and experienced Volunteers accidentally discharging weapons.[14] The dangers of soldiers confusing blank and live cartridges also resurfaced in the years immediately after the war. Several Volunteer corps complained of receiving live ammunition in the same shipping cases as blank cartridges, and in one case live ammunition was accidentally issued to Dannevirke Rifle Volunteers prior to an exercise.[15]

The surge in Volunteer corps membership that took place during the conflict proved unsustainable in the post-war period. However, the wartime encouragement given to defence rifle clubs, coupled with many veterans being allowed to retain their weapons, continued the practice of firearms being held in private ownership. In 1906 the New Zealand Defence Forces' commanding officer noted difficulties in recruiting Volunteers and considered 'less cordial relations' between employer and employee a 'serious menace' to Volunteer recruiting.[16] The following year, sports like cricket and football were also seen as contributing to the decline of the Volunteer movement.[17] In 1909, Legislative Council member John Anstey claimed it was only 'the weedy youths' who joined the Volunteers.[18] The school cadet corps did not share this decline and continued until 1970, while military-style drill in schools was an antecedent of New Zealand schoolchildren marching to martial music — a practice that continued long after the Second World War.

The influence of the South African War on New Zealanders, whether soldier or civilian, inevitably varied according to individual circumstances, but for many its impact was profound and far-reaching. New Zealand's military contribution to the South African War was significant — not because it made a difference to the war's eventual outcome, but because it showed New Zealand was capable of fulfilling what its population widely considered to be its patriotic duty to the empire. Few of the young men and women who had grown up during the South African War would have questioned New Zealand's participation in the global conflict that followed. A month after the war in South Africa ended, members of the Wanganui Collegiate School 'Parliamentary Union' debating society were asked to consider the motion 'That there is no more important fact in British History than the share which the Colonies have taken in the recent war'. Speaking in support of the motion, a student posed a rhetorical question. Noting the number of men the British colonies had contributed to the South African conflict, he asked how many would they send 'in case of Empire-shaking war?'[19] All too soon, New Zealand would learn the answer.

New Zealand Governor Lord Plunket unveils the memorial to 16 North Otago soldiers who 'gave their lives for the Empire' during the South African War. The memorial, erected on Thames Street, was unveiled on 2 February 1905. NIGEL ROBSON

NOTES

Abbreviations

AJHR	Appendix to the Journal of the House of Representatives
ANZ	Archives New Zealand
NAM NZ	National Army Museum, New Zealand
NZBDM	New Zealand Births, Deaths and Marriages
NZPD	New Zealand Parliamentary Debates
SCNZ	Statistics for the Colony of New Zealand

Preface

1. *Evening Post*, 26 September 1903, p. 5; *New Zealand Mail*, 1 March 1905, p. 53.
2. 'The Boers', URL: https://nzhistory.govt.nz/war/south-african-boer-war/the-boers, (Ministry for Culture and Heritage), updated 7 March 2018 (accessed 20 September 2020).
3. *Evening Star*, Auckland, 14 February 1879, p. 2; *New Zealand Times*, 26 April 1881, p. 2; *North Otago Times*, 13 February 1885, p. 3.
4. NZPD, Wellington: Government Printer, 108 (1899), p. 445.
5. NZPD, 110 (1899), p. 75.
6. *Otago Daily Times*, 13 October 1899, p. 5.

Chapter One: 'The flag that floats over us'

1. *Otago Daily Times*, 19 May 1900, p. 2.
2. *Marlborough Express*, 13 October 1900, p. 4.
3. Supplement to the Wanganui Collegian, August 1901.
4. *Feilding Star*, 13 July 1899, p. 2.
5. NZPD, 108 (1899), p. 446.
6. NZPD, 110 (1899), p. 77.
7. *Auckland Star*, 23 December 1899, p. 5.
8. *Chatterbox*, London: Wells Gardner, Darton & Co., 1899, Vol. XXIV, pp. 191–92.
9. *Evening Star*, 16 December 1899, p. 7.
10. *Wanganui Collegian*, No. 53, August 1900, p. 6.
11. *Otago Daily Times*, 19 May 1900, p. 2.
12. George Fowlds to Richard Seddon, telegram, 15 May, AAYS 8638 AD1/354/ay, D1900/2003, ANZ.
13. Richard Seddon to Penton, telegram, 15 May, AAYS 8638 AD1/354/ay, D1900/2003, ANZ.
14. *Otago Daily Times*, 18 May 1900, p. 5.
15. *Outlook*, 28 October 1899, p. 4. While *The Outlook* article conceded that details of the incident were unconfirmed, it nonetheless repeated grossly inaccurate newspaper reports placing Boer casualties at 1500 dead. *Nelson Evening Mail*, 21 October 1899, p. 2; *Otago Witness*, 26 October 1899, p. 22.
16. *Otago Daily Times*, 19 May 1900, p. 2.
17. *Otago Daily Times*, 19 May 1900, p. 2.
18. *Otago Daily Times*, 19 May 1900, p. 2. As contentious as this expression undoubtedly is today, its use was widespread during the period in question. 'Two niggers in a gig' is a direct quote from the newspaper, although it quite possibly referred to Pākehā in blackface, as 'minstrels' were common at the time. The term 'nigger' appeared regularly in the press well into the twentieth century and even appeared in reports tabled in Parliament.
19. *Otago Witness*, 24 May 1900, p. 25.
20. *Otago Daily Times*, 19 May 1900, pp. 2–3.
21. *Otago Daily Times*, 19 May 1900, p. 2.
22. *Otago Daily Times*, 12 May 1900, p. 8.
23. *Otago Daily Times*, 19 May 1900, p. 4.
24. *Otago Daily Times*, 19 May 1900, p. 4.
25. *Inangahua Times*, 21 May 1900, p. 2.
26. *Wanganui Chronicle*, 22 May 1900, p. 2.
27. *Hawke's Bay Herald*, 19 May 1900, p. 2.
28. *Wanganui Collegian*, No. 55, April 1901, p. 2.
29. *Mataura Ensign*, 2 June 1900, p. 2; *Ashburton Guardian*, 2 April 1900, p. 2; *Press*, 9 April 1900, p. 5.
30. *New Zealand Times*, 25 June 1900, p. 5; *New Zealand Mail*, 28 June 1900, p. 17.
31. *Evening Post*, 19 December 1899, p. 5.
32. *Otago Witness*, 1 March 1900, p. 51; *Star*, 5 August 1899, p. 6.
33. *Star*, 5 August 1899, p. 6.
34. Roll of individuals entitled to the South African Medal and clasps. British South African Police, 2 November 1901, National

Archives, Kew, UK, WO 100/238, p. 9; Roll of individuals entitled to the South African Medal and clasps. Protectorate Regt. Frontier Force, 18 February 1902, National Archives, Kew, UK, WO 100/263, p. 252; *Wanganui Collegian*, No. 53, August 1900, pp. 17, 19; *Wanganui Collegian*, No. 61, April 1903, p. 17.

35 *Wanganui Collegian*, No. 54, December 1900, pp. 13–14.

36 *Wanganui Collegian*, No. 54, December 1900, pp. 13–14.

37 W. Francis Aitken, *Baden-Powell, the Hero of Mafeking*, London: S. W. Partridge & Co., 1900, p. 145.

38 *Evening Post* (supplement), 21 July 1900, p. 3.

39 *Marlborough Express*, 18 July 1900, p. 4; Roll of individuals entitled to the South African Medal and clasps. Kimberley Light Horse, 16 July 1901, National Archives, Kew, UK, WO 100/255, p. 18.

40 Peat's wife, whom he accused of being unfaithful and wrote out of his will, claimed he had served in the Kimberley Town Guard, but his name does not appear on the corps' nominal roll. Supreme Court Affidavit, Janet Harriet Peat, 17 April 1903, C302 009 AAOM 6029 Box 152, ANZ; Last Will and Testament, Willis Harcourt Peat, 4 June 1886, C302 009 AAOM 6029 Box152, ANZ; Janet Harriet Peat to Richard John Seddon, 27 August 1900, AD1 362 I D1900/3698, ANZ.

41 *Nelson Evening Mail*, 13 June 1903, p. 1; *AJHR*, 1902, B-20B, p. 2.

42 *West Coast Times*, 11 April 1900, p. 2; Roll of individuals entitled to the South African Medal and clasps. Kimberley Town Guard, September 1901, National Archives, Kew, UK, WO 100/282, p. 89; Roll of individuals entitled to the South African Medal and clasps. Kimberley Light Horse, 16 July 1901, National Archives, Kew, UK, WO 100/255, p. 31; *West Coast Times*, 19 May 1900, p. 4.

43 Roll of individuals entitled to the South African Medal and clasps. Border Mounted Rifles, n.d., National Archives, Kew, UK, WO 100/260, p. 66.

44 *Evening Post*, 11 May 1900, p. 2.

45 *Timaru Herald*, 20 November 1899, p. 2; *Timaru Herald*, 6 February 1899, p. 2.

46 *Daily Southern Cross*, 12 April 1870, p. 3; *Timaru Herald*, 30 December 1899, p. 4; *Timaru Herald*, 16 August 1902, p. 3.

47 *Timaru Herald*, 30 December 1899, p. 4; *Evening Post*, 11 May 1900, p. 2.

48 Roll of individuals entitled to the South African Medal and clasps. Nursing Sisters, 28 September 1901, National Archives, Kew, UK, WO 100/229, p. 85; Roll of individuals entitled to the South African Medal and clasps, Locally Engaged Nursing Sisters, 30 August 1901, National Archives, Kew, UK, WO 100/229, p. 61; Roll of individuals entitled to the South African Medal and clasps. Nursing Sisters, Royal Army Medical Corps, 4 August 1901, National Archives, Kew, UK, WO 100/229, p. 57; Roll of individuals entitled to the South African Medal and clasps, 'A' Battery, Royal Horse Artillery, National Archives, Kew, UK, WO 100/139, p. 18.

49 *Press*, 17 September 1885, p. 2; *Bay of Plenty Times*, 26 October 1896, p. 3.

50 Nominal Roll of the Jameson Raiders, National Archives, Kew, UK, CO 179/193.

51 Attestation Form, n.d., Robert Jack, AABK 18805 W5515 0002781, ANZ; Nominal Roll of the Jameson Raiders, National Archives, Kew, UK, CO 179/193.

52 Roll of Officers, Warrant Officers, Non-commissioned Officers, and Men entitled to the Medal and Clasp or Clasps for Operations in Matabeleland, Rhodesia, together with those who have forfeited the Medal through misconduct. National Archives, Kew, UK, WO 100/77, p. 13; *Evening Post*, 15 October 1936, p. 10; *Waikato Argus*, 23 January 1905, p. 2.

53 *Hawera & Normanby Star*, 8 February 1900, p. 2; *Press*, 11 October 1900, p. 5.

54 *Taranaki Herald*, 12 October 1899, p. 2.

55 *Evening Star*, 6 October 1899, p. 2.

56 *Otago Witness*, 12 April 1900, p. 27.

57 *North Otago Times*, 18 December 1896, p. 4; Elsie Brown Garden, NZBDM, 1869/36412.

58 *Clutha Leader*, 10 April 1900, p. 3.

59 Attestation Form, n.d., Edgar Christophers Hazlett, AABK 18805 W5515 0006347, ANZ; Attestation Form, n.d., William Grant, AABK 18805 W5515 0002172, ANZ; Attestation Form, n.d., Amos McKegg, AABK 18805 W5515 0006407, ANZ; Attestation Form, n.d., John McConway, AABK 18805

W5515 0003369, ANZ; Walter Johnston to F. Jones, 20 September 1948; Attestation Form, n.d., Walter Johnston, AABK 18805 W5515 0002869, ANZ; Attestation Form, n.d., George Mitchell, AABK 18805 W5515 0003927, ANZ; *Otago Witness*, 12 April 1900, p. 7.

60   *AJHR*, 1902, A-2, pp. 74–75; *Otago Witness*, 11 December 1901, p. 25.

61   Louisa Hallam, 21 April 1902, Permit to Land in South Africa, ACGO 8333 IA1/849/ [9] 1902/1288, ANZ.

62   Maria Colvin, 12 June 1902, Permit to Land in South Africa, ACGO 8333 IA1/853/[64] 1902/2054, ANZ.

63   Annie Wattam, 4 June 1902, Permit to Land in South Africa, ACGO 8333 IA1/853/[11] 2/1973, ANZ; Jane Anne Nielsen, 5 June 1902, Permit to Land in South Africa, ACGO 8333 IA1/853/[15] 1902/1979, ANZ; Fanny Marsh, 21 May 1902, Permit to Land in South Africa, ACGO 8333 IA1/851/ [85] 1902/1761, ANZ.

64   *NZPD*, 110 (1899), pp. 75, 78, 81, 82.

65   *NZPD*, 110 (1899), p. 75.

66   *NZPD*, 110 (1899), p. 76.

Chapter Two: 'Rally to the call of home and country'

1   *AJHR*, 1898, H-19, p. 2.

2   *NZPD*, 108 (1899), pp. 445–46.

3   *The British Case Against the Boer Republics*, London: Imperial South African Association, n.d.

4   *NZPD*, 110 (1899), pp. 75, 78.

5   *NZPD*, 110 (1899), p. 78.

6   *NZPD*, 110 (1899), pp. 76, 83, 89, 90, 92, 93, 95.

7   Ian McIver, 'New Zealand and the South African War of 1899–1902: A Study of New Zealand's Attitudes and Motives for Involvement and of the Effect of the War on New Zealand and on Her Relations with Britain', Master of Arts thesis, Otago University, 1972, p. 32.

8   *NZPD*, 110 (1899), p. 79.

9   *NZPD*, 110 (1899), pp. 79, 81, 83, 90, 91, 92, 94, 95.

10   *NZPD*, 110 (1899), p. 92.

11   *NZPD*, 110 (1899), p. 94.

12   *Evening Post*, 30 September 1899, p. 5.

13   *NZPD*, 110 (1899), p. 96; *AJHR*, 1900, A-1, p. 8; *AJHR*, 1900, H-6K, [p. 1]; *Press*, 1 January 1900, p. 6.

14   *Wanganui Chronicle*, 19 January 1900, p. 2.

15   *New Zealand Times*, 6 October 1899, p. 8; *New Zealand Herald*, 7 October 1899, p. 5.

16   *NZPD*, 108 (1899), pp. 445–46; Keith Sinclair, *A Destiny Apart: New Zealand's Search for National Identity*, Wellington: Allen & Unwin, 1986, p. 138.

17   *Nelson Evening Mail*, 13 March 1876, p. 2.

18   *Daily Telegraph*, 16 December 1899, p. 2.

19   *Evening Post*, 28 October 1899, p. 4; *Otago Witness*, 1 February 1900, p. 28.

20   *Ohinemuri Gazette*, 3 February 1900, p. 3; *A Souvenir of the First New Zealand Contingent for South Africa, 1899–1901*, compiled by Alice St Clair Inglis, Auckland: Arthur Cleave & Co., 1902, p. 61; *Otago Witness*, 25 January 1900, p 28, 15 February 1900, p. 67.

21   Simon J. Potter, *News and the British World*, Oxford: Oxford University Press, 2003, p. 47; Allison Oosterman, 'New Zealand War Correspondence before 1915', *Pacific Journalism Review*, 16:1 (2010), p. 142.

22   *Press*, 17 January 1900, p. 6; *Otago Witness*, 25 January 1900, p. 26.

23   *Auckland Star*, 14 May 1902, p. 5; *Press*, 31 January 1901, p. 6.

24   *Press*, 5 February 1901, p. 2.

25   *Press*, 5 February 1901, p. 2.

26   *Press*, 26 December 1899, p. 5; W. H. Triggs to Major Robin, 16 February 1900, AD34 4, ANZ.

27   *NZPD*, 110 (1899), p. 77.

28   *New Zealand Tablet*, 19 October 1899, p. 17.

29   Potter, *News*, p. 47; Christchurch Press Company, *The Press, 1861–1961: The Story of a Newspaper*, Christchurch: Christchurch Press Company, 1963, pp. 122–23.

30   *Evening Post*, 2 March 1900, pp. 4–5; *Press*, 3 March 1900, pp. 7–8; *Northern Advocate*, 3 March 1900, p. 2.

31   *Daily Telegraph*, 2 March 1900, p. 8; *Otago Daily Times*, 19 May 1900, p. 3.

32   *New Zealand Official Year-Book*, Wellington: Government Printer, 1899, pp. 59–65; *New*

Zealand Official Year-Book, 1900, pp. 64–70; *New Zealand Official Year-Book*, 1901, pp. 75–81; *New Zealand Official Year-Book*, 1902, pp. 76–82.

33 *Poverty Bay Herald*, 13 June 1900, p. 2.
34 *Ashburton Guardian*, 5 March 1900, p. 2, original emphasis.
35 *Wanganui Collegian*, No. 53, August 1900, p. 19.
36 Hobart Cother Tennent to Katherine Jessie Tennent, 3 June 1901, NAM NZ 2001.1057; *Star* 24 July 1900, p. 1.
37 Mr Farquharson to William Farquharson, 3 Dec 1900, NAM NZ 1998.11.
38 *Otago Daily Times*, 23 January 1900, p. 5; *Otago Daily Times*, 24 February 1900, p. 2; Hobart Cother Tennent to Katherine Jessie Tennent, 16 April 1902, NAM NZ 2001.1057.
39 *Ashburton Guardian*, 3 April 1900, p. 2; *Timaru Herald*, 5 April 1900, p. 3; John Edward Thomas Burnett, diary, 2 July 1900, NAM NZ 2004.543.
40 Description of Discharge, 20 February 1901, Claude Lockhart Jewell, AABK 18805 W5515 0002836, ANZ; Attestation Form n.d., William Alfred Saunders, AABK 18805 W5515 0006459, ANZ; Attestation Form, April 1902, George Wilson, AABK 18805 W5515 0006076, ANZ.
41 *Ohinemuri Gazette*, 22 February 1901, p. 2.
42 *Hawke's Bay Herald*, 9 February 1901, p. 6.
43 *Colonist*, 11 April 1900, p. 3; *Star*, 17 January 1900, p. 1; *Press*, 24 July 1900, p. 6.
44 James Christie to H. D. Tuson, 9 July 1909, James Christie, AABK 18805 W5515 0001009, ANZ; Roll of individuals entitled to the South African Medal and clasps. Bushveldt Carbineers (later Pietersburg Light Horse, 10 December 1902, National Archives, Kew, UK, WO 100/263, p. 155.
45 *Clutha Leader*, 18 July 1902, p. 6; *Wanganui Chronicle*, 15 April 1902, p. 2; *Auckland Star*, 16 April 1902, p. 5; *Hastings Standard*, 24 July 1902, p. 4.
46 Malcolm McKinnon, 'Opposition to the War in New Zealand', in John Crawford and Ian McGibbon (eds), *One Flag, One Queen, One Tongue*, Auckland: Auckland University Press, 2003, p. 37; *Evening Post*, 4 January 1902, p. 5; *Evening Post*, 15 January 1902, p. 6; *Evening Post*, 18 January 1902, p. 2; *Star*, 24 February 1902, p. 1; *Hawera & Normanby Star* supplement, 1 March 1902, p. 3; *AJHR*, 1902, A-1, p. 30.
47 T. Green to Hugh Pollen, 17 January 1902, IA1 840 [1] 1902/140, ANZ.
48 *Evening Post*, 15 January 1902, p. 4; *Otago Witness*, 5 February 1902, p. 14.
49 *NZPD*, 119 (1901), pp. 312–13.
50 Crawford, 'Impact of the War', p. 210.
51 *Taranaki Daily News*, 22 August 1900, p. 3; *Wanganui Herald*, 21 August 1900, p. 2; *Timaru Herald*, 22 August 1900, p. 3; *Auckland Star*, 24 February 1900, p. 2; *Hawera & Normanby Star*, 1 March 1902, p. 2.
52 *Evening Star*, supplement, 23 September 1899, p. 3; *Evening Star*, 19 February 1901, p. 6; *Evening Star*, 23 August 1901, p. 6.
53 *Press*, 16 December 1899, p. 8; *Evening Post*, 24 February 1900, p. 2; *Otago Daily Times*, 26 February 1900, p. 5; *Grey River Argus*, 10 Oct 1899, p. 4; *Evening Post*, 4 April 1900, p. 5; *Thames Star*, 28 November 1899, p. 2; *Press*, 30 October 1899, p. 5.
54 *Evening Star*, 16 December 1899, p. 7; *Southland Times*, 16 August 1900, p. 2.
55 *Outlook*, 11 January 1902, p. 4; *Free Lance*, 12 April 1902, p. 8.
56 *Free Lance*, 12 April 1902, p. 8.
57 *Thames Star*, 1 March 1900, p. 4; *Tuapeka Times*, 3 March 1900, p. 4; *Evening Post*, 15 December 1899, p. 2.
58 *Otago Witness*, 26 July 1900, p. 25.
59 *Christ's College Register*, No. 43, August 1900, p. 396.
60 *Wanganui Collegian*, No. 52, April 1900, p. 13 (emphasis as in original).
61 *New Zealand Tablet*, 14 March 1901, p. 19; *Outlook*, 11 January 1902, p. 4.
62 *Southland Times*, 15 January 1902, p. 2; *Grey River Argus*, 10 February 1902, p. 2.
63 *Otago Witness*, 19 March 1902, p. 20.
64 Hobart Cother Tennent to Katherine Jessie Tennent, 3 June 1901, NAM NZ 2001.1057.
65 *Nelson Evening Mail*, 24 January 1902, p. 2.
66 *Southland Times*, 15 January 1902, p. 2.
67 *Wanganui Collegian*, No. 55, April 1901, p. 22.

68 *Wanganui Collegian*, No. 52, April 1900, p. 4.
69 *Wanganui Collegian*, No. 54, December 1900, p. 15.
70 Frank Perham, diary, 28, 30 January 1901, NAM NZ 2000.736; Hobart Cother Tennent to 'Oswald', 20 Jun 1901, NAM NZ 2001.1057; *Wanganui Collegian*, No. 55, Apr 1901, p. 22.
71 Charles Borland Tasker, transcript of diary, 22 June 1901, NAM NZ 1991.1956.
72 *Auckland Star*, 29 August 1900, p. 5; *Daily Telegraph*, 3 September 1900, p. 3; *New Zealand Tablet*, 20 September 1900, p. 15; *Otago Daily Times*, 10 September 1900, p. 2.
73 *Wanganui Herald*, 6 March 1901, p. 2.
74 *Ohinemuri Gazette*, 22 February 1901, p. 2.
75 *Otago Daily Times*, 25 July 1900, p. 5; *Wanganui Collegian*, No. 53, August 1900, p. 19; George Leece to Bert, 24 September 1900, Alexander Turnbull Library (ATL) MS-Papers-8464-05.
76 Luke Perham to Mrs. Perham, 28 June 1900, NAM NZ 2003.7.
77 *Otago Daily Times*, 25 July 1900, p. 5.
78 *Wanganui Collegian*, No. 53, August 1900, p. 19.
79 *Star*, 24 July 1900, p. 1.
80 *Press*, 10 August 1901, p. 5.
81 Mr Farquharson to William Farquharson, 6 July 1900, NAM NZ 1998.11.
82 *Taranaki Herald*, 13 February 1901, p. 2; *Otago Witness*, 20 March 1901, p. 55.
83 *Auckland Star*, 14 March 1902, p. 4.
84 *Thames Star*, 28 November 1901, p. 2.
85 *Evening Post*, 29 April 1901, p. 6.
86 NZPD, 113 (1900), pp. 639–40.
87 *Auckland Star*, 23 October 1899, p. 5; *Star*, 14 November 1899, p. 1.
88 *Nelson Evening Mail*, 15 November 1899, p. 2.
89 *Star*, 14 October 1899, p. 6; *Star*, 16 October 1899, p. 1.
90 *Outlook*, 11 January 1902, p. 1; Robert Lawrence to George Scott, envelope, Hocken Library, HL Misc-MS-0987.
91 Lieutenant George Leece to Miss Leece, envelope, n.d., ATL MS-Papers-8464-07.
92 *Evening Post*, 13 January 1902, p. 5.
93 *New Zealand Tablet*, 19 October 1899, p. 18.
94 *Manawatu Evening Standard*, 4 September 1901, p. 2; *Auckland Star*, 27 August 1901, p. 8; *Otago Daily Times*, 6 February 1900, p. 5.
95 *Evening Post* (supplement), 11 January 1902, p. 7.
96 *Bush Advocate*, 14 April 1902, p. 2.
97 *Evening Post*, 29 November 1901, p. 4.
98 Bert Stevens to J. H. Stevens, postcard, NAM NZ 1999.3239; *Ashburton Guardian*, 6 April 1900, p. 2.
99 Luke Perham to Theresa Perham, 28 June 1900, NAM NZ 2003.7, p. 3; *Otago Daily Times*, 11 August 1900, p. 2; Thomas Pakenham, *The Boer War*, London: Abacus, 1992, pp. 435–36. Pakenham claims that De Wet allowed British prisoners to loot the mail prior to leaving Roodewal.
100 Luke Perham to Theresa Perham, 28 June 1900, NAM NZ 2003.7, p. 3.
101 Sixth Contingent Regimental Orders, 25 May 1902, AD34 28 29 SA 8024, ANZ.
102 *AJHR*, 1902, A-1, pp. 22–23; W. P. Reeves to Richard J. Seddon, 13 December 1901, Charles Borland Tasker, AABK 18805 W5515 0005468, ANZ.
103 Sir Montagu Ommanney to W. P. Reeves, 17 December 1901, Charles Borland Tasker, AABK 18805 W5515 0005468, ANZ.
104 W. P. Reeves to Sir Montagu Ommanney, 18 December 1901, Charles Borland Tasker, AABK 18805 W5515 0005468, ANZ.
105 *Evening Post*, 19 December 1901, p. 5; *Bay of Plenty Times*, 20 December 1901, p. 2; *Auckland Star*, 19 December 1901, p. 5; *Feilding Star*, 19 December 1901, p. 2; *Evening Post*, 11 January 1902, p. 5; *Evening Post*, 29 January 1902, p. 7.
106 Proceedings of a Court of Inquiry, 29 June 1902, Levlin (Leolin) Hamar Arden, AABK 18805 W5515 0000127, ANZ.
107 Lord Ranfurly to the Acting Premier, 1 July 1902, Levlin (Leolin) Hamar Arden, AABK 18805 W5515 0000127, ANZ.
108 *Marlborough Express*, 2 July 1902, p. 2; *Wanganui Chronicle*, 3 July 1902, p. 5; *Auckland Star*, 2 July 1902, p. 5; *Press*, 3 July 1902, p. 5; Major-General Babington to Fanny Arden, 30 December 1903, Levlin (Leolin) Hamar Arden, AABK 18805 W5515 0000127, ANZ.
109 *Wanganui Collegian*, No. 59, August 1902, p. 7.

110 *Poverty Bay Herald*, 30 August 1902, p. 3; *Press*, 30 August 1902, p. 7; *Otago Witness*, 3 September 1902, p. 9.
111 *AJHR*, 1900, I-12, pp. 2–3, 5–6.
112 *AJHR*, 1900, I-12, p. 6.
113 *AJHR*, 1900, I-12, pp. 3–4.
114 *Evening Post*, 7 January 1901, p. 4.
115 Penelope Farquharson to William Farquharson, 8 Apr 1900, NAM NZ 1998.11.
116 *Auckland Star*, 10 February 1900, p. 4.
117 *Taranaki Herald*, 13 February 1901, p. 2; *Otago Witness*, 17 July 1901, p. 52.
118 *West Coast Times*, 18 April 1902, p. 3; *Auckland Star*, 27 May 1899, p. 2; *NZPD*, 111 (1900), p. 20.
119 *Evening Post*, 7 January 1901, p. 4.
120 *Nelson Evening Mail*, 1 March 1901, p. 2.
121 *NZPD*, 116 (1901), p. 294.
122 *AJHR*, 1901, A-1, p. 4.
123 *Otago Witness*, 12 February 1902, p. 29.
124 *AJHR*, 1900, I-12, pp. 2–3.
125 *Otago Witness*, 5 February 1902, p. 14.
126 *Free Lance*, 15 June 1901, p. 3.
127 *AJHR*, 1901, A-4, p. 324.
128 *Star*, 15 June 1900, p. 4; *AJHR*, 1901, A-1, pp. 5–6.
129 *AJHR*, 1901, A-3A, p. 2.
130 *AJHR*, 1901, A-1, pp. 5–6.
131 *NZPD*, 111 (1900), p. 83.
132 Colin McGeorge, 'Military Training in New Zealand Primary Schools 1900–1912', *Australia and New Zealand History of Education Society Journal*, 3:1 (1974), p. 2; Crawford and McGibbon, *One Flag, One Queen, One Tongue*, p. 206; *Nelson Evening Mail*, 16 January 1888, p. 5.
133 *NZPD*, 116 (1901), p. 286.
134 School Reports, Annual Report of the Inspector of Schools, 28 February 1901, p. 2, EB-W9 12, ANZ, EB-W9 13, ANZ.
135 *New Zealand Graphic and Ladies' Journal*, 23 November 1901, p. 985, Sir George Grey Special Collections, Auckland Libraries.
136 Walter Empson to Richard J. Seddon, 28 December 1900, AD1 372 h D1901/18; *AJHR*, 1902, E-1D, [p. 1], ANZ.
137 *Wanganui Collegian*, No. 54, December 1900, p. 16.
138 *NZPD*, 121 (1902), p. 373.
139 *Christ's College Register*, No. 42, April 1900, pp. 356–57; *Christ's College Register*, No. 44, December 1900, p. 439.
140 *AJHR*, 1901, H-6E [p. 1]; *Christ's College Register*, No. 43, August 1900, p. 370; *Christ's College Register*, No. 42, April 1900, pp. 301–2.
141 *AJHR*, 1902, E-1D, pp. [1], pp. 3–5.
142 *AJHR*, 1903, A-7, p. 37.
143 *AJHR*, 1903, H-19A, p. 1. Although a table in the parliamentary return showed 12 Volunteer mounted corps across five regions in 1898, a list naming the corps and the number of men serving in them listed 13 Volunteer mounted corps in 1898.
144 *AJHR*, 1903, H-19A, pp. 1, 7. The membership figures were calculated from the number of corps members who qualified for capitation through their attendance record.
145 J. A. B. Crawford, 'The Role and Structure of the New Zealand Volunteer Force 1885–1910', Master of Arts thesis, University of Canterbury, 1986, p. 185.
146 *AJHR*, 1905, H-19A, p. 3. One of these clubs at Stanley Brook disbanded in November 1901 but was included in the 1905 government report titled *Defence Rifle Clubs*. Similarly, the Taita club, which disbanded in April 1902, and the Parua club, which disbanded in December 1902, were also included in the totals. *AJHR*, 1905, H-19A, pp. 1–3.
147 *AJHR*, 1905, H-19A, p. 2.
148 *AJHR*, 1898, H-19, p. 8.
149 *AJHR*, 1898, H-19, pp. 7–8.
150 *Evening Post*, 2 March 1899, p. 4.
151 Simon Johnson, 'Sons of Empire: A Study of New Zealand Ideas and Opinions during the Boer War', BA Honours research exercise in history, Massey University, 1974, p. 52.
152 *Thames Star*, 13 October 1899, p. 2.
153 *Evening Star*, 14 October 1899, p. 2.
154 *Otago Daily Times*, 10 October 1899, p. 5.
155 *Wairarapa Daily Times*, 21 August 1899, p. 2; *Woodville Examiner*, 19 June 1899, p. 3; *Manawatu Herald*, 20 May 1899, p. 2; *Hawke's Bay Herald*, 29 September 1899, p. 3; *Otago Daily Times*, 29 September 1899, p. 5; *Thames Star*, 8 January 1880, p. 2;

Hawera & Normanby Star, 11 January 1884, p. 2; *Akaroa and Banks Peninsula Advertiser*, 29 March 1881, p. 2.

156 *Auckland Star*, 16 October 1899, p. 5.

157 *Marlborough Express*, 14 February 1900, p. 3; *Marlborough Express*, 31 May 1902, p. 2.

158 *Otago Daily Times*, 25 January 1900, p. 6.

159 *Otago Witness*, 1 March 1900, p. 28; *Thames Star*, 1 March 1900, p. 2; *Otago Witness*, 1 March 1900, p. 28.

160 *Otago Witness*, 1 March 1900, p. 27.

161 *Otago Daily Times*, 17 December 1900, p. 5.

162 *Evening Star*, 6 April 1901, p. 8.

163 *Evening Star*, 6 April 1901, p. 8.

164 *Otago Witness*, 1 March 1900, pp. 26–27.

165 *Star*, 13 February 1902, p. 1; *Daily Telegraph*, 17 October 1899, p. 1; *Press*, 19 February 1900, p. 7; *Otago Witness*, 3 July 1901, p. 30; *Evening Post*, 3 February 1902, p. 5; *Otago Witness*, 10 May 1900, p. 34; *Auckland Star*, 10 February 1902, p. 2.

166 *Auckland Star*, 21 October 1899, p. 3; *Colonist*, 20 June 1900, p. 2.

167 *Evening Post*, 13 January 1902, p. 7.

168 Diary of the Eighth Contingent Railway accident, Jessie Whitehead: Letters written by Corporal David Whitehead to his family & papers relating to his death, HL Misc-MS-1900.

169 F. R. W. Daw to Lieutenant-Colonel Banks, 12 February 1900, Return showing quantities of gifts forwarded to South Africa, February 1900 – August 1901, AD34 4, ANZ.

170 Ellen Ellis, 'New Zealand Women and the War', in Crawford and McGibbon, *One Flag, One Queen, One Tongue*, pp. 128–48; Anna Rogers, *While You're Away*, Auckland: Auckland University Press, 2003, pp. 12–31.

171 *Ashburton Guardian*, 6 April 1900, p. 2.

172 *Wanganui Herald*, 9 February 1900, p. 2.

173 H. L. Cameron to Major Robin, 22 March 1900, Return showing quantities of gifts forwarded to South Africa, February 1900 – August 1901, AD34 4, ANZ.

174 Alice St Claire Inglis to Major Robin, 29 April 1900, Return showing quantities of gifts forwarded to South Africa, February 1900 – August 1901, AD34 4, ANZ; Jessie Bodle to Major Robin, 8 January 1900, AD34 4, ANZ.

175 List of contents of Cases shipped per S.S. 'Monowai' to Messers. Divine, Gates & Co. Cape Town for distributing to members of the First and Second New Zealand Contingent, Return showing quantities of gifts forwarded to South Africa, February 1900 – August 1901, AD34 4, ANZ.

176 Penelope Farquharson to William Farquharson, 1 February 1900, NAM NZ 1998.11; 'Lonnie' to William Farquharson, 2 January 1901, NAM NZ 1998.11.

177 Rebecca Totman to Defence Department, 24 April 1902 (Received), Clement Totman, AABK 18805 W5515 0005634, ANZ.

178 N. G. Lyttelton to Lord Ranfurly, 30 July 1902, Clement Totman, AABK 18805 W5515 0005634, ANZ; Rebecca Totman to W. H. Webb, 12 May 1902, Clement Totman, AABK 18805 W5515 0005634, ANZ.

179 *Ohinemuri Gazette*, 27 January 1902, p. 2.

180 *New Zealand Herald*, 28 January 1902, p. 5; *Thames Star*, 29 January 1902, p. 4.

181 *Otago Witness*, 5 April 1900, p. 26.

182 Megan Hutching, '"Mothers of the World": Women, Peace and Arbitration in Early Twentieth-Century New Zealand', *NZJH*, 27:2 (1993), pp. 173–85.

183 *Otago Daily Times*, 12 May 1900, p. 6.

184 *Otago Daily Times*, 12 May 1900, p. 8.

185 *Evening Star*, 11 May 1900, p. 2.

186 *Otago Daily Times*, 12 May 1900, p. 8.

187 *Otago Daily Times*, 12 May 1900, pp. 3, 6, 8; Megan Hutching. 'Bain, Wilhelmina Sherriff — Biography', *Te Ara — the Encyclopedia of New Zealand*, updated 27 September 2010, http://www.TeAra.govt.nz/en/bibliographies/3b3

188 Hutching, 'Mothers', pp. 175–76.

189 *Otago Daily Times*, 11 May 1900, p. 2.

190 *Otago Daily Times*, 11 May 1900, p. 2; Attestation Form, 25 January 1901, Charles Borland Tasker, AABK 18805 W5515 0005468, ANZ.

191 *Evening Post*, 11 January 1902, p. 5.

192 *Evening Star*, 11 May 1900, p. 2.

193 *Evening Star*, 11 May 1900, p. 2.

194 *Otago Daily Times*, 12 May 1900, p. 6, 11 May 1900, p. 2.

195 *Otago Daily Times*, 12 May 1900, p. 8.

196 *Otago Daily Times*, 12 May 1900, p. 8; *Evening Star*, 16 July 1896, p. 4.
197 *Evening Star*, 20 April 1899, p. 4.
198 *Poverty Bay Herald*, 24 March 1900, p. 2.
199 *Poverty Bay Herald*, 13 June 1900, p. 2.
200 *Otago Daily Times*, 12 May 1900, p. 3; 14 May 1900, p. 2.
201 *Evening Star*, 19 May 1900, p. 3.
202 *Evening Star*, 12 May 1900, p. 6.
203 *Otago Witness*, 17 May 1900, p. 35.
204 *Otago Daily Times* 11 May 1900, p. 2; 12 May 1900, p. 8.
205 *Otago Daily Times*, 12 May 1900, p. 8; 14 May 1900, p. 2; 11 May 1900, p. 2.
206 *Otago Daily Times*, 26 May 1900, p. 8.
207 *Observer*, 3 March 1900, p. 15; *Evening Post*, 17 May 1900, p. 5.
208 *Hawera & Normanby Star*, 25 March 1902, p. 2; *Marlborough Express*, 27 March 1902, p. 3.
209 *Otago Witness*, 17 August 1899, p. 9; *Otago Daily Times*, 18 July 1899, p. 6; *Auckland Star*, 17 January 1902, p. 3; *Otago Daily Times*, 18 July 1899, p. 6; *Auckland Star*, 17 January 1902, p. 3.
210 *Evening Post*, 18 January 1902, p. 2.
211 *Evening Post*, 18 January 1902, p. 2.
212 *Evening Post*, 17 May 1900, p. 5.
213 *Evening Post*, 1 February 1900, p. 6; *Evening Post*, 6 February 1900, p. 6; Scott Flutey, 'Ladies at Arms for Empire: Patriotic Women's Organisations in New Zealand during the South African War, 1899–1902', BA (Hons) research essay in history, Victoria University of Wellington, 2016, p. 7.
214 *Evening Post*, 17 May 1900, p. 5.
215 *Wanganui Herald*, 30 June 1900, p. 1; *Hawke's Bay Herald*, 12 June 1900, p. 4; *New Zealand Herald*, 17 February 1900, p. 4.
216 *Thames Star*, 28 June 1900, p. 4.
217 *Wanganui Herald*, 15 October 1900, p. 1; *Mataura Ensign*, 18 August 1900, p. 1.
218 Ellis, 'New Zealand Women and the War', p. 138.
219 Edward Iveagh Lord to A. P. Douglas, 1 September 1900, AAYS 8638 AD1 459/b D1900/3427, ANZ; *Otago Witness*, 30 January 1901, pp. 26–27.
220 Edward Iveagh Lord to Under-Secretary for Defence, 14 September 1900, AAYS 8638 AD1 459/b D1900/3427, ANZ.
221 *Grey River Argus*, 27 December 1900, p. 2; James O'Sullivan to Joseph Petrie, 16 August 1901, AAYS 8638 AD1 459/b D1900/3427, ANZ; Arthur Douglas to Richard J. Seddon, 17 May 1902, AAYS 8638 AD1 459/b D1900/3427, ANZ.
222 T. F. Grey to R. J. Seddon, 13 July 1903, AAYS 8638 AD1 459/b D1900/3427, ANZ; W. Hannan to Under-Secretary for Defence, 7 October 1903, AAYS 8638 AD1 459/b D1900/3427, ANZ.
223 *Star*, 10 October 1899, p. 3; *Otago Daily Times*, 10 October 1899, p. 5; *Star*, 23 December 1899, p. 6; *Otago Witness*, 19 October 1899, p. 63.
224 *Press*, 26 December 1899, p. 3; *Ashburton Guardian*, 13 October 1899, p. 2; *Auckland Star*, 3 February 1900, p. 5; *Auckland Star*, 20 December 1933, p. 15.
225 *Press*, 12 January 1900, p. 6; *Otago Daily Times*, 5 February 1900, p. 5.
226 *Manawatu Herald*, 6 March 1900, p. 2; *Star*, 24 January 1900, p. 4; *Otago Daily Times*, 5 February 1900, p. 5.
227 *Otago Daily Times*, 1 August 1900, p. 3; *Otago Witness*, 19 July 1900, p. 8; *Evening Post*, 14 July 1900, p. 5.
228 *Otago Witness*, 22 March 1900, p. 66.
229 *Auckland Star*, 29 March 1902, p. 5.
230 *Colonist*, 19 January 1900, p. 2.
231 *Temuka Leader*, 30 January 1900, p. 1; *Press*, 24 January 1901, p. 5.
232 *Evening Post*, 3 March 1900, p. 6; *Manawatu Evening Standard*, 1 October 1900, p. 2.
233 *Press*, 13 January 1900 p. 8.
234 *Press*, 19 October 1899, pp. 4–5.
235 *Press*, 19 October 1899, pp. 4–5.
236 *Press*, 19 October 1899, pp. 4–5.
237 *Auckland Star*, 13 December 1905, p. 8.
238 *Taranaki Daily News*, 18 March 1902, p. 3.
239 *Evening Post*, 10 March 1902, p. 4; *AJHR*, 1903 A-1, p. 3.
240 *Press*, 18 March 1902, p. 2; *Observer*, 12 April 1902, p. 3.
241 *Manawatu Evening Standard*, 22 March 1902, p. 2.
242 *Marlborough Express*, 5 May 1902, p. 1;

Otago Witness, 14 May 1902, p. 41.

243   Ellen Ellis, *Teachers for South Africa: New Zealand Women at the South African War Concentration Camps*, Paekakariki: Hanorah Books, 2010, pp. 152–67.
244   *Auckland Star*, 1 November 1900, p. 3.
245   *Daily Telegraph*, 1 June 1900, p. 2; *Evening Post*, 23 February 1900, p. 6; *Press*, 3 March 1900, p. 8.
246   *Wanganui Collegian*, No. 53, August 1900, p. 1.
247   *Wanganui Collegian*, No. 53, August 1900, pp. 2–3.
248   *Evening Star*, 16 December 1899, p. 7.
249   *School Reports, Scholarship Report*, February 1900, p. 2, ADEX 16413 EB-W9 12, ANZ.
250   *School Reports, Scholarship Report*, February 1900, ADEX 16413 EB-W9 12, ANZ.
251   *School Reports, Pupil Teachers' Examination*, December 1900, ADEX 16413 EB-W9 13, ANZ.
252   *Wanganui Collegian*, No. 54, December 1900, p. 10.
253   *Otago Witness*, 1 March 1900, p. 28.
254   *Evening Post*, 1 March 1900, p. 5; *Observer*, 3 March 1900, p. 14.
255   *Mataura Ensign*, 31 May 1900, p. 2.
256   *Auckland Star*, 6 July 1900, p. 5.
257   *Otago Witness*, 12 September 1900, p. 65.
258   *Otago Witness*, 21 June 1900, pp. 61–63.
259   *Otago Witness*, 25 January 1900, p. 62; *Otago Witness*, 12 September 1900, p. 66.
260   *Star*, 16 October 1899, p. 1.
261   *Star*, 16 October 1899, p. 1.
262   *Star*, 16 October 1899, p. 1.
263   *Star*, 16 October 1899, p. 1.
264   *Otago Witness*, 22 May 1901, p. 25.
265   *Bruce Herald*, 17 October 1899, p. 3.
266   *New Zealand Illustrated Magazine*, 1 November 1899, p. 81.
267   *Hawke's Bay Herald*, 10 June, 1901, p. 4; A. Clarke to Richard J. Seddon, 21 November 1901, AD1 390 v D1901/5269, ANZ; John Hobbs to Richard J. Seddon, 11 November 1902, AD1 389 an D1901/5074, ANZ.
268   H. H. Donald to Under-Secretary for Defence, 22 March 1901, AD1 76 af D1901/1159, ANZ.
269   Johnson, 'Sons of Empire', p. 52.
270   *Outlook*, 7 October 1899, p. 4.
271   *Outlook*, 21 October 1899, p. 1; *Outlook*, 28 Oct 1899, p. 4.
272   *Outlook*, 28 October 1899, p. 4.
273   *Outlook*, 6 January 1900, p. 4.
274   *Outlook*, 21 October 1899, p. 1.
275   *Outlook*, 28 October 1899, p. 28; *Outlook*, 11 November 1899, p. 19; *Tuapeka Times*, 5 May 1900, p. 3.
276   *Outlook*, 28 October 1899, p. 4; *Outlook*, 11 November 1899, p. 19.
277   *Outlook*, 28 October 1899, p. 4.
278   *Outlook*, 4 November 1899, pp. 26, 28; *Outlook*, 18 November 1899, p. 27.
279   McKinnon, 'Opposition to the War in New Zealand', p. 29; Johnson, 'Sons of Empire', pp. 42–44.
280   *Outlook*, 21 October 1899, p. 4.
281   *Outlook*, 10 March 1900, p. 19.
282   *Otago Daily Times*, 4 November 1899, p. 6; *Outlook*, 11 November 1899, p. 18.
283   *New Zealand Tablet*, 30 November 1899, p. 6; *Outlook*, 4 November 1899, p. 19.
284   *Outlook*, 22 March 1902, p. 21.
285   *Outlook*, 4 November 1899, pp. 4–5.
286   *Outlook*, 11 January 1902, p. 1.
287   *Star*, 15 January 1902, p. 3.
288   *Star*, 15 January 1902, p. 3.
289   *Star*, 29 January 1902, p. 3.
290   *Evening Post*, 13 January 1902, p. 5.
291   *New Zealand Tablet*, 19 October 1899, pp. 2, 17.
292   *New Zealand Tablet*, 19 October 1899, p. 17.
293   *NZPD*, 110 (1899), p. 89.
294   *Otago Daily Times*, 27 February 1900, p. 5; *Otago Daily Times*, 2 March 1900, p. 2; *AJHR*, 1900, H-27, p. 2; Thomas Pakenham, *The Boer War*, London: Abacus, 1992, p. 106; Trooper named Arthur to his parents, 8 June 1900, NAM NZ 1999.2002.
295   *Otago Witness*, 28 December 1899, p. 21.
296   *New Zealand Tablet*, 29 March 1900, p. 20.
297   Attestation Form, 17 March 1900, Robert Walter Gordon Collins, AABK 18805 W5515 0006267, ANZ; Attestation Form, 6 April 1900; Certificate of Discharge, AAABK 18805 000254, ANZ; Attestation Form, 3 April

1901, Peter Fletcher; Secretary, Council of Defence to Kate Fletcher, 4 November 1907, AABK 18805 W5515 0001831, ANZ; Attestation Form, 3 April 1901, John Considine; Medical Report, 26 January 1903, AABK 18805 W5515 001135, ANZ; *Otago Witness*, 1 March 1900, p. 27; *Otago Witness*, 12 March 1902, p. 28.
298 *Outlook*, 24 March 1900, p. 5.
299 *Otago Daily Times*, 19 February 1900, p. 5.
300 Attestation Form, 15 March 1902, Percy Arthur Cohen, AABK 18805 W5515 0001064, ANZ; Attestation Form, 11 March 1900, Albert Moeller Samuel, AABK 18805 W5515 0003947, ANZ; Chief Paymaster, Cape Colony to Minister of Defence, New Zealand, 5 February 1902, Isodore Cohen, AABK 18805 W5515 0000970, ANZ.
301 *Outlook*, 21 April 1900, p. 58; *Tuapeka Times*, 29 September 1900, p. 3.
302 *Otago Daily Times*, 19 June 1900, p. 2.
303 *Auckland Star*, 22 March 1900, p. 2; *Auckland Star*, 9 March 1900, p. 3.
304 *Auckland Star*, 10 March 1900, p. 2; *Auckland Star*, 27 March 1900, p. 2.
305 *New Zealand Herald*, 31 January 1901, p. 6; *Evening Star*, 18 January 1897, p. 2; *Otago Witness*, 22 March 1900, p. 25.
306 *Outlook*, 17 February 1900, p. 5.
307 *Mataura Ensign*, 7 June 1900, p. 2; *New Zealand Tablet*, 24 April 1902, p. 4.
308 *AJHR*, 1900, H-6, p. 29; John Edward Thomas Burnett, transcript of diary, 1 April & 9 June 1900, NAM NZ 2009.549.
309 Frank Perham, diary, 18–19 May 1901, NAM NZ 2000.736.
310 *Outlook*, 19 April 1902, p. 29.
311 *Outlook*, 19 April 1902, p. 29.
312 *Northern Advocate*, 9 February 1901, p. 6.
313 James Egan, 6th Contingent crime and offence reports, March 1901 – March 1902, AD34 7 7 8032, ANZ.
314 *Otago Witness*, 15 February 1900, p. 11; *Auckland Star*, 7 March 1900, p. 5; *Outlook*, 24 March 1900, p. 5.
315 *Otago Witness*, 15 February 1900, p. 11; *Otago Daily Times*, 7 March 1900, p. 3.
316 *Auckland Star*, 7 March 1900, p. 5; *Wanganui Chronicle*, 15 January 1902, p. 3; *New Zealand Herald*, 28 January 1902, p. 5.

317 *Outlook*, 1 February 1902, p. 35.
318 Application for the New Zealand Service Medal, Arthur William Henry Compton, AABK 18805 W5515 0006268, p. 17; *AJHR*, 1902, H-6, p. 1.
319 *Auckland Star*, 22 June 1909, p. 4; *AJHR*, 1902, H-6A, pp. 1, 15; *AJHR*, 1909, C-12, p. 2; *Outlook*, 8 March 1902, p. 1; *AJHR*, 1902, H-6B, pp. 1, 14.
320 Jared Davidson, *Dead Letters: Censorship and Subversion in New Zealand 1914–1920*, Dunedin: Otago University Press, 2019, p. 215.
321 John Williamson to Colonel Penton, 18 April 1900, AAYS 8638 AD1 Box 354/ag D1900/1967, ANZ.
322 Harrison Bros. to Colonel Penton, 17 May 1900, AAYS 8638 AD1 Box 354/ag D1900/1967, ANZ.
323 *Otago Daily Times*, 25 October 1899, p. 2.
324 *Evening Post*, 11 October 1899, p. 6.
325 *Colonist*, 12 October 1899, p. 3.
326 *Otago Witness*, 28 December 1899, p. 20.
327 *Wanganui Herald*, 9 February 1900, p. 2. Although the term 'rough riders' appeared in the New Zealand press as early as 1856 in reference to experienced riders, it was the nickname given to a volunteer United States Cavalry unit commanded by Theodore Roosevelt during the 1898 Spanish-American War. As Roosevelt's force was often mentioned in New Zealand newspapers, it seems likely the US Cavalry unit influenced the popular name given to both the Third and Fourth contingents (see *AJHR*, 1900, H-6, pp. 11, 16).
328 *Otago Witness*, 22 March 1900, p. 66.
329 *Timaru Herald*, 3 March 1900, p. 3; *Otago Daily Times*, 19 May 1900, p. 2.
330 *Otago Daily Times*, 19 May 1900, p. 2.
331 *New Zealand Herald*, supplement, 13 April 1900, p. 3.
332 Davidson, *Dead Letters*, p. 88.
333 W. Morton to Joseph Ward, 7 January 1901, AAYS 8638, AD1 372/bt D1901/199, ANZ.
334 *Evening Post*, 20 December 1901, p. 5.
335 *Evening Post*, 20 December 1901, p. 5.
336 *Evening Post*, 10 January 1902, p. 2.
337 *Evening Post*, 20 December 1901, p. 5;

Evening Post, 8 January 1902, p. 5; *Evening Post*, 10 January 1902, p. 2.
338 *Evening Post*, 10 January 1902, p. 2.
339 *Evening Post*, 20 January 1902, p. 5.
340 *Evening Post*, 13 January 1902, p. 5.
341 *Wanganui Herald*, 24 January 1902, p. 2.
342 *Hawke's Bay Herald*, 31 May 1900, p. 4.
343 *Star*, 15 January 1902, p. 3.
344 *Grey River Argus*, 28 January 1902, p. 4.
345 Hillside Workshops illuminated address, AAAA 21953 D579 Box 1, ANZ.
346 *Nelson Evening Mail*, 16 January 1902, p. 2; *Marlborough Express*, 28 September 1901, p. 4.
347 NZPD, 111 (1900), p. 20.
348 NZPD, 110 (1899), pp. 80, 96; *Wanganui Chronicle*, 2 December 1899, p. 2.
349 *Wanganui Chronicle*, 2 December 1899, p. 2.
350 AJHR, 1900, H-26, p. 2.
351 *Evening Post*, 20 December 1899, p. 3.
352 AJHR, 1903, H-26, p. 2.
353 NZPD, 110 (1899), pp. 80, 87–88, 96; AJHR, 1906, H-14, pp. 3, 10; 1903, AJHR, H-26, p. 3.
354 AJHR, 1900, H-26, p. 2.
355 AJHR, 1900, H-26, p. 3.
356 *Colonist*, 18 March 1902, p. 2; McKinnon, 'Opposition to the War in New Zealand', p. 37.
357 *Colonist*, 18 March 1902, p. 2.
358 Margaret Buist, 'Reaction in New Zealand to the Boer War with Particular Reference to Canterbury and Marlborough', Master of Arts thesis, University of Otago, 1970, pp. 31–32.
359 *New Zealand Tablet*, 8 May 1902, p. 2.
360 *Nelson Evening Mail*, 1 March 1901, p. 2.
361 AJHR, 1900, H-6, [p. 1]; AJHR, 1901, H-6D, [p. 1]; AJHR, 1901, H-6A, p. 9; *Observer*, 18 January 1902, p. 20.
362 *Otago Witness*, 16 July 1902, p. 82.
363 *Tuapeka Times*, 3 March 1900, p. 3.
364 *Akaroa Mail and Banks Peninsula Advertiser*, 24 January 1902, p. 2.
365 *Otago Witness*, 19 March 1902, pp. 26–29.
366 McKinnon, 'Opposition to the War in New Zealand', pp. 28–37.
367 NZPD, 110 (1899), pp. 81–82.
368 NZPD, 110 (1899), p. 82.
369 *Evening Post*, 30 September 1899, p. 5.
370 *Otago Witness*, 5 October 1899, p. 19; *Daily Telegraph*, 14 July 1900, p. 8.
371 *Evening Star*, 16 May 1900, p. 4; *Evening Star*, 22 May 1900, p. 5; *Evening Star*, 27 July 1900, p. 1.
372 Barclay, Alfred Richard, *The Origin of Wealth: Being the Theory of Karl Marx in Simple Form*, Dunedin: S. Lister, 1899, ATL Pam 1899 BAR 2285; *Evening Star*, 16 May 1900, p. 4; *Evening Star*, 22 May 1900, p. 5; *Evening Star*, 27 July 1900, p. 1; *Evening Star*, 18 May 1900, p. 7.
373 NZPD, 112 (1900), p. 215.
374 NZPD, 112 (1900), p. 215; *Evening Star*, 27 July 1900, p. 1.
375 *Southland Times*, 27 July 1900, p. 3.
376 *Auckland Star*, 28 May 1900, p. 4; *Auckland Star*, 29 May 1900, p. 2.
377 *Auckland Star*, 28 May 1900, p. 4; *Auckland Star*, 29 May 1900, p. 2.
378 *Auckland Star*, 15 June 1900, p. 2.
379 *Auckland Star*, 27 May 1899, p. 2; *Auckland Star*, 8 June 1900, p. 6; *Hawke's Bay Herald*, 6 October 1899, p. 3.
380 Auckland Peace Association Minutes of Committee Meetings, 1899–1906, 26 October 1899, ATL MS-Papers-2530; *Evening Post*, 5 October 1899, p. 5; *Poverty Bay Herald*, 5 October 1899, p. 3; *West Coast Times*, 5 October 1899, p. 3; *Advertiser*, Adelaide, Australia, 6 October 1899, p. 5.
381 Auckland Peace Association Minutes of Committee Meetings, 1899–1906, 26 October 1899, ATL MS-Papers-2530; *Auckland Star*, 5 September 1899, p. 8; *New Zealand Herald*, 22 June 1900, p. 3.
382 Auckland Peace Association Minutes of Committee Meetings, 1899–1906, 26 October 1899, ATL MS-Papers-2530; *Auckland Star*, 8 June 1900, p. 6.
383 *Auckland Star*, 8 June 1900, p. 6.
384 *New Zealand Herald*, 12 June 1900, p. 4.
385 *New Zealand Herald*, 12 June 1900, p. 4.
386 *Evening Post*, 28 February 1902, p. 4.
387 *Evening Star*, 4 March 1902, p. 4.
388 *Free Lance*, 12 April 1902, p. 22.
389 *Free Lance*, 15 March 1902, p. 22.
390 *Daily Telegraph*, 7 March 1900, p. 5.

391 *Evening Post*, 1 September 1899, p. 5; *Evening Post*, 22 April 1911, p. 9; William Cowper Smith, 6 February 1891, ACGS 16211 J1 489/cg 1891/141, ANZ; William Cowper Smith, 2 February 1892, 16211 J1 489/cg 1892/80, ANZ; S. Menzies, ACGS 16211 J1 669/x 1901/1325, ANZ.

392 *Evening Post*, 7 August 1899, p. 6; *Evening Post*, 9 August 1899, p. 4; *Evening Post*, 16 August 1899, p. 7; *Evening Post*, 25 August 1899, p. 7.

393 *Evening Post*, 16 August 1899, p. 7.

394 *Evening Post*, 10 August 1899, p. 4.

395 *Press*, 23 November 1899, p. 6.

396 Vilhelm Jensen to Francis Henry Dillon Bell, 2 April 1901, ACGO 8333 IA1 821/[34] 1901/1090, ANZ; Evidence of Harold Eastcott Seccombe, 5 April 1901, ACGO 8333 IA1 821/[34] 1901/1090, ANZ.

397 Evidence of Harold Eastcott Seccombe, 5 April 1901, ACGO 8333 IA1 821/[34] 1901/1090, ANZ; *Evening Post*, 19 April 1901, p. 6; Norman Smith to Commandant of New Zealand Forces, 5 April 1901, ACGO 8333 IA1 821/[34] 1901/1090, ANZ.

398 Francis Henry Dillon Bell to Colonial Secretary, 16 April 1901, ACGO 8333 IA1 821/[34] 1901/1090, ANZ; *London Gazette*, 14 December 1900, p. 8457.

399 *Evening Post*, 19 April 1901, p. 6; Joseph Ward to Francis Henry Dillon Bell, 6 April 1901, ACGO 8333 IA1 821/[34] 1901/1090, ANZ.

400 R. J. Seddon to Colonial Secretary, 6 April 1901, ACGO 8333 IA1 821/[34] 1901/1090, ANZ.

401 *Evening Post*, 16 August 1899, p. 7; *Evening Post*, 5 October 1901, p. 7.

402 *Wanganui Chronicle*, 25 November 1903, p. 7.

403 *Wanganui Chronicle*, 25 September 1899, p. 1; *Wanganui Chronicle*, 25 November 1903, p. 7.

404 *Wanganui Chronicle*, 25 September 1899, p. 1; *Wanganui Chronicle*, 9 October 1899, p. 2.

405 *Wanganui Chronicle*, 28 October 1899, p. 2.

406 *Daily Telegraph*, 8 March 1900, p. 5.

407 *Press*, 27 March 1900, p. 4; Johnson, 'Sons of Empire', p. 47.

408 *Feilding Star*, 25 January 1900, p. 2; *Evening Post*, 27 January 1900, p. 5.

409 *Manawatu Evening Standard*, 19 January 1900, p. 2; Johnson, 'Sons of Empire', p. 47.

410 New Zealand Ensign Act 1901 (1 EDW VII 1901 No 74).

411 *New Zealand Herald*, 28 November 1901, p. 6.

412 *New Zealand Herald*, 28 November 1901, p. 6.

413 *New Zealand Herald*, 28 November 1901, p. 6.

414 *AJHR*, 1902, E-1, p. 15; *New Zealand Herald*, 16 January 1902, p. 7.

415 *New Zealand Herald*, 16 January 1902, p. 7.

416 *Press*, 17 January 1902, p. 3; *Auckland Star*, 17 January 1902, p. 3.

417 *New Zealand Herald*, 16 January 1902, p. 7; *New Zealand Herald*, 2 December 1901, p. 5.

418 *New Zealand Herald*, 28 November 1901, p. 6.

419 *NZPD*, 116 (1901), pp. 283, 303.

420 State-School Children Compulsory Drill Bill 1900 (60-1).

421 State-School Children Compulsory Drill Bill 1900 (60-1).

422 *Marlborough Express*, 16 October 1901, p. 2.

423 *NZPD*, 116 (1901), p. 283; *Hastings Standard*, 12 July 1901, p. 3.

424 *Press*, 19 July 1901, p. 3.

425 *Otago Daily Times*, 31 October 1900, p. 5.

426 *Otago Daily Times*, 2 November 1900, p. 6.

427 *Otago Daily Times*, 12 November 1900, p. 3.

428 *Otago Daily Times*, 12 November 1900, p. 3.

429 *Otago Daily Times*, 8 November 1900, p. 6.

430 *Otago Daily Times*, 8 November 1900, p. 6.

431 *Auckland Star*, 27 October 1899, p. 2.

432 *Auckland Star*, 28 October – 28 December 1899, passim; *Observer*, 9 December 1899 – 27 January 1900, passim.

433 *Observer*, 14 April 1900, p. 20.

434 *Observer*, 23 December 1899, p. 10; *Observer*, 30 December 1899, p. 2; *Observer*, 27 January 1900, p. 20.

435 *Observer*, 25 November 1899, p. 16.

436 *West Australian Sunday Times*, (Aus.), 24 December 1899, p. 1. *The Observer* reported that Bewicke was also mentioned in the Sydney *Truth*. *Observer*, 30 December 1899, p. 8.

437 *Temuka Leader*, 6 February 1900, p. 3.
438 *Nelson Evening Mail*, 28 July 1900, p. 3; *NZPD*, 112 (1900), p. 216.
439 Keith Sinclair, *William Pember Reeves: New Zealand Fabian*, Oxford: Clarendon Press, 1965, p. 258. Sinclair notes that while Reeves saw the empire as 'a force for good in the world', he was ambivalent towards the monarchy.
440 *Evening Star*, 8 January 1900, p. 3.
441 *NZPD*, 111 (1900), p. 210; *Otago Daily Times*, 28 February 1900, p. 2.
442 *Daily Telegraph*, 9 March 1900, p. 8.
443 *NZPD*, 111 (1900), p. 210.
444 *NZPD*, 111 (1900), p. 210.
445 *Tuapeka Times*, 21 July 1900, p. 2.
446 *Otago Daily Times*, 24 November 1900, p. 2.
447 *Otago Witness*, 26 July 1900, p. 30.
448 *New York Times*, New York, 26 November 1899, p. 21.
449 *New York Times*, New York, 26 November 1899, p. 21.
450 *AJHR*, 1900, H-29, p. 2.
451 *NZPD*, 111 (1900), p. 159.
452 *AJHR*, 1900, I-8, pp. 2–3.
453 *NZPD*, 112 (1900), pp. 12–50.
454 Johnson, pp. 50–51.
455 *NZPD*, 110 (1899), p. 96; *NZPD*, 112 (1900), p. 53.
456 *NZPD*, 112 (1900), p. 23.
457 *NZPD*, 112 (1900), p. 53.
458 *New York Times*, New York, 10 June 1900, p. 3.
459 *New York Times*, New York, 2 September 1900, p. 15.
460 *New York Times*, New York, 2 September 1900, p. 15; *Waikato Argus*, 16 April 1912, p. 3.
461 *Manawatu Herald*, 20 March 1900, p. 2; *Grey River Argus*, 23 March 1900, p. 2; *Evening Post*, 11 January 1902, p. 5.
462 *New York Times*, New York, 2 September 1900, p. 15.
463 *Timaru Herald*, 10 May 1901, p. 3.
464 *Ashburton Guardian*, 11 February 1902, p. 2.
465 *Observer*, 8 September 1900, p. 22.
466 *New Zealand Herald*, 25 October 1900, p. 4.

Chapter Three: 'An especially fine lot of fighting-men'

1 *AJHR*, 1900, H-6A, p. 2; *Evening Post*, 17 January 1901, p. 5; *AJHR*, 1900, H-6G, p. 1; *Thames Star*, 17 December 1901, p. 4.
2 Sir Ian Hamilton to William Hall-Jones, 30 January 1901, ATL MS-Papers-5755-59.
3 *AJHR*, 1900, H-6A, p. 7; *Manawatu Evening Standard*, 23 February 1900, p. 2.
4 *Evening Post*, 22 February 1900, p. 6.
5 *Times* (London), 14 February 1900, p. 9; *Times* (London), 5 May. 1900, p. 15.
6 *Evening Post*, 18 January 1900, p. 5.
7 *Times* (London), 14 February 1900, p. 12.
8 *Times* (London), 17 January 1900, p. 3; *Times* (London), 14 February 1900, p. 9.
9 *Waikato Argus*, 13 July 1901, p. 3, *AJHR*, 1896, H-5, [p. 1].
10 *New Zealand Herald*, 15 March 1900, p. 6; *Otago Witness*, 1 March 1900, p. 12.
11 *Otago Daily Times*, 22 February 1900, p. 2; *Otago Witness*, 1 March 1900, p. 12.
12 *Otago Daily Times*, 22 February 1900, p. 2; *Otago Witness*, 1 March 1900, p. 12.
13 *Evening Post*, 18 January 1900, p. 6.
14 Arthur Pole Penton to Hugh Gourley, 19 January 1900, Samuel Gourley, AABK 18805 W5515 0002143, ANZ.
15 *Evening Post*, 18 January 1900, p. 5; *Otago Daily Times*, 6 February 1900, p. 5.
16 *NZPD*, 1900, Vol. 111, p. 18.
17 *Evening Post*, 18 January 1900, p. 5; *Times* (London), 18 January 1900, p. 5.
18 *Ohinemuri Gazette*, 21 March 1900, p. 3.
19 *Otago Witness*, 10 September 1902, p. 41.
20 *New Zealand Herald*, 28 February 1910, p. 12.
21 *Tuapeka Times*, 7 April 1900, p 3; *Evening Post*, 3 August 1900, p. 2; Prosper Raine Berland to Robin, n.d., Prosper Rain Berland, AABK 18805 W5515 0000359, ANZ.
22 Prosper Raine Berland to Robin, n.d., Prosper Rain Berland, AABK 18805 W5515 0000359, ANZ.
23 Prosper Raine Berland to Robin, n.d., Prosper Rain Berland, AABK 18805 W5515 0000359, ANZ.
24 *Marlborough Express*, 23 August 1900, p. 3; *AJHR*, 1903, H-6A, pp. 5, 11.
25 *Times* (London), 17 September 1900, p. 9.

26 *AJHR*, 1901 H-6E, p. 2.
27 Summary of data regarding Private M. Canavan, 4th Contingent, n.d., Michael Canavan, AABK 18805 W5515 0000921, ANZ; Proceedings of a Medical Board, 15 October 1900, Robert Walter Gordon Collins, AABK 18805 W5515 0006267, ANZ; Medical History, n.d., William Henry Vinsen, AABK 18805 W5515 0005736, ANZ.
28 *NZPD*, 1900, Vol. 113, p. 389; *Otago Witness*, 26 September 1900, pp. 25–26.
29 *Wanganui Collegian*, No. 54, December 1900, p. 14.
30 *Wanganui Collegian*, No. 55, April 1901, p. 23.
31 *Auckland Star*, 15 February 1901, p. 8; *Wanganui Collegian*, No. 55, April 1901, p. 23.
32 *Wanganui Collegian*, No. 55, April 1901, p. 23.
33 *AJHR*, 1903, H-6A, pp. 2, 4, 6–7.
34 Proceedings of the Medical Board, 10 March 1901, Albert Marr Beath, AABK 18805 W5515 0000308, ANZ; Proceedings of the Medical Board, 10 March 1909, Albert Marr Beath, AABK 18805 W5515 0000308, ANZ; Edward Walter Clervaux Chaytor to General Officer Commanding N.Z. Forces, 14 February 1905, Sidney Charles Godfray, AABK 18805 W5515 0006325, ANZ.
35 *Star*, 10 January 1901, p. 1.
36 *Dominion*, 24 February 1910, p. 5.
37 Proceedings of an Enquiry by Col. Porter, Officer Commanding VII N. Z. Regiment, 19 February 1902, James Kirkwood, AABK 18805 W5515 0003054, ANZ.
38 Proceedings of an Enquiry by Col. Porter, Officer Commanding VII N. Z. Regiment, 19 February 1902, James Kirkwood, ABK 18805 W5515 0003054, ANZ.
39 *Manawatu Evening Standard*, 3 June 1902, p. 3; *New Zealand Times*, 28 February 1902, p. 5.
40 *AJHR*, 1903, H-6A, p. 11.
41 *Times* (London), 27 February 1902, p. 5.
42 *Times* (London), 24 February 1902, p. 5; *Times* (London), 1 March 1902, p. 11; Kenneth Gordon Malcolm, *The Seventh N.Z. Contingent: Its Record on the Field*, Wellington: Deslandes & Lewis (Printers), 1903, p. 39.
43 *Evening Post*, 14 April 1902, p. 5.
44 *Dominion*, 24 February 1910, p. 5.
45 *Auckland Star*, 18 April 1902, p. 5.
46 *Times* (London), 27 February 1902, p. 5; *Times* (London), 28 February 1902, p. 3: *West Australian*, Perth, 1 March 1902, p. 7; *Maitland Daily Mercury* (Aus.), 14 March 1902, p. 3; *Register*, Adelaide, 1 March 1902, p. 7.
47 Statement of Case of Lieutenant C. O. Phair, 6 August 1902, Charles Oakshot Phair, AABK 18805 W5515 0004472, ANZ; *New Zealand Herald*, 19 December 1902, p. 6; *Otago Witness*, 11 March 1903, p. 35.
48 *Feilding Star*, 19 April 1902, p. 2.
49 *Times* (London), 28 February 1902, p. 3; *Times* (London), 1 March 1902, p. 11.
50 *Auckland Star*, 6 May 1902, p. 1; *Otago Witness*, 5 March 1902, p. 27.
51 *Southland Times*, 14 April 1902, p. 3.
52 *Times* (London), 27 February 1902, p. 5; *Times* (London), 28 February 1902, p. 3; *Evening Post*, 3 July 1903, p. 4.
53 *Press*, 25 April 1902, p. 5; *Dominion*, 24 February 1910, p. 5; Proceedings of Medical Board, 14 June 1902, Charles Frederick Normanby Minifie, AABK 18805 W5515 0003917, ANZ.
54 *Auckland Star*, 26 April 1902, p. 5.
55 *Auckland Star*, 18 April 1902, p. 5.
56 *AJHR*, 1903, H-6A, pp. 4–5, 9, 11; Malcolm, pp. 40, 41.
57 *Press*, 25 April 1902, p. 5.
58 History-Sheet, Lytton Alphonse Ditely; Description of Lytton Alphonse Ditely on Enlistment, 15 August 1914, AABK 18805 W5515 0001484, ANZ.
59 *Otago Daily Times*, 3 May 1902, p. 3.
60 Proceedings of the Medical Board, 9 February 1904; Kenneth Gordon Malcolm to James Allen, 22 February 1916, Kenneth Gordon Malcolm, AABK 18805 W5515 0003695, ANZ.
61 Attestation-Form, 3 April 1901, Duncan Anderson, AABK 18805 W5515 0000082, ANZ; Attestation-Form, 3 April 1901, Percy Nation, 18805 W5515 0004139, ANZ; *Hawera & Normanby Star*, 13 March 1902, p. 2.
62 S. A. Field Force: Nominal Roll of Officers,

63. Warrant Officers, Non-Commissioned Officers and Men Admitted into or Discharged from the No. 1 General Hospital Wynberg, 11 July 1902, Stapylton Cotton Caulton; James C. Collins to Under Secretary for Defence, 4 November 1903, AABK 18805 W5515 0000878, ANZ.
63. Richard John Seddon to N. Z. Agent General, London, 19 September 1904, Stapylton Cotton Caulton; Proceedings of the Medical Board, 27 February 1908, AABK 18805 W5515 0000878, ANZ; NZBDM, Guy Stapylton Caulton, 1867/19012; Florence Isabella Caulton, 1871/22251; Alexander Stapylton Caulton, 1872/19043; Ada Emily Caulton,1874/42240; Lena Eliza Stapylton Caulton, 1877/9711; Hilda Rose Caulton, 1882/19199.
64. N. Z. War Medal Roll, pt. 1, Roll A, 1871; *New Zealand Gazette*, No. 40, 29 July 1886, p. 889.
65. *Times* (London), 27 February 1902, p. 5; *Times* (London), 1 March 1902, p. 11; *Sydney Morning Herald*, 28 April 1902, p. 7; *Albany Advertiser* (Aus.), 29 April 1902, p. 3; *Kalgoorlie Western Argus*, 29 April 1902, p. 37; *Barrier Miner* (Aus.), 28 April 1902, p. 2.
66. Secretary of State for the Colonies to Lord Ranfurly, received 19 March 1902, Telegrams to and from the Secretary of State — 2 January 1902 – 11 April 1903, ACHK 16561, G5 box 6, ANZ.
67. *AJHR*, 1903, E-1B, p. 42.
68. *Southland Times*, 3 March 1902, p. 2.
69. *AJHR*, 1903, H-6A, pp. 10, 12; *Cromwell Argus*, 22 April 1902, p. 5.
70. *Press*, 16 April 1902, p. 7.
71. *Press*, 20 May 1902, p. 5.
72. *Southland Times*, 22 May 1902, p. 2.
73. *Southland Times*, 22 May 1902, p. 2.
74. *Hawera & Normanby Star*, 19 February 1900, p. 2.
75. *Supplement to the London Gazette*, 29 November 1917, p. 2356; Casualty Form – Active Service, Duncan Barrie Blair, AABK 18805 W5515 0000415, ANZ.
76. *Wanganui Collegian*, No. 54, December 1900, p. 13.
77. *Wanganui Collegian*, No. 54, December 1900, p. 14.
78. *AJHR*, 1903, H-6A, pp. 7–8; *Wanganui Collegian*, No. 54, December 1900, p. 14.
79. *Wanganui Collegian*, No. 54, December 1900, p. 14; Extract from Casualty Roll Third Contingent, John Joseph Heasley, AABK 18805 W5515 0002462, ANZ; *London Gazette*, 7 May 1901, p. 3114.
80. *Wanganui Collegian*, No. 54, December 1900, p. 14; Proceedings of Medical Board, 15 March 1901, Robert Hooper Aldworth, AABK 18805 W5515 0000038, ANZ.
81. *AJHR*, 1903, H-6A, p. 7.
82. William Hall-Jones to Richard J. Seddon, telegram, 12 March 1902, p.1, ATL MS-Papers-5755-57.
83. William Hall-Jones to Richard J. Seddon, telegram, 12 March 1902, pp.1–2, ATL MS-Papers-5755-57.
84. *Otago Witness*, 23 April 1902, pp. 37, 40, 41.
85. *Otago Daily Times*, 7 April 1900, p. 6.
86. *Times* (London), 28 February 1902, p. 3.
87. *Nelson Evening Mail*, 15 March 1902, p. 4.
88. *Nelson Evening Mail*, 18 March 1902, p. 2.
89. *New Zealand Herald*, 18 March 1902, p. 5; *Auckland Star*, 18 March 1902, p. 1.
90. *Otago Witness*, 4 January 1900, p. 27; *Otago Witness*, 25 January 1900, p. 26.
91. *AJHR*, 1900, H-6A, p. 12.
92. *Press*, 13 August 1900, p. 2.
93. *NZPD*, 1900, Vol. 112, pp. 145–46; *AJHR*, 1899, H-6, p. 3.
94. *Wanganui Collegian*, No. 53, August 1900, p. 12.
95. *NZPD*, 1900, Vol. 112, p. 146.
96. *Wanganui Collegian*, No. 53, August 1900, p. 16.
97. *NZPD*, 1900, Vol. 112, p. 56.
98. *NZPD*, 1900, Vol. 112, p. 145.
99. Frederick Harcourt to Richard J. Seddon, 22 October 1900, Frederick Harcourt, AABK 18805 W5515 0002336, ANZ.
100. *Mataura Ensign*, 5 April 1900, p. 2.
101. Luke Perham to Theresa Perham, 28 June 1900, NAM NZ 2003.7.
102. *Wanganui Collegian*, No. 53, August 1900, p. 17.
103. *Press*, 17 October 1900, p. 5.
104. *West Coast Times*, 21 August 1900, p. 4.

105 *Grey River Argus*, 24 August 1900, p. 4.
106 Alfred Robin to Arthur Pole Penton, n.d., William Alexander Leslie, AABK 18805 W5515 0003197, ANZ.
107 *Grey River Argus*, 8 September 1900, p. 3.
108 *Ashburton Guardian*, 18 January 1901, p. 2; *Thames Star*, 18 January 1901, p. 2; *North Otago Times*, 19 January 1901, p. 1.
109 *Ashburton Guardian*, 6 March 1901, p. 3; *Star*, 22 May 1899, p. 3.
110 *Otago Witness*, 22 March 1900, p. 34.
111 *Evening Post*, 26 April 1901, p. 5.
112 *Evening Post*, 2 April 1901, p. 8.
113 *Press*, 20 April 1901, p. 1; *Timaru Herald*, 22 April 1901, p. 1; *Evening Post*, 25 April 1901, p. 1.
114 *Star*, 22 April 1901, p. 3.
115 *Evening Post*, 26 April 1901, p. 5.
116 *Evening Post*, 26 April 1901, p. 5.
117 *Free Lance*, 4 May 1901, p. 3.
118 *Evening Post*, 26 April 1901, p. 4.
119 *New Zealand Herald*, 27 April 1901, p. 5; *Evening Post*, 26 April 1901, p. 4.
120 *Thames Star*, 9 May 1901, p. 3.
121 *New Zealand Herald*, 27 April 1901, p. 5.
122 *Star*, 9 August 1901, p. 1.
123 *Otago Daily Times*, 10 August 1901, p. 8.
124 Mrs Blanche Ethel Maude Hughes, Application to land in South Africa, 29 January 1902, ACGO 8333 IA1/841/[1], 1902/300, ANZ.
125 History Sheet, Frederick Boulton Hughes, AABK 18805 W5541 31/ 0057546, ANZ.
126 *Colonist*, 19 January 1900, p. 2.
127 *AJHR*, 1902, H-6A, pp.1–8, 10, 13.
128 *Press*, 4 January 1900, p. 6.
129 Richard John Seddon to Albert Andrew C456 773 SEDDON1 3 8/47, ANZ.
130 Thomas Jowsey to Commandant New Zealand Forces, 24 March 1900, AD1 354, D1900/1980, ANZ.
131 Diary of Major Thomas Jowsey, 20 February 1900, AD1 354, D1900/1980, ANZ.
132 Diary of Major Thomas Jowsey, 19–20 February 1900, AD1 354, D1900/1980, ANZ.
133 Third Contingent Diary (Major Jowsey), February 1900 – May 1901, 3, 11–12 March 1900, AD1 345, 8020, ANZ.
134 Third Contingent Diary (Major Jowsey), February 1900 – May 1901, 13 March 1900, AD1 345, 8020, ANZ.
135 Third Contingent Diary (Major Jowsey), February 1900 – May 1901, 10 March 1900, AD1 345, 8020, ANZ.
136 Third Contingent Diary (Major Jowsey), February 1900 – May 1901, 6, 17 March 1900, AD1 345, 8020, ANZ.
137 Thomas Jowsey to Commandant New Zealand Forces, 24 March 1900, AD1 354, D1900/1980, ANZ.
138 *Wanganui Collegian*, No. 53, August 1900, p. 16.
139 *Wanganui Collegian*, No. 53, August 1900, p. 16.
140 *Wanganui Collegian*, No. 52, April 1900, p. 12.
141 *Wanganui Collegian*, No. 54, December 1900, p. 13.
142 *Wanganui Collegian*, No. 53, August 1900, p. 15.
143 Extract from 1900/4762, AD1 380 cf D1901/2335, ANZ.
144 Frank Perham, diary, 21 March 1901, NAM NZ 2000.736. Perham incorrectly gives the date as 'Wednesday 21st March' in his 1901 diary. In 1901, 21 March was a Thursday. Presumably the incident he recorded occurred on 20 March 1901; *Wanganui Collegian*, No. 60, December 1902, p. 12.
145 *Clutha Leader*, 19 January 1900, p. 5.
146 *Clutha Leader*, 19 January 1900, p. 5.
147 Attestation Form, n.d., Hugh Smith, AABK 18805 W5515 0005177, ANZ.
148 Attestation Form, 3 April 1901, Charles M. Baré, AABK 18805 W5515 0000234, ANZ; *Otago Witness*, 24 May 1900, p. 35.
149 *New Zealand Herald*, 28 February 1898, p. 5; Roll of individuals entitled to the South African Medal and clasps. 1st Imperial Light Horse, n.d., National Archives, Kew, UK, WO 100/250, p. 69; Roll of individuals entitled to the South African Medal and clasps. 1st Imperial Light Horse, 10 August 1904, National Archives, Kew, UK, WO 100/250, p. 204; Roll of individuals entitled to the South African Medal and clasps. South African Light Horse, 15 July 1901, National Archives, Kew, UK, WO 100/274, p. 90; *Otago Witness*, 24 May 1900, p. 35.

150 *Timaru Herald*, 9 July 1900, p. 3.
151 *Otago Witness*, 14 August 1901, p. 27.
152 *Otago Witness*, 14 August 1901, p. 27; Mrs. J. O'Hagan to Joseph Ward, 18 February 1908, AAYS 8636 AD1/518/bn, D1908/594, ANZ.
153 *Timaru Herald*, 12 August 1901, p. 2.
154 *New Zealand Herald*, 8 August 1901, p. 5.
155 *AJHR*, 1901, H-6, pp. 1–14; Attestation Form, 11 March 1900, William Thomson, AABK 18805 W5515 0005583, ANZ; Attestation Form, 26 February 1902, William Thomson, AABK 18805 W5515 0005583, ANZ.
156 *AJHR*, 1902, H-6A, pp. 2, 4, 5, 9, 11.
157 *AJHR*, 1902, H-6A, p. 11.
158 Attestation Form, 14 March 1900, Leo Sam Matthews, AABK 18805 W5573 0356312, ANZ.
159 *Free Lance*, 11 August 1900, p. 16; *AJHR*, 1903, H-6A, p. 5.
160 *Wanganui Collegian*, No. 60, December 1902, p. 12.
161 *Clutha Leader*, 19 January 1900, p. 5.
162 *Wanganui Herald*, 8 February 1900, p. 2.
163 *AJHR*, 1905, H-19A, p. 3.
164 *Evening Post*, 2 January 1900, p. 4.
165 *NZPD*, 1899, Vol. 110, p. 185.
166 *Evening Post*, 17 January 1901, p. 5.
167 *NZPD*, 1899, Vol. 110, pp. 892–93.
168 *NZPD*, 1899, Vol. 110, pp. 892–93.
169 *NZPD*, 1900, Vol. 111, p. 366.
170 *NZPD*, 1900, Vol. 111, p. 366.
171 *NZPD*, 1900, Vol. 111, p. 366.
172 *NZPD*, 1900, Vol. 111, p. 366.
173 *AJHR*, 1900, H-6A, p. 1.
174 *AJHR*, 1900, H-6A, p. 4.
175 *AJHR*, 1900, H-6A, p. 3.
176 *AJHR*, 1900, H-6A, pp. 3–4.
177 Sixth Contingent, Crime and Offence Reports, March 1901 – March 1902, AD34 7 7 8032, ANZ.
178 Captain Fred W. Abbott, diary, 23 October 1901, NAM NZ 1997.1841; *AJHR*, 1903, H-6A, p. 9.
179 Captain Fred W. Abbott, diary, 23 October 1901, NAM NZ 1997.1841; Evidence of Trooper A. J. Whitney, Proceedings of Regimental Enquiry, 23 October 1901, William James Byrne, AABK 18805 W5515 0000776, ANZ.
180 Evidence of Trooper A. J. Whitney, Proceedings of Regimental Enquiry, 23 October 1901, William James Byrne, AABK 18805 W5515 0000776, ANZ.
181 *New Zealand Herald*, 19 December 1901, p. 5.
182 Attestation Form, 3 April 1901, Alfred John Whitney, AABK 18805 W5515 0005970, ANZ.
183 Evidence of Captain Dickinson, William James Byrne, AABK 18805 W5515 0000776, ANZ; *AJHR*, 1903, H-6A, p. 9.
184 *Pelorus Guardian and Miners' Advocate*, 18 April 1902, p. 4.
185 *Pelorus Guardian and Miners' Advocate*, 18 April 1902, p. 4.
186 6th N.Z.M.R. Casualty Return, 2 October 1901, William Frederick Raynes, AABK 18805 W5515 0004655, ANZ.
187 *Evening Post*, 20 November 1901, p. 6; *New Zealand Herald*, 18 November 1901, p. 5.
188 Finding of Lieutenant-Colonel Porter, Proceedings of Regimental Enquiry, 23 October 1901, William James Byrne, AABK 18805 W5515 0000776, ANZ.
189 Captain Frederick W. Abbott, diary, 24 October 1901, NAM NZ 1997.1841; Medical Report, 27 January 1903, AABK 18805 W5515 0001263, ANZ.
190 *AJHR*, 1901, H-6C, p. 2; Copy of cablegram received by His Excellency the Governor from Casualty, Cape Town, 29 August 1901, AABK 18805 W5515 0005017, ANZ.
191 *AJHR*, 1901, H-6C, p. 2; Proceedings on Re-examination of a Recruit, 31 August 1917, Henry Houchen, AABK 18805 W5515 0002660, ANZ.
192 Herbert Ernest Hart, diary, 5 June 1902, NAM NZ 1990.1024; Proceedings of a Court of Enquiry, 24 May 1902, Horace Baker, AABK 18805 W5515 0000215, ANZ.
193 Coroner's Inquest into the death of Percy Mathew Crawford, 20 January 1902, J46 Box 235 COR1902/76, ANZ; *Wanganui Herald*, 20 January 1902, p. 2.
194 Evidence of Edward Fisher and J. E. Barltrop, Coroner's Inquest into the death of Percy Mathew Crawford, 20 January 1902, J46 Box 235 COR1902/76, pp. 6–7, ANZ.
195 *Hawke's Bay Herald*, 27 January 1900, p. 2;

Notes

Hawke's Bay Herald, 24 February 1900, p. 3; *Hawke's Bay Herald*, 13 April 1900, p. 3; *Hawke's Bay Herald*, 30 March 1900, p. 2.
196 *Wanganui Collegian*, No. 54, December 1900, p. 16.
197 *Wanganui Collegian*, No. 54, December 1900, p. 16.
198 *New Zealand Herald*, 20 January 1900, pp. 5–6.
199 *Hawke's Bay Herald*, 8 February 1900, p. 3.
200 *AJHR*, 1903, H-6A, p. 10.
201 *Otago Daily Times*, 30 January 1900, p. 5.
202 *NZPD*, 1902, Vol. 121, pp. 368–69.
203 *AJHR*, 1901, H-6B, p. 3; Sydney Charles A'Court to Defence Department, 18 December 1901, Sydney Charles A'Court, AABK 18805 W5515 0000005, ANZ.
204 Sydney Charles A'Court to Defence Department, 18 December 1901, Sydney Charles A'Court, AABK 18805 W5515 0000005, ANZ.
205 Sydney Charles A'Court to Arthur Douglas, 23 December 1901, Sydney Charles A'Court, AABK 18805 W5515 0000005, ANZ.
206 *Press*, 20 August 1902, p. 7.
207 *AJHR*, 1903, H-6A, p. 14.
208 F. Perham to Under-Secretary for Defence, 30 January 1902, AABK 18805 W5515 0004448, ANZ; F. W. Bezar to Under-Secretary for Defence, 14 May 1902, AABK 18805 W5515 0000379, ANZ; John Joseph O'Reilly to Under-Secretary for Defence, 26 October 1902, AABK 18805 W5515 0004283, ANZ; Annotation by Defence Storekeeper J. O'Sullivan dated 14 November 1902 on John Joseph O'Reilly to Under-Secretary for Defence, 26 October 1902, AABK 18805 W5515 0004283, ANZ.
209 Archibald Craig Hutton to Richard John Seddon, 27 September 1901, Archibald Craig Hutton AABK 18805 W5515 0002739, ANZ.
210 Arthur Douglas to Archibald Craig Hutton, 17 October 1901, Archibald Craig Hutton 18805 W5515 0002739, ANZ.
211 Frank Perham to Arthur Douglas, 30 January 1902, Frank Perham, AABK 18805 W5515 0004448, ANZ.
212 Charles Edward Nurse to Arthur Douglas, 19 April 1902, Charles Edward Nurse, AABK 18805 W5515 0004226, ANZ.
213 *NZPD*, 1900, Vol. 115, p. 548.
214 *AJHR*, 1902, H-6A, p. 15; Attestation Form, 20 March 1902, Robert Hall Bakewell, AABK 18805 W5515 0000220, ANZ.
215 *Auckland Star*, supplement, 11 May 1904, p. 2.
216 *Manawatu Evening Standard*, 26 March 1902, p. 3.
217 *Auckland Star*, supplement, 11 May 1904, p. 2; Proceedings of the Medical Board, 6 February 1904, Robert Hall Bakewell; Attorney General, Proposed Grant of Military Pension to R. H. Bakewell, 10 March 1904, Robert Hall Bakewell, AABK 18805 W5515 0000220, ANZ.
218 *Auckland Star*, supplement, 11 May 1904, p. 2.
219 *Auckland Star*, supplement, 11 May 1904, p. 2.
220 *Wanganui Herald*, 14 October 1903, p. 2.
221 *Otago Witness*, 26 December 1900, p. 31; John C. Adams to Agent-General for New Zealand, 23 October 1900, Hugh McDonagh, AABK 18805 W5515 0003382, ANZ.
222 *Otago Witness*, 26 December 1900, p. 31; John C. Adams to Agent-General for New Zealand, 23 October 1900, Hugh McDonagh, AABK 18805 W5515 0003382, ANZ.
223 John C. Adams to Agent-General for New Zealand, 23 October 1900, Hugh McDonagh, AABK 18805 W5515 0003382, ANZ.
224 *Auckland Star*, 16 February 1900, p. 2.
225 Record of Military Service, James Nathaniel Hamer, 12 February 1901, James Nathaniel Hamer, AABK 18805 W5515 0002307, ANZ.
226 Record of Military Service, James Nathaniel Hamer, 12 February 1901, James Nathaniel Hamer, AABK 18805 W5515 0002307, ANZ; James Nathaniel Hamer to William Pember Reeves, 23 February 1905, James Nathaniel Hamer, AABK 18805 W5515 0002307, ANZ.
227 E. N. Senior to James Nathaniel Hamer, 4 February 1901, James Nathaniel Hamer, AABK 18805 W5515 0002307, ANZ.
228 James Nathaniel Hamer to William Pember Reeves, 23 February 1905, James Nathaniel Hamer, AABK 18805 W5515 0002307, ANZ.
229 R. H. Brade to William Pember Reeves, 11 March 1905, James Nathaniel Hamer; William Pember Reeves to Richard J.

Seddon, 14 March 1905, AABK 18805 W5515 0002307, ANZ.

230 Attestation-Form, William Edward Mahood, 21 October 1899; William Edward Mahood to Lieutenant-Colonel White, 9 June 1901; Certificate of Discharge, 1 August 1902, William Edward Mahood, AABK 18805 W5515 0003689, ANZ; Service of Major H. H. Browne, M.B.E. in the New Zealand Permanent Force, AABK 18805 W5515 0000648, ANZ.

231 Henry Harwood Browne to Lieutenant-Colonel Gordon, 2 July 1900, AABK 18805 W5515 0000648, ANZ.

232 Record of Active Service in Campaigns Prior to August, 1914, Henry Harwood Browne; Statement of Military Service, n.d. AABK 18805 W5515 0000648, ANZ.

233 *New Zealand Herald*, 30 July 1943, p. 5; *Press*, 12 February 1932, p. 11; *Evening Post*, 29 July 1943, p. 7.

234 Attestation Form, Arthur Robert Johnstone Dewar, 31 March 1900, AABK 18805 W5515 0001466, ANZ; *Hastings Standard*, 25 May 1897, p 2.

235 *Nelson Evening Mail*, 2 June 1897, p. 3; *New Zealand Herald*, 13 April 1896, p. 6; *Colonist*, 20 June 1900, p. 2; Nominal Roll of the Jameson Raiders, National Archives, Kew, UK, CO 179/193.

236 *Nelson Evening Mail*, 2 June 1897, p. 3; *Press*, 25 April 1900. p. 5; *Wanganui Chronicle*, 26 October 1899, p. 2.

237 *Wanganui Collegian*, No. 52, April 1900, p. 16; *Wanganui Collegian*, No. 59, August 1902, p. 6.

238 *Daily Telegraph*, 17 September 1900, p. 5.

239 Attestation-Form, 28 March 1900, Edward Ward Lascelles, AABK 18805 W5515 0003127, ANZ; *Daily Telegraph*, 17 September 1900, p. 5.

240 Attestation-Form, 31 March 1900, Thomas Errington Tanner, AABK 18805 W5515 0005458, ANZ; *Otago Witness*, 5 April 1900, p. 27.

241 *Otago Witness*, 5 April 1900, p. 27.

242 *Manawatu Evening Standard*, 28 July 1900, p. 2; Abstract, 5th N.Z. Contingent, Thomas Errington Tanner, AABK 18805 W5515 0005458, ANZ.

243 *Daily Telegraph*, 17 September 1900, p. 5.

244 *Feilding Star*, 2 August 1900, p. 2; *Daily Telegraph*, 17 September 1900, p. 5; Attestation-Form, 31 March 1900, Thomas Errington Tanner, AABK 18805 W5515 0005458, ANZ.

245 Certificate of Discharge, n.d., Thomas Errington Tanner, AABK 18805 W5515 0005458, ANZ; Roll of Officers, Non-commissioned Officers, and Men entitled to the Medal for operations in Matabeleland, Rhodesia, 1893, National Archives, Kew, UK, WO 100/77, p. 13.

246 *Otago Witness*, 5 April 1900, p. 27; Roll of Officers, Non-commissioned Officers, and Men entitled to the Medal for operations in Matabeleland, Rhodesia, 1893, National Archives, Kew, UK, WO 100/77, p. 13.

247 *Hawke's Bay Herald*, 1 May 1884, p. 3.

248 *Otago Witness*, 5 April 1900, p. 27; Roll of Officers, Non-commissioned Officers, and Men entitled to the Medal for operations in Matabeleland, Rhodesia, 1893, National Archives, Kew, UK, WO 100/77, p. 13.

249 *Otago Witness*, 5 April 1900, p. 27; Roll of individuals entitled to the South African Medal and clasps: 27 Compy, 7th Bn. Impl. Yeoy, 13 July 1901, WO 100/123, p. 76.

250 *New Zealand Herald*, 8 May 1901, p. 6; *Otago Daily Times*, 15 August 1914, p. 12.

251 N.Z.R. Staff Return, William Lancelot Miles, n.d., Douglas Miles collection; *Press*, 7 July 1899, p. 6; *AJHR*, 1881, H-2, p. 23.

252 *AJHR*, 1893, D-18, p. 1; *AJHR*, 1894, D-15, p. 5; N.Z.R. Staff Return, William Lancelot Miles, n.d., Douglas Miles collection.

253 *Press*, 7 July 1899, p. 6; *Star*, 5 August 1899, p. 6.

254 *Press*, 7 July 1899, p. 6; *Star* 5 August 1899, p. 6; *Auckland Star*, 21 November 1899, p. 2.

255 Form of Passenger List, *Umtata*, 19 October 1899, National Archives, Kew, UK, BT26.

256 Roll of individuals entitled to the South African Medal and clasps: Natal Corps of Guides, 1 April 1902, WO 100/301, p. 28; Roll of individuals entitled to the South African Medal and clasps: Intelligence Department, 11 October 1901, WO 100/301, p. 8; Roll of individuals entitled to the South African Medal and clasps: Colonial Scouts, n.d., WO 100/242, pp. 160, 178.

257 Military Pass, W. L. Miles, Intelligence Department, 29 March 1900; Douglas Miles collection; Permanent Pass, Field Intelligence Department, W. L. Miles, n.d., Douglas Miles collection; Roll of individuals entitled to the South African Medal and clasps: Colonial Scouts, n.d., WO 100/242, pp. 160, 178.

258 *Star*, 25 January 1901, p. 1.

259 *Auckland Star*, 13 November 1931, p. 8.

260 *Auckland Star*, 13 November 1931, p. 8; Roll of individuals entitled to the South African Medal and clasps: Mashonaland Division British South African Police, 22 August 1901, WO 100/238, p. 95; Roll of individuals entitled to the South African Medal and clasps: Rhodesian Coronation Contingent, 30 June 1902, WO 100/267, p. 89.

261 Roll of individuals entitled to the South African Medal and clasps: Rhodesian Coronation Contingent, 30 June 1902, WO 100/267, p. 89; Roll of individuals entitled to the King's South African Medal and clasps: British South African Police, 21 May 1903, WO 100/358, p. 171; *Evening Post*, 26 September 1904, p. 5.

262 *Auckland Star*, 13 November 1931, p. 8.

263 *NZPD*, 1899, Vol. 108, p. 446; *Manawatu Evening Standard*, 27 March 1900, p. 2.

264 *Manawatu Evening Standard*, 27 March 1900, p. 2.

265 Bert Stevens to James Stevens, 27 January 1902, pp. 4–5, NAM NZ 1999.3239.

266 *Evening Post*, 19 March 1900, p. 5.

267 *AJHR*, 1900, H-6, p. 21; *AJHR*, 1902, H-6, p. 2; Attestation-Form, n.d., George Ralph Miller, AABK 18805 W5515 0003911, ANZ; *Press*, 11 April 1900, p. 6; Hugh Gordon Bonar, NZBDM 1882/15595; Archibald James Merle Bonar, NZBDM 1876/8993; Attestation-Form, n.d., Archibald James Merle Bonar, AABK 18805 W5515 0000461, ANZ; Attestation-Form, 5 January 1902, Hugh Gordon Bonar, AABK 18805 W5515 0000462, ANZ.

268 Roll of individuals entitled to the South African Medal and clasps. Peninsula Horse, 25 September 1904, National Archives, Kew, UK, WO 100/263, p.122; Roll of individuals entitled to the South African Medal and clasps. Scott's Railway Guards, January 1903, National Archives, Kew, UK, WO 100/270, n.p.; Roll of individuals entitled to the South African Medal and clasps. Commander in Chief's Body Guard, 9 August 1905, National Archives, Kew, UK, WO 100/243, p. 183.

269 *AJHR*, 1906, Session II, H-14, p. 16; Attestation-Form, 12 February 1900, Bertie William Willis, AABK 18805 W5515 0006049, ANZ; Attestation-Form, 17 March 1902, Ernest Raymond Willis, AABK 18805 W5515 0006051, ANZ.

270 *Evening Post*, 22 November 1900, p. 5; *AJHR*, 1900, H-6, p. 16; *AJHR*, 1906, H-14, p. 13; *AJHR*, 1900, H-26, p. 2.

271 *AJHR*, 1901, H-6D, p. 1; Thomas Francis Grey to George Hector Fitzgerald Rolleston, 23 March 1903, George Hector Fitzgerald Rolleston, AABK 18805 W5515 0004826, ANZ.

272 Roll of individuals entitled to the South African Medal and clasps. Kaffrarian Rifles, 15 July 1901, National Archives, Kew, UK, WO 100/254, p. 24; Francis Marion Bates Fisher to Defence Minister, 12 March 1906, George Henry Fisher, AABK 18805 W5515 0001806, ANZ; *AJHR*, 1902, H-6B, p. 14.

273 *AJHR*, 1903, H-6A, p. 6; Principal Medical Officer to Pilcher, telegram, 28 August 1900, William Charles Colvin, AABK 18805 W5515 0001104, ANZ; *Marlborough Express*, 15 April 1904, p. 1.

274 *AJHR*, 1900, H-6I, p. 5.

275 *New Zealand Times*, 22 July 1901, p. 5; Roll of individuals entitled to the South African Medal and clasps. 1st Kitchener's Fighting Scouts, 20 January 1906, National Archives, Kew, UK, WO 100/256, p. 292; Roll of individuals entitled to the South African Medal and clasps. Army Pay Department (Imp. Yeo. Branch), n.d., National Archives, Kew, UK, WO 100/230, p. 30.

276 *AJHR*, 1906, H-14, p. 12; *Evening Post*, 7 December 1899, p. 2; *AJHR*, 1902, H-6, p. 20.

277 *Press*, 6 March 1901, p. 6.

278 Attestation-Form, 29 January 1902, Robert Heaton Rhodes, AABK 18805 W5515 0004783, ANZ; *Evening Post*, 23 December 1899, p. 5.

279 *Press*, 5 February 1902, p. 8.

280 *Akaroa Mail and Banks Peninsula Advertiser*, 11 February 1902, p. 2.

281 *Press*, 10 September 1902, p. 8.

282 *New Zealand Herald*, 31 October 1931, p. 12; *AJHR*, 1906, H-14, p. 12; *AJHR*, 1900, H-6, p. 36.

283 Richard Seddon to Major Alfred Robin, 12 April 1900, Robert Witheford, AABK 18805 W5901 0006498, ANZ.

284 Attestation-Form, 26 April 1900, Robert Witheford, AABK 18805 W5901 0006498, ANZ.

285 Arthur Pole Penton to Robert Witheford, 29 January 1901, Robert Witheford; Robert Witheford to R. J. Seddon, 3 February 1901, AABK 18805 W5901 0006498, ANZ.

286 Notification of Death, 16 January 1936, Robert Witheford, AABK 18805 W5901 0006498, ANZ.

287 Attestation-Form, 8 May 1900, Frank Willis, AABK 18805 W5515 0006052, ANZ.

288 Attestation-Form, 8 May 1900, Frank Willis, AABK 18805 W5515 0006052, ANZ; *Auckland Star*, 9 July 1900, p. 5; *AJHR*, 1900, H-6, p. 36.

289 Attestation Form, Frank Willis, 12 May 1900, AABK 18805 W5515 0006052, ANZ; *AJHR*, 1901, H-6A, p. 6; *AJHR*, 1902, H-6B, p. 4; Commission, Frank Willis, AABK 18805 W5515 0006052, ANZ. There is some confusion surrounding the contingents in which Willis served. He appears on the nominal rolls of the Seventh and Tenth contingents, but his South African reference papers record service in the First, Fifth, Seventh and Tenth. He also signed attestation forms for the First and Fourth contingents, and in the First World War indicated service in the First, Seventh, Eight and Tenth contingents.

290 *NZPD*, 1899, Vol. 110, pp. 80, 96, 891–93.

291 *Press*, 18 May 1900, p. 3.

292 Tom Brooking, *Richard Seddon: King of God's Own — The Life and Times of New Zealand's Longest-serving Prime Minister*, Auckland: Penguin Books, 2014, p. 182.

293 *NZPD*, 1900, Vol. 112, p. 562.

294 High Court Deposition, Richard John Seddon, n.d., William Hutchison, AABK 18805 W5515 0002735, ANZ.

295 *Press*, 13 August 1900, p. 2.

296 William Hutchison to William Massey, 9 January 1915, William Hutchison, AABK 18805 W5515 0002735, ANZ; *Press*, 11 August 1900, p. 8.

297 *NZPD*, 1900, Vol. 112, pp. 560–62.

298 William Hutchison to James Allen, 9 January 1915, William Hutchison, AABK 18805 W5515 0002735, ANZ.

299 *Press*, 11 August 1900, p. 8.

300 *NZPD*, 1900, Vol. 112, p. 562.

301 *NZPD*, 1900, Vol. 112, p. 561.

302 *Evening Post*, 10 August 1900, p. 4.

303 Richard John Seddon, sworn affidavit, n.d., William Hutchison, AABK 18805 W5515 0002735, ANZ.

304 William Hutchison to William Massey, 9 January 1915; William Hutchison to James Allen, 9 January 1915, AABK 18805 W5515 0002735, ANZ.

305 William Hutchison to James Allen, 9 January 1915, AABK 18805 W5515 0002735, ANZ.

306 Roll of individuals entitled to the South African Medal and clasps. Rhodesian Field Force Artillery, 16 October 1906, National Archives, Kew, UK, WO 100/266, p. 224.

307 William Hutchison, 14 April 1902, Permit to Land in South Africa, ANZ Wellington, ACGO 8333 IA1 848/ [22] 1902/1167; *Hawera & Normanby Star*, 1 March 1902, p. 2.

308 William Hutchison to James Allen, 9 January 1915; Alfred William Robin to James Allen, 7 April 1915, AABK 18805 W5515 0002735, ANZ.

309 *AJHR*, 1903, H-6A, p. 9; *Star*, 10 January 1901, p. 1.

310 *Otago Witness*, 9 January 1901, p. 31.

311 *London Gazette*, UK, No. 27307, 23 April 1901, p. 2776; 9 July 1901, p. 4558; *AJHR*, 1901, H-6, p. 9.

312 *AJHR*, 1901 H-6E, p. 3; *AJHR*, 1903 H-6A, p. 9.

313 *Feilding Star*, 11 June 1901, p. 2; Extract from telegram dated 4 April 1901 from O/C 'C' Squadron, Kimberley to O/C 5th N.Z. Regt, Krugersdorp, AABK 18805 W5515 0006158, ANZ.

314 Extract from telegram dated 4 April 1901 from O/C 'C' Squadron, Kimberley to O/C 5th N.Z. Regt, Krugersdorp; Births in the District of Hobart, 1879, AABK 18805 W5515 0006158, ANZ; *Oamaru Mail*, 24 May 1901, p. 2.

315 *Colonist*, 18 December 1900, p. 3.

316 *Manawatu Evening Standard*, 24 August 1900, p. 2; *Feilding Star*, 12 June 1900, p. 2; Roll of individuals entitled to the South African Medal and clasps. N.S.W. Citizens Bushmen's Contingent, 5 March 1904, National Archives, Kew, UK, WO 100/288, p. 311.

317 *Manawatu Evening Standard*, 24 August 1900, p. 2.

318 *Manawatu Evening Standard*, 12 June 1900, p. 2.

319 *Feilding Star*, 12 June 1900, p. 2.

320 *Feilding Star*, 18 May 1900, p. 2; *Feilding Star* 12 June 1900, p. 2; Roll of individuals entitled to the South African Medal and clasps. New South Wales Mounted Infantry, n.d., National Archives, Kew, UK, WO 100/288, p. 64; Roll of individuals entitled to the South African Medal and clasps. New South Wales Mounted Infantry, 2nd Contingent, 1 August 1904, National Archives, Kew, UK, WO 100/289, p. 184; *New Zealand Herald*, 15 January 1900, p. 5; *New Zealand Herald*, 6 July 1901, p. 5.

321 *New Zealand Herald*, 1 November 1915, p. 7; Roll of individuals entitled to the South African Medal and clasps. 5th Victorian Mounted Rifles, n.d., National Archives, Kew, UK, WO 100/291, p. 175; Roll of individuals entitled to the South African Medal and clasps. 5th Victorian Mounted Rifles, 31 December 1903, National Archives, Kew, UK, WO 100/291, p. 217.

322 *Wanganui Collegian*, No. 61, April 1903, p. 20.

323 *Otago Witness*, 3 May 1900, p. 30.

324 *Press*, 28 April 1900, p. 7; *Manawatu Evening Standard*, 3 May 1900, p. 2; *Hawke's Bay Herald*, 1 May 1900, p. 2.

325 John Taylor Marshall to Paymaster-General, 9 February 1905, AAYS 8638 AD1/486/a, D1907/316, ANZ; *AJHR*, 1902, H-28, p. 2; *Nelson Evening Mail*, 7 September 1900, p. 3.

326 *AJHR*, 1902, H-28, p. 2; *Nelson Evening Mail*, 7 September 1900, p. 3; Roll of individuals entitled to the South African Medal and clasps. 1st Brabant's Horse, 28 June 1905, National Archives, Kew, UK, WO 100/237, p. 146; Roll of individuals entitled to the South African Medal and clasps. 1st Brabant's Horse (later Brabant's Horse), 9 April 1902, National Archives, Kew, UK, WO 100/237, p. 52; Roll of individuals entitled to the South African Medal and clasps. 1st Brabant's Horse (later Brabant's Horse), 9 April 1902, National Archives, Kew, UK, WO 100/237, p. 12; Roll of individuals entitled to the South African Medal and clasps. 1st Brabant's Horse, 30 November 1903, National Archives, Kew, UK, WO 100/237, p. 128; Roll of individuals entitled to the South African Medal and clasps. 1st Brabant's Horse (later Brabant's Horse), 9 April 1902, National Archives, Kew, UK, WO 100/237, p. 3.

327 Attestation-Form, 6 January 1902, Kerr Andrew Maxwell, AABK 18805 W5515 0003808, ANZ; Attestation-Form, 21 January 1902, Walter William Glass, AABK 18805 W5515 0002076, ANZ; L. Grey to Archibald Craig Hutton, 7 October 1903, AABK 18805 W5515 0002739, ANZ.

328 Roll of individuals entitled to the South African Medal and clasps. 1st Brabant's Horse (later Brabant's Horse), 9 April 1902, National Archives, Kew, UK, WO 100/237, p. 60.

329 John Taylor Marshall to Officer Commanding Militia and Volunteer District, Wellington, 24 March 1902, AABK 18805 W5515 0003743, ANZ.

330 John Taylor Marshall to Officer Commanding Militia and Volunteer District, Wellington, 24 March 1902, AABK 18805 W5515 0003743, ANZ; *Colonial Militia and Volunteers, New Zealand Militia, 1878*, GBM Army List, p. 854; *AJHR*, 1888, I-2, p. 9; *Colonist*, 26 September 1874, p. 4.

331 *AJHR*, 1888, I-2, p. 9; *Evening Post*, 1 December 1877, p. 2; *New Zealand Gazette*, No. 87, 18 October 1877, p. 1029; *New Zealand Gazette*, No. 96, 29 November 1877, p. 1140; *New Zealand Gazette*, No. 8, 17 January 1878, p. 79; *New Zealand Gazette*, No. 12, 10 February 1881, p. 198; *Hawke's Bay Herald*, 27 September 1899, p. 3; *New Zealand Gazette*, No. 40, 29 July 1886, p. 889; *Evening Post*, 31 August 1886, p. 3.

332 Account of Henry Fernie & Sons, 6 December 1900, AD1 380 cf D1901/2335, ANZ; Men going on S.S. *Ormazon* as grooms, n.d., AD1 380 cf D1901/2335, ANZ. Primary records refer to the vessel as both

the *Ormazan* and the *Ormazon*, though the former is correct.

333 E. Bezar to Richard J. Seddon, 31 July 1901, AABK 18805 W5515 0000378, ANZ; *Evening Post*, 16 September 1901, p. 5.

334 *AJHR*, 1902, Session I, H-6C, p. 66; Otto R. Cook to E. W. C. Chaytor, 18 December 1902, AABK 18805 W5901 0001151, ANZ.

335 *Observer*, 2 June 1900, p. 6; *Newcastle Morning Herald* (Aus.), 17 March 1900, p. 4; Roll of individuals entitled to the South African Medal and clasps. 2nd Brabant's Horse, 16 March 1902, National Archives, Kew, UK, WO 100/237, pp. 186, 209, 240.

336 *Hawera & Normanby Star*, 25 August 1900, p. 2; Roll of individuals entitled to the South African Medal and clasps. 2nd Imperial Horse, 13 July 1904, National Archives, Kew, UK, WO 100/251, p. 177; Roll of individuals entitled to the South African Medal and clasps. Utrecht & Vryheid Mounted Police, n.d., National Archives, Kew, UK, WO 100/261, p. 192; Roll of individuals entitled to the South African Medal and clasps. 1st Scottish Horse, 30 April 1903, National Archives, Kew, UK, WO 100/263, p. 188.

337 *New Zealand Herald*, 29 December 1902, p. 6.

338 *Advertiser* (Adelaide), 1 February 1900, p. 5; *Oamaru Mail*, 1 February 1900, p. 2.

339 *Auckland Star*, 16 February 1900, p. 2; *Press*, 14 February 1900, p. 4; Form of Enrolment of Volunteers, 13 January 1900, William Henry White, AABK 18805 W5515 0005960, ANZ.

340 Attestation Form, 6 January 1902, Leonard Edward John Worthington, AABK 18805 W5515 0006170, ANZ.

341 *Temuka Leader*, 4 January 1900, p. 2; *Temuka Leader*, 16 January 1900, p. 2.

342 Roll of individuals entitled to the South African Medal and clasps: 2nd Brabant's Horse, n.d., WO 100/237, p. 70; *Otago Daily Times*, 13 February 1900, p. 6.

343 Crime and Offence Report, Leonard Edward John Worthington, AABK 18805 W5515 0006170, ANZ.

344 Casualty Form – Active Service, Leonard Edward John Worthington, AABK 18805 W5515 0006170, ANZ.

345 Casualty Form – Active Service, Leonard Edward John Worthington, AABK 18805 W5515 0006170, ANZ.

346 *Otago Daily Times*, 15 February 1900, p. 4.

347 *Press*, 9 May 1902, p. 5.

348 *Press*, 9 May 1902, p. 5.

349 *Press*, 25 April 1900, p. 6; *Star*, 26 April 1900, p. 1; Herbert Ernest Hart, diary, 21 March 1902, NAM NZ 1990.1024; *AJHR*, 1902, H-6A, pp. 25–26.

350 *Star*, 26 April 1900, p. 1.

351 *Wairarapa Daily Times*, 3 April 1900, p. 3.

352 *Star*, 13 June 1900, p. 4.

353 *NZPD*, 1899, Vol. 110, p. 185; *Observer*, 3 March 1900, p. 16.

354 *Wairarapa Daily Times*, 14 April 1902, p. 3. Text as in original.

355 *Wairarapa Daily Times*, 14 April 1902, p. 3; *Oamaru Mail*, 26 March 1902, p. 3; *Marlborough Express*, 27 January 1902, p. 3.

356 *Otago Witness*, 5 February 1902, p. 14. The trooper was possibly either Cyril Morris or Henry Fenton Brown. Morris, who served in the Heretaunga Mounted Rifles and the Eighth Contingent, was 23 years old and 5'4½", while 19-year-old Brown, who also served in both corps, was 5'5¾".

357 Attestation Form, 13 January 1901, Alan Bruce Saunders, AABK 18805 W5515 0004951, ANZ.

358 *Manawatu Evening Standard*, 22 October 1900, p. 2; *Colonist*, 11 April 1900, p. 3.

359 *Manawatu Evening Standard*, 6 April 1900, p. 2.

360 Certificate of Discharge, May 1902, Alan Bruce Saunders, AABK 18805 W5515 0004951, ANZ.

361 *Press*, 28 July 1900, p. 5.

362 *Press*, 28 July 1900, p. 5.

363 *Manawatu Evening Standard*, 6 July 1901, p. 2; *AJHR*, 1902, B-20B, p. 2.

364 Harcourt Eugene Louis Peat, NZBDM 1884001588.

365 John Evelyn Duigan, NZBDM, 1883006669; *Wanganui Collegian*, No. 59, August 1902, p. 15.

366 Attestation Form, Henry Ray Vercoe, 3 April 1901, Resignation, Transfer to the Reserve of Officers, or Retirement of an Officer of Territorial Force, 8 May. 1946, AABK 18805 W5901 Box 7 0006493, ANZ.

367 Certificate of Discharge, 26 August

1902; New Zealand Expeditionary Force Attestation Form, 27 January 1915, AABK 18805 W5901 Box 7 0006493, ANZ.

368 *Wanganui Collegian*, No. 59, August 1902, p. 15; *Nelson Evening Mail*, 18 April 1900, p. 2; *Wanganui Chronicle*, 19 August 1903, p. 5.

369 *Wanganui Collegian*, No. 59, August 1902, p. 15; *Wanganui Herald*, 14 June 1900, p. 2; *Wanganui Herald*, 25 June 1900, p. 2; *Wanganui Herald*, 21 December 1900, p. 3.

370 *Wanganui Herald*, 17 November 1900, p. 2; *Wanganui Herald*, 21 December 1900, p. 3.

371 *Wanganui Collegian*, No. 59, August 1902, p. 15; *Timaru Herald*, 24 January 1901, p. 1; *Auckland Star*, 12 January 1901, p. 4.

372 *Timaru Herald*, 9 April 1901, p. 2; *Bush Advocate*, 2 April 1902, p. 2; Roll of individuals entitled to the South African Medal and clasps: 2nd Kitchener's Fighting Scouts, 14 September 1904, WO 100/257, p. 172.

373 *Wanganui Collegian*, No. 59, August 1902, p. 15.

374 *Bush Advocate*, 2 April 1902, p. 2; *AJHR*, 1902, H-6B, p. 9.

375 *NZPD*, 1900, Vol. 111, p. 88.

376 *NZPD*, 1900, Vol. 111, pp. 88–89.

377 *NZPD*, 1900, Vol. 111, p. 375; *Nelson Evening Mail*, 12 July 1900, p. 2.

378 *NZPD*, 1900, Vol. 112, p. 247.

379 *Press*, 24 March 1902, p. 5; *Otago Witness*, 17 July 1901, p. 85; *Auckland Star*, 27 August 1902, p. 2.

380 *London Gazette*, No. 27362, 4 October 1901, p. 6481.

381 *London Gazette*, No. 27377, 15 November 1901, p. 7383.

382 *Poverty Bay Herald*, 11 October 1901, p. 4; *AJHR*, 1903, H-6A, pp. 3, 5; Proceedings of Medical Board, 14 August 1902, John Henry Helm, AABK 18805 W5515 0002484, ANZ; Extract from Quarterly Report by Lieut. Colonel Porter, 14 October 1901; Proceedings of Medical Board, 23 August 1901, AABK 18805 W5515 0003249, ANZ.

383 *Wanganui Chronicle*, 3 December 1901, p. 2; Extract from Army Orders, South Africa, 13 September 1901, Ernest Barnett Lockett, AABK 18805 W5515 0003249, ANZ; *London Gazette*, No. 27364, 11 October 1901, p. 6643.

384 Ivanhoe Edward Baigent to Richard John Seddon, 20 June 1902, Ivanhoe Edward Baigent, AABK 18805 W5515 0000189, ANZ.

385 *Auckland Star*, 31 January 1903, p. 3; *Bush Advocate*, 13 November 1901, p. 2.

386 *AJHR*, 1903, H6A, p. 5; *Auckland Star*, 31 January 1903, p. 3; *Bush Advocate*, 13 November 1901, p. 2; *Evening Post*, 20 November 1901, p. 6.

387 *Bush Advocate*, 13 November 1901, p. 2.

388 *Timaru Herald*, 19 December 1900, p. 3. Hugh Charles Grahame served as a sergeant in the Ninth Contingent before receiving a lieutenant's commission.

389 *AJHR*, 1901, H-6E, p. 3; *AJHR*, 1903, H-6A, pp. 3, 5.

390 *Ashburton Guardian*, 12 October 1900, p. 2; *Lyttelton Times*, 11 October 1900, p. 3.

391 *Timaru Herald*, 19 December 1900, p. 3.

392 *AJHR*, 1900, H-6, p. 11; *Lyttelton Times*, 28 November 1900, p. 5.

393 Arthur Bauchop to Thomas William Porter, n.d., AABK 18805 W5515 0003571, ANZ.

394 Arthur Bauchop to Thomas William Porter, n.d., AABK 18805 W5515 0003571, ANZ.

395 *Star*, 10 January 1901, p. 1.

396 *London Gazette*, No. 27490, 31 October 1902, p. 6909.

397 *London Gazette*, No. 27331, 9 July 1901, p. 4556.

398 *London Gazette*, No. 27331, 9 July 1901, p. 4556; J. W. Poynton to Arthur Douglas, 26 September 1901, AABK 18805 W5515 0006197, ANZ; *AJHR*, 1901, H-6E, p. 3; *AJHR*, 1903, H-6A, p. 3.

399 M. E. Warmington to Officer in Command, Wellington Military District, 12 June 1901, Appointment of Nurses, March 1900 – February 1957, AD34 2/2, ANZ.

400 *Ashburton Guardian*, 14 September 1901, p. 3; Roll of individuals entitled to the South African Medal and clasps: Civilian Nurses, 22 August 1901, WO 100/229, p. 207; Roll of individuals entitled to the South African Medal and clasps: Colonial Nursing Sister (New Zealand) No. 5, 1901, WO 100/229, p. 311.

401 *Taranaki Herald*, 19 December 1901, p. 3.

402 *Daily Telegraph*, 22 August 1900, p. 8;

Times (London), 10 April 1900, p. 7; Roll of individuals entitled to the South African Medal and clasps: 2nd Battalion Durham Light Infantry Company Burma Mounted Infantry, n.d., WO 100/201, p. 166.

403 *Hawera & Normanby Star*, 18 October 1900, p. 2.
404 Joseph Chamberlain to Lord Ranfurly, 27 April 1901, AJHR, 1902, A-2, pp. 27–28; A. W. Andrew to Station Paymaster 'S' Branch, 9 February 1902, Owen Thomas Baigent, AABK 18805 W5515 0000190, p. 52.
405 *Daily Telegraph*, 4 November 1901, p. 3; Roll of individuals entitled to the South African Medal and clasps: Prince of Wales' Light Horse, 14 November 1901, WO 100/264, p. 162; Roll of individuals entitled to the South African Medal and clasps: Prince of Wales' Light Horse, 25 August 1904, WO 100/264, p. 188.
406 *Wanganui Herald, Bay of Plenty Times*, 12 February 1902, p. 2; *Grey River Argus*, 23 May 1904, p. 4.
407 Arthur James Vogan to Joseph Ward, 14 May 1902, AABK 18805 W5515 0005740, ANZ; *Evening Post*, 11 February 1902, p. 6.
408 NZPD, 1902, Vol. 122, p. 124.
409 General Officer Commanding Transvaal and Orange River Colonies to Commandant New Zealand Forces, 19 September 1902, AABK 18805 W5515 0005740, ANZ.
410 Arthur James Vogan to Norman B. Smith, 10 November 1902; Arthur J. Vogan to the Editor, *The Empire*, n.d., AABK 18805 W5515 0005740, ANZ.
411 Roll of individuals entitled to the South African Medal and clasps: Prince of Wales' Light Horse, 14 November 1901, WO 100/264, p. 162.
412 Roll of individuals entitled to the South African Medal and clasps: 1st Brabant's Horse, 11 May. 1904, WO 100/237, p. 131; Roll of individuals entitled to the South African Medal and clasps: Prince of Wales Light Horse, 14 November 1901, WO 100/264, p. 148; *Star*, 9 August 1901, p. 1; Roll of individuals entitled to the South African Medal and clasps: Bushveldt Carbineers. Later Pietersburg Light Horse, 10 December 1902, WO 100/263, p. 155.
413 Petition of John James Clark, AJHR, 1903, I-1C, pp. 1–2. Clark found that a total of 117 contingent members were eligible for the King's South Africa Medal.
414 Ivanhoe Edward Baigent to Richard John Seddon, 20 June 1902; Richard John Spotswood Seddon to Ivanhoe Edward Baigent, 22 March 1909; H. D. Tuson to Commissioner of Police, 3 March 1909; Donald Shaw to Inspector of Police, 14 March 1909, AABK 18805 W5515 0000189, ANZ.
415 Ivanhoe Edward Baigent to Richard John Spotswood Seddon, n.d., AABK 18805 W5515 0000189, ANZ.
416 Ivanhoe Edward Baigent, History Sheet, n.d., AABK 18805 W5515 0000189, ANZ.
417 Thomas Byrne to Joseph Ward, 1906; Joseph Ward to Thomas Byrne, 1906, AABK 18805 W5515 0000777, ANZ; AJHR, 1903, H-6A, p. 7; Catherine Bruce to Lord Ranfurly, 25 August 1902, AABK 18805 W5515 0000657, ANZ.
418 NZPD, 1902, Vol. 120, p. 139.
419 *Auckland Star*, 12 April 1904, p. 2.
420 Frank Hemphill to Under-Secretary for Defence, 26 March 1903, AABK 18805 W5515 0002487, ANZ.
421 *Auckland Star*, 31 January 1903, p. 3.
422 NZPD, 1901, Vol. 116, p. 401.
423 John Oswald Murray to Commandant, Defence Department, 1 September 1902, AABK 18805 W5515 0004126, ANZ.
424 AJHR, 1900, H-6N, p. 1; AJHR, 1901, H-6E, p. 1; AJHR, 1903, H-6A, p. 7; *Otago Daily Times*, 7 June 1902, p. 2.
425 *Manawatu Evening Standard*, 3 June 1902, p. 3.
426 *Manawatu Evening Standard*, 1 July 1901, p. 2.
427 AJHR, 1900, H-6A, p. 6; *Times* (London), 5 May 1900, p. 15; Roll of individuals entitled to the South African Medal and clasps. 1st Battalion, The Princess of Wales Own (Yorkshire Regiment), 11 June 1902, National Archives, Kew, UK, WO 100/178, p. 64.
428 AJHR, 1900, H-6A, p. 6.

Chapter Four: 'Loyalty to the British Empire'

1 AJHR, 1868, D-21, p. 7.
2 Military Pensions Act 1866 (30 Victoriae 1866 No. 6).

3   *Otago Daily Times*, 21 October 1881, p. 2.
4   *Taranaki Herald*, 23 August 1892, p. 2; W. Stuart to Under-Secretary, 6 September 1892, J1 490 z 1892-879, ANZ.
5   *New Zealand Herald*, 21 September 1889, p. 5; Commissioner of Police Walter Gudgeon to Under-Secretary of Defence Colonel Humphrey, 9 September 1889, AAAL 22514 W5741 Box 4 P.M. 89/326, ANZ.
6   *Evening Post*, 6 January 1920, p. 5; Walter Gudgeon to Colonel Humphrey, 9 September 1889, AAAL 22514 W5741 Box 4 P.M. 89/326, ANZ.
7   *Nelson Evening Mail*, 24 July 1901, p. 4.
8   John Edward Hume to Under-Secretary for Defence, 22 January 1896, AD1 296/ag PM1896/73, ANZ.
9   Petition of Tenetahi, No. 161, ACIH 16036 MA1/1029 1910/4648, ANZ.
10 Little Barrier Island Purchase Act 1894 (Local) (58 Vict 1894 No. 27); John Edward Hume to Arthur Douglas, 22 January 1896, AD1 292/ag PM1896/279, ANZ; Petition No. 136/19110 Tenetahi re Hauturu, Little Barrier Island. Referred to Government for favourable consideration, ACIH 16036 MA1 1029 / 1910/4648.
11 Alexander Barron to Arthur Douglas, 20 November 1896, AD1 292/ag PM1896/279, ANZ.
12 *Press*, 24 May 1897, p. 6; *Press*, 1 September 1897, p. 5.
13 Report of Constable McGilp, 25 August 1897, ACIS 17627 P1 Box 251 1897/1041, ANZ.
14 Hōne Tōia & Rameka Ngamanu, petition, J1 578 ah 1897/1037, ANZ.
15 *Evening Post*, 2 May 1898, p. 4; *Hawke's Bay Herald*, 3 May 1898, p. 3; *New Zealand Herald*, 13 May 1898, p. 11.
16 *Otago Daily Times*, 10 May 1898, p. 3; Criminal Code Act 1893 (57 VICT 1893 No. 56); Justice Conolly Report, 7 December 1898, ACGS 16211 J1 625 a 1898/1178, ANZ.
17 *AJHR*, 1900, E-2, p. 6.
18 *Auckland Star*, 24 August 1901, p. 4; *New Zealand Herald*, 24 August 1901, p. 4.
19 *NZPD*, 116 (1901), p. 35.
20 National Archive, UK, CO 209/260, p. 215.
21 *Northern Advocate*, 3 March 1900, p. 4; *New Zealand Herald*, 2 February 1928, p. 12.
22 *Northern Advocate*, 3 March 1900, p. 2.
23 *Northern Advocate*, 3 March 1900, p. 4.
24 *Oamaru Mail*, 23 March 1899, p. 4; *Northern Advocate*, 8 April 1899, p. 2.
25 *Poverty Bay Herald*, 11 April 1899, p. 2; *North Otago Times*, 27 March 1899, p. 2; *New Zealand Times*, 6 April 1899, p. 3.
26 *Great Britain Parliamentary Papers*, 1862 93040, New Zealand 13, Vol. XXXVII, p. 25.
27 *Observer*, 26 December 1891, p. 14.
28 *Wanganui Herald*, 7 March 1900, p. 2.
29 *Evening Post*, 29 December 1900, p. 2.
30 *Evening Post*, 27 December 1900, p. 4.
31 *Wanganui Chronicle*, 1 December 1900, p. 2.
32 *New Zealand Times*, 6 September 1901, p. 3; *Auckland Star*, 30 October 1901, p. 4.
33 *Auckland Star*, 30 October 1901, p. 4.
34 Nigel Robson, 'Counting the Cost: The Impact of the South African War 1899–1902 on New Zealand Society', Master of Arts in History thesis, Massey University, 2012, p. 151; Blair Nicholson, 'Viewpoints on the Veldt: Attitudes and Opinions of New Zealand Soldiers during the South African War, 1899–1902', Master of Arts in History thesis, University of Waikato, 2011, p. 50.
35 *Colonist*, 25 September 1863, p. 3; *Wellington Independent*, 12 September 1846, p. 4; *Otago Daily Times*, 19 May 1900, p. 2; *AJHR*, 1906, A-1, p. 2.
36 *Otago Witness*, 19 December 1900, p. 47.
37 *Wanganui Collegian*, No. 55, April 1901, p. 22; Frank Perham, diary, 11 December 1900, NAM NZ 2000.736.
38 Richard Davis to Joseph Ward, ANZ, Mangere, BBAO, 5544 A78 Box 74/a 1902/687.
39 *Wanganui Collegian*, No. 60, December 1902, p. 12.
40 *Wanganui Collegian*, No. 60, December 1902, p. 12.
41 *Evening Post*, 22 February 1901, p. 7.
42 *Wairarapa Daily Times*, 22 June 1900, p. 3; Kingsley Field, *Soldier Boy: A Young New Zealander Writes Home from the Boer War*, Auckland: New Holland Publishers (NZ), 2007, p. 31.
43 *NZPD*, 110 (1899), p. 95.
44 *NZPD*, 110 (1899), pp. 79–80.
45 *NZPD*, 110 (1899), p. 96.

46 *AJHR*, 1897, H-14, pp. 3–5.
47 *AJHR*, 1897, H-14, p. 3.
48 *NZPD*, 110 (1899), pp. 90–91.
49 *AJHR*, 1900, H-6K, p. 2; *AJHR*, 1900, H-27B, pp. [1]–2, 6, 8.
50 *AJHR*, 1900, H-27B, pp. [1], 5–6.
51 *AJHR*, 1900, H-6K, p. 2.
52 *Hawke's Bay Herald*, 26 February 1900, p. 3.
53 *Star*, 19 February 1900, p. 2.
54 *Evening Post*, 31 March 1900, p. 5.
55 Lord Ranfurly to Secretary of State for the Colonies, 26 December 1900, ACHK 16561, G5/4, ANZ.
56 *Otago Daily Times*, 28 December 1900, p. 6; *Evening Post*, 29 December 1900, p. 6; *Manawatu Evening Standard*, 31 December 1900, p. 5; *Free Lance*, 5 January 1901, p. 4; *Free Lance*, 12 January 1901, p. 9.
57 *New Zealand Herald*, supplement, 9 February 1901, p. 5.
58 *Clutha Leader*, 28 December 1900, p. 5; *Evening Post*, 27 December 1900, p. 4; *Manawatu Evening Standard*, 29 December 1900, p. 4.
59 *Press*, 19 October 1899, p. 4.
60 *Evening Post*, 27 December 1900, p. 4.
61 *Timaru Herald*, 11 April 1902, p. 2.
62 *Manawatu Evening Standard*, 29 December 1900, p. 4.
63 *Poverty Bay Herald*, 29 December 1900, p. 2.
64 *Times* (London), 27 December 1900, p. 7.
65 *Evening Post*, 29 December 1900, p. 5.
66 Colonial Office minute, 28 December 1900, National Archives, Kew, UK, CO209/261.
67 Capitation Rolls, Papawai Native Rifle Volunteers, AAYS 8790 ARM41 Box 35, 1890/2i Capitation Roll, Ngati Porou Rifle Volunteers, ARM41 31 1888/2aa, ANZ.
68 *Press*, 31 December 1900, p. 8; *Poverty Bay Herald*, 31 December 1900, p. 3.
69 *Free Lance*, 5 January 1901, p. 4; *Free Lance*, 12 January 1901, p. 9.
70 *Press*, 31 December 1900, p. 4; *Otago Daily Times*, 31 December 1900, p. 6.
71 *Otago Daily Times*, 31 December 1900, p. 6.
72 *Star*, 18 May 1901, p. 4.
73 *Star*, 8 January 1901, p. 2.
74 *Otago Daily Times*, 11 February 1902, p. 2.
75 *Times* (London), 24 February 1902, p. 5; *Auckland Star*, 21 January 1902, p. 5; *Auckland Star*, 22 February 1902, p. 5.
76 Lord Ranfurly to Secretary of State for the Colonies, 1 February 1902, ACHK 16561 G5/6, ANZ.
77 Lord Ranfurly to Secretary of State for the Colonies, n.d., ACHK 16561 G5/6, ANZ.
78 Lord Ranfurly to Secretary of State for the Colonies, 9 February 1902, ACHK 16561 G5/6, ANZ.
79 *Auckland Star*, 22 February 1902, p. 5.
80 *Auckland Star*, 22 February 1902, p. 5.
81 Secretary of State for the Colonies to Lord Ranfurly, 20 February 1902, ACHK 16561 G5/6, ANZ.
82 Secretary of State for the Colonies to Lord Ranfurly, 20 February 1902, ACHK 16561 G5/6, ANZ.
83 *Auckland Star*, 31 March 1902, p. 4.
84 *Poverty Bay Herald*, 5 April 1902, p. 4; *Evening Post*, 5 April 1902, p. 5.
85 *Star*, 30 January 1902, p. 1.
86 *Evening Post*, 5 April 1902, p. 4.
87 *Poverty Bay Herald*, 9 April 1902, p. 4.
88 *Examiner* (Launceston, Tas.), 21 April 1902, p. 4; *Sunday Times* (Sydney), 13 April 1902, p. 8; *Western Mail* (Perth), 19 April 1902, p. 59.
89 *Evening Post*, 6 May 1902, p. 6.
90 *Timaru Herald*, 7 April 1902, p. 2.
91 *Timaru Herald*, 7 April 1902, p. 2.
92 *Timaru Herald*, 11 April 1902, p. 2.
93 *AJHR*, 1901, H-26B, p. 14.
94 John McLellan, 'Maori and the South African War, 1899–1902: An Overlooked Presence in a "White Man's War"', research essay, Victoria University of Wellington, 2015.
95 Capitation Rolls, Tauranga Mounted Rifles Volunteers, 30 March 1901, AAYS 8790 ARM41 Box 68 1910/3x, ANZ; *AJHR*, 1901, H-6A, p. 4; NZ Expeditionary Force Attestation Form, 27 January 1915, Henry Ray Vercoe, AABK 18805 W5901 7/ 0006493, ANZ.
96 Chiefs and Arawa to Governor, 15 January 1896, G15/1, ANZ.
97 Governor to Chiefs of the Arawa tribe, 22 January 1896, G15/1, ANZ.
98 *Hawera & Normanby Star*, 11 February 1902, p. 2.

99 *Wanganui Chronicle*, 2 December 1899, p. 2.
100 *NZPD*, 110 (1899), p. 185.
101 Napier Guards Rifle Volunteers 1899 Capitation Roll, ARM41 128 1911/59a, ANZ; *New Zealand Times*, 8 February 1900, p. 7; *Otago Witness*, 15 February 1900, p. 46.
102 *Thames Star*, 9 February 1900, p. 1; *Wanganui Herald*, 8 February 1900, p. 2; *Otago Witness*, 15 February 1900, p. 46; *Te Puke ki Hikurangi*, 30 June 1900, pp. 1–2.
103 *Te Puke ki Hikurangi*, 30 June 1900, pp. 1–2; *Otago Witness*, 15 February 1900, pp. 46–47.
104 Statement of Service of Arthur Joseph in the South African War of 1899–1902, 1 December 1936, AABK 18805 W5515 0002931, ANZ; *Star*, 29 April 1897, p. 4; *AJHR* 1897, H-14, p. 5.
105 *Bay of Plenty Times*, 11 June 1900, p. 4.
106 Ahere Te Koari Hohepa, New Zealand Expeditionary Force Attestation Form, 18805 W5541 0055707, ANZ.
107 Capitation Rolls – Waikato Mounted Rifle Volunteers No 3 Company, ARM41 148 1911/790, ANZ.
108 *Taranaki Herald*, 20 October 1899, p. 2; Mike Dwight, *Walter Callaway: A Māori Warrior of the Boer War*, Mike Dwight, 2010, p. 91.
109 *Te Puke ki Hikurangi*, 30 June 1900, pp. 1–2; Capitation Rolls, Napier Guards Rifle Volunteers, 1899, ARM41 128 1911/59a, ANZ; Capitation Rolls, No. 3 Company, Waikato Mounted Rifles, 28 February 1900, ARM41 Box 148, 1911/790, ANZ.
110 *AJHR*, 1900, H-6, p. 19; *AJHR*, 1901, H-6, p. 2; *AJHR*, 1901, H-6A, p. 3.
111 *Hastings Standard*, 22 November 1916, p. 2.
112 *New Zealand Herald*, 3 March 1900, p. 5.
113 *AJHR*, 1901, H-26B, p. 8.
114 Capitation Rolls, Kawakawa Rifle Volunteers, 30 March 1901, ARM41 58 1905/1a, ANZ; Capitation Rolls, Hokianga Mounted Rifle Volunteers, 16 April 1901, ARM41 61 1907/1f, ANZ.
115 *New Zealand Herald*, 17 April 1901, p. 4; *Auckland Star*, 17 April 1901, p. 1; *Thames Star*, 18 April 1901, p. 4.
116 *Auckland Weekly News*, 5 July 1900, p. 8.
117 *Auckland Weekly News*, 5 July 1900, p. 8; Sandra Coney, *Standing in the Sunshine: A History of New Zealand Women Since They Won the Vote*, Auckland: Penguin Books, 1993, p. 93.
118 *Evening Post*, 11 May 1900, p. 2.
119 *Daily Telegraph* (Launceston, Tas.), 2 May 1901, p. 2.
120 *Otago Daily Times*, 16 March 1900, p. 5; *Weekly Press*, 14 March 1900, p. 62.
121 *Star*, 20 May 1901, p. 3.
122 *Auckland Star*, 18 February 1901, p. 5; *Wanganui Herald*, 23 February 1901, p. 2; *Wanganui Herald*, 18 February 1901, p. 2.
123 *Weekly Press*, 14 March 1900, p. 62.
124 *NZPD*, 110 (1899), p. 96; *Bush Advocate*, 12 October 1899, p. 2; *New Zealand Times*, 12 October 1899, p. 6; *Evening Post*, 22 December 1899, p. 5.
125 *Evening Post*, 12 October 1899, p. 5; S. M. Chrisp. 'Te Tau, Taiawhio Tikawenga', *Dictionary of New Zealand Biography. Te Ara — the Encyclopedia of New Zealand*, updated 4 March 2014.
126 *Evening Post*, 10 September 1897, p. 5.
127 *Evening Post*, 10 September 1897, p. 5.
128 Lachy Paterson, 'Identity and Discourse: Te Pipiwharauroa and the South African War, 1899–1902', *South African Historical Journal*, 65:3 (2013): 444–62, DOI:10.1080/02 582473.2013.770063, at p. 448.
129 *Wanganui Herald*, 4 December 1897, p. 2; *AJHR*, 1898, H-19A, p. 3.
130 *AJHR*, 1901, E-1, p. xv; John Thornton to Minister of Defence, 26 February 1897, Te Aute Native College Cadets, AD1 309/av D1898/765, ANZ.
131 Te Aute Native College Cadets, AD1 309/av D1898/765, ANZ.
132 Capitation Roll, Ngati Porou Rifle Volunteers, 31 December 1887, ARM41 31 1888/2aa, ANZ.
133 *Evening Post*, 1898, 2 June, p. 5.
134 *Exeter and Plymouth Gazette*, 11 July 1898, p. 5; *Morning Post* (UK), 9 July 1898, p. 2; *Star*, 23 December 1899, p. 6; *Otago Witness*, 28 December 1899, p. 25; Tuta Nihoniho, *Narrative of the Fighting on the East Coast (Nga Pakanga Ki Te Tai Rawhiti) 1865–71, with a Monograph on Bush Fighting (Me Nga Korero Mo Uenuku)*, Wellington: John MacKay, 1913, p. 45.
135 Transcription of the history of the Maori

mere, Porourangi, given to Field Marshal Earl Roberts by Tuta Nihoniho, 1901, National Army Museum United Kingdom (NAM U.K.), 1971-01-25-5-2; *Sheffield Daily Telegraph* (UK), 31 January 1901, p. 9.

136 Field Marshal Earl Roberts to M. H. Tuta Nihoniho, 5 February 1901, NAM U.K., 1971-01-25-5-1.

137 *Reading Mercury* (UK), 16 February 1901, p. 10; *Times* (London), 30 January 1901, p. 6; *Times* (London), 9 February 1901, p. 14; *Sheffield Daily Telegraph* (UK), 31 January 1901, p. 9.

138 *Otago Witness*, 5 February 1902, p. 73.

139 *Western Times* (Exeter), 3 April 1902, p. 6.

140 *Otago Witness*, 5 February 1902, p. 73.

141 *Poverty Bay Herald*, 26 May 1900, p. 3.

142 *Poverty Bay Herald*, 9 April 1901, p. 2.

143 *Poverty Bay Herald*, 22 May 1900, p. 4.

144 *AJHR*, 1900, H-6I, p. 9.

145 Claudia Orange (ed.), *Dictionary of New Zealand Biography. Volume Two, 1870–1900*, Wellington: Bridget Williams Books & Department of Internal Affairs, 1993, p. 306; *Evening Post*, 29 September 1899, p. 5.

146 *Evening Post*, 30 March 1900, p. 5.

147 *Evening Post*, 29 March 1900, p. 5; *Hawke's Bay Herald*, 2 April 1900, p. 4. The *Evening Post* claimed that '[n]ever in the history of the Basin Reserve had there been a crowd of equal proportions'.

148 *Evening Post*, 28 March 1900, p. 6; *Evening Post*, 29 March 1900, p. 5.

149 Angela Ballara, 'Parata, Katherine Te Rongokahira – Biography', *Dictionary of New Zealand Biography. Te Ara — the Encyclopedia of New Zealand*, updated 1 September 2010.

150 *Poverty Bay Herald*, 5 June 1900, p. 4.

151 *Hawke's Bay Herald*, 2 April 1900, p. 4.

152 *Auckland Star*, 6 July 1900, p. 5; *Otago Witness*, 21 June 1900, p. 8.

153 *Auckland Star*, 3 July 1900, p. 4; *Evening Star*, 25 September 1902, p. 2.

154 *Evening Post*, 15 October 1900, p. 5.

155 A-080-025, ATL.

156 *New Zealand Herald*, 21 January 1901, p. 4.

157 *AJHR*, 1900, H-19A, pp. 1–2.

158 *AJHR*, 1900, H-6I, p. 3.

159 *Taranaki Herald*, 20 October 1899, p. 2; John Crawford with Ellen Ellis, *To Fight for the Empire*, Auckland: Reed Books, 1999, p. 26; John Edward Thomas Burnett, diary, 31 March 1900, NAM NZ 2004.549.

160 *Ashburton Guardian*, 19 January 1900, p. 2.

161 Jim Christie to Jack Christie, 10 May 1903, NAM NZ 1991.2451.

162 *Wanganui Collegian*, No. 53, August 1900, p. 18.

163 *New Zealand Herald*, 1 March 1900, p. 5.

164 *Nelson Evening Mail*, 4 January 1902, p. 2.

165 *Timaru Herald*, 3 February 1902, p. 2.

166 *AJHR*, 1902, A-1, p. 30.

167 *Auckland Star*, 3 November 1899, p. 4.

168 *Bay of Plenty Times*, 22 January 1902, p. 2; *Auckland Star*, 21 January 1902, p. 5.

169 *Bay of Plenty Times*, 22 January 1902, p. 2; *Auckland Star*, 21 January 1902, p. 5.

170 Letters from J. Carroll received in London, reports of speeches, particulars of Maori Coronation Contingent, SEDDON 1 7 29, ANZ. The Coronation Contingent also included men from South Island iwi Ngāti Mamoe, as well as North Island iwi Ngāti Ata, Ngāi Te Ūpokoiri, Tūwharetoa, Te Whānau-ā-Apanui, Ngāiterangi, Ngāti Whatua, Ngāti Awa and Ngāti Hau.

171 *Evening Post*, 7 January 1903, p. 5; *Taranaki Herald*, 15 November 1881; William Bazire Messenger to Arthur Douglas, 15 May 1902, AD1 404 bo D1902/3229, ANZ.

172 *Evening Post*, 20 March 1902, p. 4; Roll of the Māori Coronation Contingent, SEDDON1 7 29, ANZ.

173 *Thames Star*, 29 May 1901, p. 2.

174 *AJHR*, 1907, G-1B, pp. 5–6.

175 *Nelson Evening Mail*, 3 June 1902, p. 3.

176 Angela Ballara, 'Te Waharoa, Tupu Atanatiu Taingakawa – Biography', *Dictionary of New Zealand Biography, Te Ara — the Encyclopedia of New Zealand*, updated 1 September 2010; *Auckland Star*, 3 November 1899, p. 4.

177 *Manawatu Evening Standard*, 23 January 1902, p. 2.

178 *Akaroa Mail and Banks Peninsula Advertiser*, 24 January 1902, p. 2; *Lyttelton Times*, 22 January 1902, p. 7.

179 *New Zealand Herald*, 22 January 1902, p. 5.
180 South African War Reference Papers, n.d., William Pitt; Attestation Form, 21 January 1902, William Pitt, AABK 18805 W5515 0004511, ANZ.
181 Attestation Form, 12 January 1917, Bernard Reed, AABK 18805 W5515 0004667, ANZ.
182 Attestation Form, 6 January 1902, Bernard Reed, AABK 18805 W5515 0004667, ANZ.
183 *AJHR*, 1902, A-7, p. 34.
184 *Ashburton Guardian*, 29 December 1900, p. 2.

Chapter Five: 'Yelling yahoos in yellow'

1 *Otago Daily Times*, 20 May 1901, p. 5.
2 *Timaru Herald*, 13 July 1899, p. 2.
3 *Observer*, 7 October 1899, p. 7; *Feilding Star*, 3 October 1988, p. 2.
4 *Evening Post*, 20 June 1901, p. 5; *AJHR*, 1901, H-19A, pp. 17, 19, 42. Other estimates of the number of Volunteers involved ranged from four to 'sixty or seventy'.
5 *Evening Post*, 20 June 1901, p. 5; *AJHR*, 1901, H-19A, pp. 9, 17, 42.
6 *AJHR*, 1901, H-19A, p. 16.
7 *AJHR*, 1901, H-19A, p.16.
8 *AJHR*, 1901, H-19A, p. 40.
9 *NZPD*, 116 (1901), pp. 33–34.
10 *NZPD*, 116 (1901), p. 34.
11 *Star*, 30 January 1902, p. 1.
12 *Hawke's Bay Herald*, 28 January 1902, p. 3; *Wanganui Herald*, 29 January 1902, p. 2; *Patea Mail*, 31 January 1902, p. 4.
13 *Clutha Leader*, 31 January 1902, p. 6.
14 *Wanganui Herald*, 29 January 1902, p. 2.
15 *New Zealand Herald*, 21 March 1902, p. 6.
16 *Observer*, 29 March 1902, p. 2.
17 *Wanganui Herald*, 29 January 1902, p. 2; *Wanganui Herald*, 10 October 1902, p. 2; Michael Sullivan, Attestation-Form, 2 January 1901, AABK 18805 0005408, ANZ.
18 *New Zealand Herald*, 14 August 1902, p. 7.
19 *Tuapeka Times*, 3 March 1900, p. 2; *Otago Daily Times*, 23 February 1900, p.6; *Clutha Leader*, 2 March 1900, p. 5.
20 *Otago Witness*, 26 September 1900, p. 25.
21 *Otago Witness*, 26 September 1900, p. 25; *Otago Daily Times*, 18 September 1900, p. 7.
22 Asiatic Restriction Act 1896 (60 Victoriae 1896 No. 64), p. 233; *Post*, 29 October 1900, p. 5.
23 *Otago Witness*, 26 September 1900, p. 25.
24 Opium Prohibition Act 1901 (1 EDW VII 1901 No. 26).
25 *Evening Post*, 2 April 1901, p. 4.
26 *Evening Post*, 4 April 1901, p. 4.
27 *Bush Advocate*, 4 April 1901, p. 3.
28 *Evening Post*, 27 January 1902, p. 5.
29 *Evening Star*, 27 January 1902, p. 4.
30 *Evening Post*, 27 January 1902, p. 5; *Nelson Evening Mail*, 30 January 1902, p. 2; *Evening Post*, 3 February 1902, p. 6; Leonard Noly Jacobs, Attestation-Form, 28 March 1900, AABK 18805 W5515 0002799, ANZ.
31 *Colonist*, 31 January 1902, p. 4; *Evening Star*, 1 February 1902, p. 6.
32 *Evening Post*, 27 January 1902, p. 5; *Evening Star*, 27 January 1902, p. 4.
33 *Nelson Evening Mail*, 30 January 1902, p. 2; *Evening Post*, 2 May 1902, p. 5; Thomas Page, Attestation-Form, 14 January 1902, AABK 18805 W5515 0004314, ANZ; Frances Raikes, Attestation-Form, AABK 18805 W5515 0004631, ANZ.
34 *Evening Post*, 1 May 1902, p. 6.
35 *Evening Post*, 1 May 1902, pp. 5–6; 2 May 1902, p. 5.
36 *Evening Post*, 2 May 1902, p. 5.
37 *Free Lance*, 10 May 1902, p. 13.
38 *Evening Post*, 1 May 1902, p. 5.
39 *Daily Telegraph* (UK), 26 July 1900, p. 8; *Feilding Star*, 6 June 1900, p. 2; *Auckland Star*, 2 June 1900, p. 13; *Press*, 10 July 1900, p. 6; *Auckland Star*, 17 July 1900, p. 5.
40 Frank Perham, diary, 1900–1901, 5 December 1900, NAM NZ, 2000.736; Herbert Ernest Hart, diary, 8 February 1902 – 5 January 1903, 23 July 1902, NAM NZ, 1990.1024.
41 Percy Mowlem, Sixth Contingent Crime and Offence Reports, March 1901 – March 1902, AD34 7, ANZ.
42 William Christian Laurie, 6th Contingent crime and offence reports, March 1901 – March 1902, AD34 7 7 8032, ANZ.
43 *Natal Witness*, 23 June 1902, p. 5; *Natal*

Advertiser, 2 July 1902, p. 5; *Natal Advertiser*, 8 July 1902, p. 5.

44 Roll of Individuals Entitled to the South Africa Medal and Clasps, 69th Company, 14th Battalion Imperial Yeomanry, National Archives, Kew, WO 100/126, pp. 85, 92; Statement of charges, Abdol Salih, Circuit Court for the District of Worcester, together with the Districts of Robertson, Montagu, Ceres, Tulbagh, Beaufort West, Fraserburg, Sutherland and Prince Albert, Cape Archive Repository, Cape Supreme Court, CSC 1/2/1/125 Ref 1; *Cape Argus*, 5 June 1901, p. 5. Reid's name was incorrectly spelt as 'Reed' in both the trial documents and the newspaper accounts. The soldier injured at Worcester was Saddler Robert Samuel Reid, No. 1678, Fifth Contingent. The Kew medal rolls indicate that Corporal A. Luck, No. 16473, was 'left at Worcester', and list H. Dixie, No. 16438, as 'deceased 2, 6, 01'.

45 *Cape Argus*, 5 June 1901, p. 5; Proceedings of Medical Board, 14 December 1901, Robert Samuel Reid, AABK 18805 W5515 0004690, ANZ.

46 *West Coast Times*, 25 July 1901, p. 4.

47 *West Coast Times*, 25 July 1901, p. 4; *Feilding Star*, 22 July 1901, p. 2; *Manawatu Evening Standard*, 23 July 1901, p. 4.

48 *Evening Telegraph* (Dundee), 7 June 1901, p. 4; *Gloucester Citizen* (UK), 7 June 1901, p. 3; *Sheffield Daily* (UK), 7 June 1901, p. 5; *Sunderland Daily Echo and Shipping Gazette* (UK), 7 June 1901, p. 6.

49 *Times* (London), 5 June 1901, p. 10; *Times* (London), 7 June 1901, p. 5.

50 Frank Perham, diary, 2 June 1901, NAM NZ, 2000.736.

51 *Nelson Evening Mail*, 25 July 1901, p. 2.

52 NZ Camps at Worcester, orders, 3 June 1901, AD34 7 7 8030, ANZ.

53 Registrar of the Circuit Court's report on Salih's case, n.d., Cape Archive Repository, Cape Supreme Court, CSC 1/2/1/125 Ref 1.

54 *Cape Argus*, 5 June 1901, p. 5; *Feilding Star*, 22 July 1901, p. 2; *Wanganui Chronicle*, 23 July 1901, p. 3; *Nelson Evening Mail*, 25 July 1901, p. 2.

55 Resident Magistrate of Worcester to the Inspector of Public Works, 1 August 1901, Cape Archive Repository, Cape Town, Public Works Department, PWD 1/2/38 Ref. B25; A. S. Howard to Secretary for Public Works, 13 August 1901, PWD 1/2/38 Ref. B25.

56 Medical certificate, 3 March 1902, 18805 W5515 0004690, ANZ.

57 Cecil Street, Leslie O'Callaghan & Victor Kelsall to Montague Lewin, 23 January 1902, AABK 18805 W5515 0005698, ANZ.

58 Montague Lewin to Brigade Major, n.d., AABK 18805 W5515 0005698, ANZ.

59 John Rose, Attestation-Form, 29 January 1900, AABK 18805 W5515 0004834, ANZ; *Press*, 19 February 1900, p. 5; *AJHR*, 1900 H-6, p. 11.

60 *Star*, 19 March 1900, p. 3.

61 *Western Mail* (Perth), 10 March 1900, p. 35; *Otago Witness*, 3 May 1900, p. 30; Extract from Diary of Major Thomas Jowsey, 4 March 1900, AABK 18805 W5515 0004834, ANZ.

62 *Marlborough Express*, 30 April 1900, p. 3.

63 *Poverty Bay Herald*, 28 April 1900, p. 2; *Marlborough Express*, 30 April 1900, p. 3; Extract from Diary of Major Thomas Jowsey, 4 March 1900, AABK 18805 W5515 0004834, ANZ.

64 *Timaru Herald*, 1 May 1900, p. 3; *Marlborough Express*, 30 April 1900, p. 3.

65 Third Contingent Diary (Major Jowsey), February 1900 – May 1901, 12 March 1900, AD1 345, 8020, ANZ.

66 Third Contingent Diary (Major Jowsey), February 1900 – May 1901, 29 March 1900, AD1 345, 8020, ANZ.

67 *Auckland Star*, 24 July 1900, p. 5.

68 *Otago Witness*, 5 April 1900, p. 26.

69 *Auckland Star*, 15 October 1901, p. 5.

70 *Bruce Herald*, 11 April 1902, p. 3.

71 *Wanganui Herald*, 7 April 1902, p. 2.

72 Isidore M. Cohen to Richard J. Seddon, 12 October 1901, AABK 18805 W5515 0000970, ANZ.

73 *Manawatu Times*, 16 June 1903, p. 2; Isidore M. Cohen (alias J. M. Cantos), 23 January 1904, AAWC 17671 W6105 Box 3 [1225], ANZ.

74 Sixth Contingent Crime and Offence Reports, March 1901 – March 1902, AD34 7, ANZ.

75 Certificates of Discharge, May 1902,

Notes

76 AABK 18805 W5515 0005071, ANZ; Sixth Contingent Crime and Offence Reports, March 1901 – March 1902, AD34 7, ANZ.

76 Sixth Contingent Crime and Offence Reports, March 1901 – March 1902, AD34 7, ANZ.

77 Charles Morris, Attestation-Form, 11 January 1901; J. H. Banks to Commandant, Defence Forces, 16 February 1901; J. H. Banks to Defence Headquarters Office, 17 February 1901, AABK 18805 W5515 0004030, ANZ.

78 Charles Morris, Attestation-Form, 11 January 1901; R. J. S. Seddon to C. W. Morris, 30 May 1907; Charles Morris to Under Secretary for Defence, 9 November 1903, AABK 18805 W5515 0004030, ANZ.

79 Sixth Contingent Crime and Offence Reports, March 1901 – March 1902, AD34 7, ANZ; Roll of Individuals Entitled to the Queen's South Africa Medal and Clasps, 6th New Zealand Mounted Rifles, n.d., National Archives, Kew, UK, WO 100/294, p. 245.

80 Thomas Jowsey to Commandant New Zealand Forces, 9 May 1900, AD1 365, D1900/4368, ANZ.

81 Diary of Major Thomas Jowsey, 12 April 1900 – 9 May 1900, AD1 365, D1900/4368, ANZ.

82 Arthur Douglas to Officer Commanding District, Dunedin, 20 August 1902; Arthur Douglas to Officer Commanding Militia and Volunteer District, Dunedin, 9 September 1902, AABK 18805 W5515 0000754, ANZ.

83 *Wanganui Collegian*, No. 52, April 1900, p. 13.

84 *Evening Post* 11 October 1901, p. 5; Statement of Mr. Hurst Davies [sic] late of the 6th Contingent, n.d., AABK 18805 W5515 0001401, ANZ.

85 J. H. Banks to Commandant of the Forces, 5 August 1901, AABK 18805 W5515 0004791, ANZ; Roll of Individuals Entitled to the South African Medal and Clasps: Sixth Contingent, n.d., National Archives, Kew, UK, WO 100/294, p. 242; Sixth Contingent Crime and Offence Reports, March 1901 – March 1902, AD34 7, ANZ.

86 *Evening Post* 11 October 1901, p. 5; N. Hurst Davis to Under Secretary for Defence, 5 April 1902, AABK 18805 W5515 0001401, ANZ.

87 Memorandum from the Premier's Office to Lord Ranfurly, 21 January 1902; Under-Secretary Defence to the chairman of the Public Petitions Committee, 30 September 1903, AABK 18805 W5515 0002670, ANZ; Samuel Haslett to Commander in Chief British Army, 27 September 1902, AD1 421 by D1903-1086, ANZ.

88 Onesimus Howe to the Secretary of Defence, n.d., AABK 18805 W5515 0002670, ANZ.

89 Memorandum from the Premier's Office to Lord Ranfurly, 21 January 1902; Onesimus Howe, Attestation-Form dated 14 January 1901, AABK 18805 W5515 0002670, ANZ.

90 Sixth Contingent Crime and Offence Reports, March 1901 – March 1902, AD34 7, ANZ; Lord Ranfurly to Joseph Ward, 19 June 1902, AABK 18805 W5515 0002670, ANZ.

91 *Auckland Star*, 17 March 1902, p. 5; *Manawatu Evening Standard*, 5 February 1902, p. 3; *Wanganui Herald*, 6 February 1902, p. 2; *Hawera & Normanby Star*, 13 February 1902, p. 2; *Bruce Herald*, 18 February 1902, p. 1.

92 *Manawatu Evening Standard*, 5 February 1902, p. 3.

93 Frank Perham, diary, 30 May 1901, NAM NZ, 2000.736.

94 Frank Perham, diary, 31 May 1901, NAM NZ, 2000.736.

95 Frank Perham, diary, 22 January 1901; 24 January 1901, NAM NZ, 2000.736.

96 Charles Borland Tasker, Reg. No. 1882/10727, Department of Internal Affairs, NZ Births, Deaths and Marriages; Claudia Orange (ed.), *The Dictionary of New Zealand Biography. Volume Two, 1870–1900*, Wellington: Bridget Williams Books & Department of Internal Affairs, 2003, pp. 504–5.

97 J. Babington to P. O. O., Woolwich, 14 April 1904, AABK 18805 W5515 0005468, ANZ; Sixth Contingent Crime and Offence Reports, March 1901 – March 1902, AD34 7 ANZ; Charles Borland Tasker, diary, 9 July 1901, NAM NZ, 1991.1956.

98 Charles Borland Tasker, diary, 17 July 1901; letter, author unknown, 28 July 1919, AABK 18805 W5515 0005468, ANZ; *AJHR*, 1902, A-1, p. 23; *Bay of Plenty Times*, 20 December 1901, p. 2.

99 *Evening Post*, supplement, 14 December 1901, p. 7.

100 *Evening Post*, 1 November 1901, p. 5; *Manawatu Evening Standard*, 5 November 1901, p. 4; *Feilding Star*, 4 November 1901, p. 2.
101 *Evening Post*, supplement, 9 November 1901, p. 6.
102 *AJHR*, 1902, A-1, p. 23.
103 E. W. D. Ward to Under Secretary of State, Colonial Office, 4 December 1901, ACHK 8604 G1/135, 1902/54, ANZ.
104 Joseph Chamberlain to Lord Ranfurly, 11 December 1901, ACHK 8604 G1/135, 1902/54, ANZ.
105 W. P. Reeves to Richard J. Seddon, 13 December 1901, AABK 18805 W5515 0005468 ANZ; *Bush Advocate*, 10 February 1902, p. 2.
106 *Evening Post*, 11 January 1902, p. 5.
107 *Evening Post*, supplement, 9 November 1901, p. 6.
108 Francis Evenson Beamish to Adjutant General, 5 July 1919, AABK 18805 W5515 0005468, ANZ.
109 Francis Evenson Beamish to Adjutant General, 5 July 1919; C. E. Andrews to C. B. Tasker, 12 December 1919, AABK 18805 W5515 0005468, ANZ; Roll of Individuals Entitled to the Queen's South Africa Medal and Clasps, 6th New Zealand Mounted Rifles, n.d., National Archives, Kew, UK, WO 100/294, p. 198.
110 *Evening Post*, 4 September 1900, p. 5.
111 *Tuapeka Times*, 16 November 1901, p. 3.
112 *Otago Witness*, 12 March 1902, p. 28.
113 Russell Family Papers, 23 September 1901, Alexander Turnbull Library (ATL) MS Papers 3854, p. 5; Leonard George Armstrong, diary, 20 August 1901, NAM NZ 2004.530.
114 *Thames Advertiser*, 12 October 1899, p. 1; *West Coast Times*, 28 April 1900, p. 3; *Manawatu Daily Times*, 29 January 1901, p. 4; *Nelson Evening Mail*, 18 January 1902, p. 2.
115 John Edward Thomas Burnett, diary, 17 October 1900, NAM NZ 2004.549.
116 *Press*, 17 January 1900, p. 6; *Press*, 24 January 1900, p. 5.
117 *Press*, 24 January 1900, p. 5.
118 *Oamaru Mail*, 31 May 1901, p. 4.
119 *Wanganui Collegian*, No. 55, April 1901, p. 23.
120 *Press*, 17 May 1900, p. 5; *Star*, 17 May 1900, p. 3; *Auckland Star*, 18 August 1900, p. 5; *Wanganui Collegian*, No. 54, December 1900, p. 14; *Press*, 28 February 1900, p. 4; *Hastings Standard*, 21 October 1901, p. 3; *Manawatu Evening Standard*, 28 November 1901, p. 4.
121 *Wanganui Collegian*, No. 54, December 1900, p. 13.
122 *Otago Witness*, 28 November 1900, p. 29.
123 *Otago Daily Times*, 11 January 1902, p. 10.
124 *Clutha Leader*, 19 January 1900, p. 5.
125 *Wanganui Collegian*, No. 53, August 1900, p. 18.
126 *Mataura Ensign*, 5 May 1900, p. 2.
127 Frank Perham, diary, 12 January 1901, NAM NZ, 2000.736.
128 Frank Perham, diary, 30 January 1901; 3 February 1901, NAM NZ, 2000.736.
129 Frank Perham, diary, 1 February 1901, NAM NZ, 2000.736.
130 Luke Perham to Theresa Perham, 28 June 1900, National Army Museum, New Zealand (NAM NZ), 2003.7, p. 3; Leonard George Armstrong, diary, 18 July 1901, NAM NZ, 2004.530. *Star*, 24 July 1900, p. 1.
131 Frank Perham, diary, 7 March 1901, NAM NZ, 2000.736; Attestation-Form, 16 Jan 1901, AABK 18805 W5515 0002368, ANZ; Sixth Contingent Crime and Offence Reports, March 1901 – March 1902, AD34 7, ANZ.
132 *Evening Post*, 22 January 1900, p. 5; William Hall-Jones to Lord Ranfurly, 24 September 1902, AABK 18805 W5515 0001475, ANZ.
133 James Clarke to Katie Clarke, transcript of letter, 15 June 1902, in J. N. Clarke, *Still Jogging Along: Being the Diary and Some Letters of Pte. J. N. Clarke, 9th Contingent N.Z.M.R.*, Dunedin: Otago Military Museum [1989], p. 27.
134 *Wanganui Collegian*, No. 54, December 1900, pp. 14–15.
135 Secretary for the Interior to Adjutant General, 8 August 1930, AAYS 8663 AD34 Box 2-2 8006, ANZ.
136 P. J. Mulhern to General Headquarters [4 January 1929], AAYS 8663 AD34 Box 2-2 8006 [p. 1], ANZ.
137 *Taranaki Daily News*, 2 October 1900, p. 2; *Southland Times*, 26 September 1903, p. 3.

138 *Tuapeka Times*, 28 April 1900, p. 3.
139 *NZPD*, 113 (1900), p. 148.
140 *Ohinemuri Gazette and Upper Thames Warden*, 10 August 1903, p. 2.
141 *Ohinemuri Gazette and Upper Thames Warden*, 10 August 1903, p. 2.
142 *Manawatu Evening Standard*, 13 October 1903, p. 8; *Hawera & Normanby Star*, 27 October 1903, p. 2.
143 *Clutha Leader*, 28 December 1900, p. 4.
144 *Bay of Plenty Times*, 13 July 1903, p. 2; *Marlborough Express*, 11 July 1903, p. 2; *Taranaki Herald*, 11 July 1903, p. 2; *Grey River Argus*, 11 July 1903, p. 3; *Feilding Star*, 11 July 1903, p. 2; *Bruce Herald*, 8 April 1904, p. 2.
145 *Otago Daily Times*, 18 June 1902, p. 8.
146 *Evening Star*, 11 July 1903, p. 6; *Wanganui Chronicle*, 11 July 1903, p. 5; *New Zealand Gazette*, No. 69, 3 September 1903, p. 1913.
147 *Evening Post*, 14 September 1903, p. 7.
148 *Manawatu Evening Standard*, 13 October 1903, p. 8.
149 Edward Chaytor to William Alexander, 2 March 1906, AD1 429 bu D1903/2770 ANZ; William Alexander to Edward Chaytor, 23 July 1906, AD1 429 bu D1903/2770, ANZ.
150 *Manawatu Evening Standard*, 23 August 1921, p. 5.
151 *Evening Post*, 3 April 1926, p. 4.
152 *Auckland Star*, 22 October 1936, p. 6.
153 Sidney Skerman to Under-Secretary for Defence, 21 December 1901, AD1 394 z D1902/426, ANZ.
154 J. S. Purdy to Richard J. Seddon, 21 June 1901, AD1 394 z D1902/426, ANZ; Walter Empson to Richard J. Seddon, 11 August 1901, AD1 394 z D1902/426, ANZ.
155 Herbert Pilcher to Under-Secretary for Defence, 18 February 1902, AD1 394 z D1902/426, ANZ.
156 Lord Ranfurly to Richard J. Seddon, 11 January 1902, AD1 394 z D1902/426, ANZ.
157 Amos McKegg to his brother, 1 July 1900, Auckland War Memorial Museum (AWMM), McKegg, Amos, papers, MS 934.
158 *Clutha Leader*, 19 Jan 1900, p. 5.
159 *New Zealand Tablet*, 1 Oct 1903, p. 18.
160 *Auckland Star*, 28 September 1901, p. 5; Staff Officer to Emily Markham, 10 April 1902, AABK 18805 W5515 0003727, ANZ.
161 *Press*, 22 May 1901, p. 5.
162 Albert Andrew to Commandant New Zealand Forces, 13 January 1902, AABK 18805 W5515 0003727, ANZ.
163 *Taranaki Daily News*, 15 February 1905, p. 3.
164 *Observer*, 7 March 1908, p. 4.
165 Kaffrarian Rifles, Roll of Individuals Entitled to the South African Medal and Clasps, under the Army Order granting the Medal on 1st April, 1901, 15 July 1901, WO 100/254, p. 24; Albert Andrew to Commanding Officer New Zealand Forces, 15 December 1903, ATL MS-Papers-1619-036.
166 Albert Andrew to Commanding Officer New Zealand Forces, 15 December 1903, ATL MS-Papers-1619-036.
167 James B. Heywood to William Pitt, 26 June 1905, AABK 18805 W5515 0004511, ANZ.
168 Richard J. Seddon to High Commissioner, 31 July 1905, AABK 18805 W5515 0004511, ANZ, pp. 1–2; James B. Heywood to William Pitt, 14 June 1906, AABK 18805 W5515 0004511, ANZ.
169 Richard J. Seddon to General Officer Commanding, Cape Town, 24 December 1902, AABK 18805 W5515 0005765, ANZ.
170 Henry Laing to Richard J. Seddon, 18 December 1902, AABK 18805 W5515 0005765, ANZ.
171 *Wanganui Chronicle*, 30 May 1903, p. 4; *Taranaki Herald*, 3 June 1903, p. 3.
172 *NZPD*, 124 (1903), p. 266.
173 Officer Commanding, South Africa: 12 February 1904, AD1 438 D1904/737, ANZ.
174 Prime Minister's office to Richard J. Seddon, 12 February 1904, AD1 438 D1904/737, ANZ.
175 *Wairarapa Daily Times*, 14 April 1902, p. 3.
176 *Auckland Star*, 24 September 1902, p. 4.
177 *Star*, 4 September 1902, p. 3.
178 Staff Diary, Natal Command, May 1902, UK National Archives, WO 32/8114, pp. 12, 14; Staff Diary, Natal Command, June 1902, UK National Archives, WO 32-8114, p. 24.
179 Staff Diary, Natal Command, June 1902, UK National Archives, WO 32-8114, pp. 4, 24.
180 *North Otago Times*, 22 July 1902, p. 1.
181 Herbert Ernest Hart, diary, 20 June 1902,

NAM NZ, 1990.1024.
182 Proceedings of a Court of Enquiry, 2 July 1902, Opinion, ATL MS-Papers-5755-57 [pp. 1-2]; J. F. Jaques to R. S. Charleston, 5 July 1902, ATL MS-Papers-5755-57.
183 Proceedings of a Court of Enquiry, 2 July 1902, ATL MS-Papers-5755-57, pp. 2–3.
184 Proceedings of a Court of Enquiry, 2 July 1902, ATL MS-Papers-5755-57, pp. 6–7.
185 Proceedings of a Court of Enquiry, 2 July 1902, ATL MS-Papers-5755-57, p. 5.
186 Herbert Ernest Hart, diary, 21 July 1902, NAM NZ 1990.1024.
187 J. F. Jaques to R. S. Charleston, 5 July 1902, ATL MS-Papers-5755-57.
188 Claims against 8th N.Z. Contingent, Looting, 13 July 1902; Proceedings of a Court of Enquiry, 2 July 1902, Opinion, ATL MS-Papers-5755-57.
189 Proceedings of a Court of Enquiry, 2 June 1902, Abstract of Claims, ATL MS-Papers-5755-57; *AJHR*, 1904, Session I, B-1, p. 69.
190 N. B. Fetherstonhaugh to Governor of New Zealand, 13 July 1902, ATL MS-Papers-5755-57.
191 *Wanganui Chronicle*, 4 September 1902, p. 5; *Ashburton Guardian*, 4 September 1902, p. 2; *Manawatu Daily Times*, 4 September 1902, p. 2; *Bush Advocate*, 4 September 1902, p. 2; *West Coast Times*, 4 September 1902, p. 2; *Marlborough Express*, 4 September 1902, p. 4; *Colonist*, 4 September 1902, p. 2.
192 *Wanganui Chronicle*, 4 September 1902, p. 5.
193 Herbert Ernest Hart, diary, 11 July 1902, NAM New Zealand, 1990.1024.
194 *NZPD*, 122 (1902), pp. 110–12.
195 *Star*, 4 September 1902, p. 3.
196 *Evening Post*, 5 September 1902, p. 5.
197 Staff Diary, Natal Command, June 1902, WO 32/8114, p. 24; South African Contingents: (Deaths of Members of) In South Africa and Since Leaving South Africa, and Particulars as to Locality, etc., of Graves, *AJHR*, 1903, H-6A p. 10; Proceedings of a Court of Inquiry, 16 June 1902, AABK 18805 W5515 0001098, ANZ.
198 *Evening Post*, 5 September 1902, p. 5.
199 Staff Diary, Natal Command, June 1902, WO 32/8114, p. 24.
200 *Sydney Morning Herald*, (Aus.), 9 September 1902, p. 5.
201 *Evening Post*, 5 September 1902, p. 5.
202 *Ashburton Guardian*, 5 September 1902, p. 2.
203 *Ashburton Guardian*, 4 September 1902, p. 2; *Marlborough Express*, 4 September 1902, p. 4; *Manawatu Daily Times*, 4 September 1902, p. 2; *Taranaki Herald*, 4 September 1902, p. 2.
204 *Wanganui Herald*, 4 September 1902, p. 3; *Feilding Star*, 4 September 1902, p. 2; *West Coast Times*, 5 September 1902, p. 2.
205 *Manawatu Daily Times*, 8 September 1902, p. 2.
206 *Otago Witness*, 10 September 1902, p. 27.
207 *North Western Advocate and the Emu Bay Times* (Aus.), 10 September 1902, p. 4.
208 *North Western Advocate and the Emu Bay Times* (Aus.), 10 September 1902, p. 4. 'Jinrickshas' (rickshaws) were a common form of conveyance in Cape Town.
209 *Western Australian*, 5 September 1902, p. 6.
210 *Manawatu Daily Times*, 8 September 1902, p. 2.
211 *Natal Advertiser*, Natal, 12 July 1902, p. 3.
212 *Natal Advertiser*, Natal, 12 July 1902, p. 3.
213 *Wanganui Herald*, 4 September 1902, p. 3; *Taranaki Herald*, 4 September 1902, p. 2; *Otago Witness*, 10 September 1902, p. 27.
214 *Wanganui Collegian*, No. 60, December 1902, p. 12.
215 *Wanganui Herald*, 9 September 1902, p. 3.
216 Herbert Ernest Hart, diary, 12–13 July 1902, NAM New Zealand, 1990.1024.
217 *AJHR*, 1902, H-6C, p. 278.
218 *New Zealand Herald*, 20 August 1902, p. 5.
219 Albert Andrew to Commandant New Zealand Forces, telegram, 24 June 1902, AD1 405 D1902/3177.
220 Albert Andrew to Commandant New Zealand Forces, telegram, 12 July 1902, AD1 405 D1902/3177.
221 *AJHR*, 1902, H-6C, p. xii.
222 *Wanganui Chronicle*, 25 November 1903, p. 7.
223 *Auckland Star*, 16 September 1907, p. 5.
224 *Observer*, 7 June 1902, p. 2.

Chapter Six: 'Maimed, crippled and completely broken'

1. *Wanganui Collegian*, No. 54, December 1900, p. 14.
2. *AJHR*, 1903, H-6A, pp. 1–15.
3. *NZPD* 110 (1899), p. 79; *AJHR*, 1903, H-6A, pp. 1–15.
4. *Press*, 11 December 1899, p. 5; *Wanganui Collegian*, No. 53, August 1900, p. 7; *Auckland Star*, 17 December 1900, p. 3; *Auckland Star*, 11 April 1907, p. 3; Roll of individuals entitled to the South African Medal and clasps. R.A.M.C. No. 3 Stationary Hospital, 24 September 1901, National Archives, Kew, UK, WO 100/223, p. 4.
5. Extract from D02/156, John Thomas, AABK 18805 W5515 0005536, ANZ.
6. H. R. Cummins to Walter Armstrong, 28 March 1900, AAOG W3559 23650 Box 668 19004, ANZ.
7. *Wanganui Chronicle*, 5 May 1900, p. 3.
8. Deputy Assistant Director Army Medical Service to Walter Armstrong, 3 May 1900, AAOG W3559 23650 Box 668 1900/4, ANZ.
9. Will, Walter Douglas Armstrong, 3 January 1900, AAOG W3559 23650 Box 668 1900/4, ANZ.
10. H. Cummins to Walter Armstrong, 28 April 1900, AAOG W3559 23650 Box 668 1900/4, ANZ; Roll of individuals entitled to the South African Medal and clasps. R.A.M.C. 14th Brigade Field Hospital, November 1901, National Archives, Kew, UK, WO 100/223, p. 107; *Wanganui Collegian*, No. 53, August 1900, p. 7.
11. *Wanganui Chronicle*, 5 May 1900, p. 3.
12. *AJHR*, 1903, H-6A, pp. 1–15.
13. Richard J. Seddon to C. W. Harvey, 20 August 1900, AABK 18805 W5515 0002406, ANZ.
14. C. W. Harvey to Richard Seddon, 23 August 1900, John Allen Harvey, AABK 18805 W5515 0002406, ANZ.
15. *Evening Star*, 20 August 1900, p. 3; *Clutha Leader*, 24 August 1900, p. 5.
16. *Oamaru Mail*, 24 August 1900, p. 3.
17. *Otago Witness*, 14 August 1901, p. 10; *Otago Witness*, 23 October 1901, p. 35.
18. *Otago Witness*, 28 August 1901, p. 35.
19. *Clutha Leader*, 1 November 1901, p. 5.
20. *Press*, 23 December 1899, p. 5.
21. *Marlborough Express*, 21 December 1899, p. 1.
22. *Newcastle Morning Herald and Miners' Advocate* (Aus.), 29 December 1899, p. 5; *Australian Town and Country Journal*, 13 January 1900, p. 14.
23. *Bunbury Herald* (Aus.), 21 December 1899, p. 3; *Kalgoorlie Miner*, 21 December 1899, p. 5.
24. *Wairarapa Daily Times*, 23 January 1900, p. 3.
25. *Nelson Evening Mail*, 6 January 1900, p. 3; *Nelson Evening Mail*, 17 January 1900, p. 3.
26. *New Zealand Herald*, 22 May 1901, p. 7.
27. *Ohinemuri Gazette*, 18 May 1903, p. 2.
28. *National Advocate* (Aus.), 27 January 1900, p. 2.
29. *Auckland Star*, 23 August 1900, p. 5; *Manawatu Evening Standard*, 23 August 1900, p. 2; *Taranaki Herald*, 22 August 1900, p. 3. The *Hawera & Normanby Star* reported that '[t]he Premier has also received a correction to the effect that it was Private Bolton, of Nelson, not Private Bottom, of Denniston, who was dangerously wounded at Ottoshoop' (22 August 1900, p. 2.). However, the *Colonist* reported in November that Bolton had written a letter home in which he made no mention of being wounded (7 November 1900, p. 3).
30. Certificate of Discharge, 25 October 1902, AABK 18805 W5515 0000479, ANZ.
31. *Marlborough Express*, 23 August 1900, pp. 3–4.
32. *Otago Witness*, 30 August 1900, p. 29.
33. Attestation-Form, Henry Benjamin Bolton, 22 March 1900; Proceedings of Medical Board, 15 August 1901, AABK 18805 W5515 0000459, ANZ.
34. *Marlborough Express*, 6 July 1901, p. 2; *Southland Times*, 6 July 1901, p. 2; *Northern Advocate*, 13 July 1901, p. 8.
35. Kaffrarian Rifles, Roll of Individuals Entitled to the South African Medal and Clasps, under the Army Order granting the Medal on 1st April, 1901, 15 July 1901, WO 100254, p. 24.
36. *Marlborough Express*, 20 July 1901, p. 2.
37. Francis Marion Bates Fisher to Richard

J. Seddon, 12 March 1906, George Henry Fisher, AABK 18805 W5515 0001806, ANZ; *Marlborough Express*, 15 April 1904, p. 1.
38 *NZPD* 110 (1899), p. 85.
39 *AJHR*, 1902, H-6B, p. 14; *Evening Post*, 15 March 1905, p. 7; *AJHR*, 1906, H-14, p. 5.
40 *New Zealand Herald*, 14 June 1900, p. 6; *Otago Daily Times*, 18 June 1900, p. 5.
41 Proceedings of the Medical Board, 8 April 1904, James Vincent Fahey, AABK 18805 W5515 0001734, ANZ.
42 *NZPD* 111 (1900), p. 421.
43 *Otago Witness*, 12 July 1900, p. 31.
44 *Manawatu Evening Standard*, 28 August 1900, p. 4.
45 *Daily Telegraph* (UK), 16 June 1900, p. 8; *Manawatu Evening Standard*, 4 October 1900, p. 4.
46 *Manawatu Evening Standard*, 4 October 1900, p. 4.
47 *Hawke's Bay Herald*, 3 October 1900, p. 3; *Wanganui Herald*, 4 October 1900, p. 2.
48 *Press*, 20 April 1901, p. 9.
49 *Feilding Star*, 27 August 1900, p. 2.
50 *Manawatu Times*, 7 June 1901, p. 2.
51 *Feilding Star*, 19 July 1900, p. 2.
52 *NZPD*, 111 (1900), p. 218.
53 Pilcher was a New Zealander employed in South Africa by the South British Insurance Company. *NZPD*, 111 (1900), p. 219.
54 *Evening Post*, 27 March 1902, p. 6.
55 Alex Willis to Richard J. Seddon, 17 July 1902, AABK 18805 W5515 0005149, ANZ.
56 *Pelorus Guardian and Miners' Advocate*, 11 April 1902, p. 5.
57 Colin McGeorge, 'The Social and Geographical Composition of the New Zealand Contingents', in John Crawford and Ian McGibbon (eds), *One Flag, One Queen, One Tongue*, Auckland: Auckland University Press, 2003, pp. 114–15. Attestation Form, 11 January 1901, AABK 18805 W5515 0002487, ANZ; Attestation-Form, 5 January 1902, AABK 18805 W5515 0006348, ANZ; Attestation-Form, 3 April 1901, AABK 18805 W5515 0002488, ANZ; Attestation-Form, 16 March 1900, AABK 18805 W5515 0004775, ANZ; Attestation-Form, 3 April 1901, AABK 18805 W5515 0004800, ANZ; Attestation Form, 3 April 1901, AABK 18805 W5515 0004776, ANZ; Attestation-Form, 3 April 1901, AABK 18805 W5515 0003062, ANZ; Attestation-Form, 3 April 1901, AABK 18805 W5515 0003063, ANZ; Attestation-Form, 3 April 1901, AABK 18805 W5515 0006382, ANZ.
58 *Otago Daily Times*, 3 March 1902, p. 5; *AJHR*, 1903, H-6A, p. 5.
59 Certificate of Discharge, 6 April 1901, Peter Cooper Donnelly, AABK 18805 W5515 0001519, ANZ.
60 J. Gethin Hughes to James Hemphill, n.d.; James Hemphill to Defence Department, 6 March 1902, AABK 18805 W5515 0002487, ANZ.
61 Casualty Form, John Malone Hemphill, AABK 18805 W5515 0002488, ANZ.
62 *Manawatu Evening Standard*, 4 August 1904, p. 8.
63 *Star*, 21 April 1874, p. 2; Registrar of Births Deaths and Marriages, New Zealand, Leslie Seton Melville, 28 Aug 1874, 1874015008; Alec Melville, 8 October 1875, 187500494; Hugh Melville, 8 October 1875, 1875004995; Ida Melville, 21 May 1877, 1877013152; Hamilton Melville, 20 March 1879, 1879013775; Rowan Melville 1 November 1881, 1881013146; 10 May 1882, 1882002717. Hamilton Melville's birth certificate indicates Louisa Alicia Melville's maiden name was French, and that she was born in Dublin. In 1879, Louisa was 38 while her husband was 43.
64 *Wanganui Collegian*, No. 53, 1900, p. 7.
65 *Star*, 18 April 1891, p. 3; *Poverty Bay Herald*, 15 April 1891, p. 3; Names and Descriptions of Passengers, S.S. *Rimutaka*, UK Archives, BT26, piece 17, item 13.
66 *Star*, 31 December 1891, p. 3.
67 *South Australian Register*, 18 April 1900, p. 5.
68 D. C. Aitken to J. B. K. Farrier, 21 June 1900, Natal Archives, NT52 T1332-1900 Death of L. S. Melville.
69 *Wanganui Herald*, 15 December 1899, p. 3; D. C. Aitken to J. B. K. Farrier, 21 June 1900, Natal Archives, NT52 T1332-1900 Death of L. S. Melville; *Wanganui Collegian*, No. 53, August 1900, p. 18; *Star*, 9 December 1899, p. 5; *Wanganui Herald*, 15 December 1899, p. 3.

70 *Wanganui Collegian*, No. 53, August 1900, p. 7, *Wanganui Herald*, 15 December 1899, p. 3.
71 *Wanganui Collegian*, No. 54, December 1900, p. 10.
72 *Wanganui Collegian*, No. 53, August 1900, pp. 7, 18.
73 *Wanganui Collegian*, No. 53, August 1900, pp. 17–18.
74 *Evening Post*, 11 May 1900, p. 2.
75 *Evening Post*, 11 May 1900, p. 2.
76 D. C. Aitken to J. B. K. Farrier, 21 June 1900, Natal Archives, NT52 T1332-1900 Death of L. S. Melville.
77 *AJHR*, 1900, H-6, p. 9; *AJHR*, 1901, H-6, p. 8; *AJHR*, 1902, H-6, p. 8.
78 *AJHR*, 1897, H-38, p. 3; *AJHR*, 1897, H-38B, p. 2.
79 *Hawera & Normanby Star*, 28 September 1901, p. 2.
80 *Ohinemuri Gazette*, 15 August 1902, p. 2.
81 *Feilding Star*, 21 November 1912, p. 2; *Patea Mail*, 25 October 1882, p. 2.
82 *Wanganui Collegian*, No. 53, 1900, p. 17; *Wanganui Collegian*, No. 52, 1900, p. 13.
83 *Wanganui Collegian*, No. 52, 1900, p. 13.
84 *Tuapeka Times*, 3 July 1901, p. 3; *AJHR*, 1900 H-6A, p. 11.
85 *Wanganui Collegian*, No. 52, 1900, p. 13.
86 *AJHR*, 1900, H-6A, p. 3.
87 *AJHR*, 1900, H-6A, p. 1.
88 *AJHR*, 1900, H-6A, p. 12.
89 *Wanganui Collegian*, No. 55, April 1901, p. 23.
90 *Otago Daily Times*, 11 January 1901, p. 2.
91 *Timaru Herald*, 3 August 1900, p. 2.
92 D. H. Webb to Charles Nation, 18 October 1907, Percy Nation, AABK 18805 W5515 0004139, ANZ; *Evening Post*, 20 January 1902, p. 5.
93 *Otago Witness*, 6 March 1901, p. 38; *Otago Witness*, 18 June 1902, p. 29.
94 *Otago Witness*, 18 June 1902, p. 29.
95 Proceedings of Medical Board, 26 July 1901, AABK 18805 W5515 0005318, ANZ.
96 Attestation-Form, 11 March 1900, Henry Alexander Stephens; Proceedings of Medical Board, 26 July 1901; Henry Stephens to Major Owen, 22 April 1900, AABK 18805 W5515 0005318, ANZ.
97 *NZPD*, 122 (1902), p. 970.
98 Certificate of Discharge, 15 September 1902, Arthur George Coleman; NZ Expeditionary Force medical treatment authorisation, 11 November 1921; NZ Expeditionary Force medical treatment authorisation, 8 December 1921, Arthur George Coleman, AABK 18805 W5515 0001072.
99 *New Zealand Herald*, 18 November 1901, p. 5; *Evening Post*, 20 November 1901, p. 6; *Poverty Bay Herald*, 22 November 1901, p. 3; *Press*, 23 November 1901, p. 5.
100 Norman Smith to William Hosking, 9 February 1903, AD1 347 0 D1900/476, ANZ.
101 *Star*, 3 May 1900, p. 1; *New Zealand Herald*, 3 May 1900, p. 5.
102 Proceedings of a Court of Enquiry, 16 June 1902, Hugh Edward Collison, AABK 18805 W5515 0001098, pp. 15–17, *AJHR*, 1903, H-6A, p. 10.
103 James Bennet to Richard J. Seddon, 3 February 1902; A. W. Robin, 1 February 1902, Robert McKeich, AABK 18805 W5515 0003519, ANZ.
104 Lord Ranfurly to Joseph Ward, 9 June 1902, Robert McKeich, AABK 18805 W5515 0003519, ANZ.
105 *AJHR*, 1902, H-6A, p. 2; *AJHR*, 1902, H-6B, p. 23; *Thames Star*, 28 July 1902, p. 4.
106 Herbert Hart, diary, 5 June 1902, NAM NZ 1990.1024.
107 William John McFarlane to Jessie McFarlane, 23 May 1902, NAM 1991.2451. Hart and McFarlane's accounts differ regarding the nature of the injury McKeich sustained. Hart claimed McKeich was shot in the head, while McFarlane claimed he was shot from side to side through the chest 'cutting the large artery near the heart'.
108 *NZPD*, 110 (1899), p. 75.
109 *NZPD*, 110 (1899), p. 726.
110 *NZPD*, 111 (1900), pp. 322–23.
111 *NZPD*, 111 (1900), p. 323.
112 Military Pensions Extension to Contingents Act 1900 (64 VICT 1900 No 62).
113 Military Pensions Act 1901 (1 EDW VII 1901 No 53); Military Pensions Act 1902 (2 EDW VII 1902 No 54); Military Pensions Amendment Act 1903 (3 EDW VII 1903 No 31).

114 Lord Ranfurly to Richard J. Seddon, 16 April 1900, Appointment of Nurses, AD34 2/2 8008, ANZ.
115 *NZPD*, 111 (1900), p. 218.
116 John Taylor Marshall to Secretary of Lloyds' Patriotic Fund, AAYS 8638 AD1/486/a D1907/316, ANZ.
117 Joseph Ward to John Taylor Marshall, 21 March 1901, AAYS 8638 AD1/486/a D1907/316, ANZ.
118 John Taylor Marshall to Secretary of Lloyd's Patriotic Fund, 26 May 1903, AAYS 8638 AD1/486/a D1907/316, ANZ.
119 Military Pensions Extension to Contingents Act 1900 (64 Vict. 1900 No. 62); *AJHR*, 1901, B-1, p. 37.
120 Attestation-Form, 10 February 1900, William John Berry; A. Penton to Under Secretary of State for War, 7 September 1900, AABK 18805 W5515 0000368, ANZ.
121 *NZPD*, 110 (1899), p. 185; Attestation-Form 10 February 1900, William John Berry, AABK 18805 W5515 0000368, ANZ; William Fletcher Berry, 1898/14155, NZBDM; Mona Gwendoline Mabel Berry, 1896/2479, NZBDM.
122 Attestation-Form, 20 March 1902, Robert Hall Bakewell, AABK 18805 W5515 0000220, ANZ.
123 *Hawke's Bay Herald*, 20 September 1900, p. 2; Arthur Douglas, Order Paper, 9 October 1900, Reply to Alfred Levavasour Durell Fraser's question in the House of Representatives, William John Berry, AABK 18805 W5515 0000368, ANZ.
124 *Feilding Star*, 26 September 1900, p. 2.
125 *NZPD*, 114 (1900), p. 191.
126 *NZPD*, 115 (1900), pp. 554–55.
127 *NZPD*, 114 (1900), p. 191.
128 *Daily Telegraph*, 3 November 1900, p. 4.
129 *Otago Daily Times*, 1 April 1901, p. 5.
130 *AJHR*, 1903, H-6, p. 3.
131 *AJHR*, 1903, H-6, p. 4.
132 Military Pensions Act 1866 (30 Victoriae 1866 No 6).
133 *SCNZ*, 1902, Wellington: John McKay, 1903, p. 356.
134 *Observer*, 7 June 1902, p. 2.
135 *Wanganui Chronicle*, 10 April 1902, p. 3.

The paper claimed the issue of pension disparities was raised in the First New Zealand Mounted Rifles Association *Bulletin*.
136 Proceedings of Medical Board, 15 October 1900; Attestation-Form, 17 March 1900; Robert W. Collins to Paymaster General, n.d., Robert Walter Gordon Collins, AABK 18805 W5515 0006267, ANZ; *Evening Post*, 25 October 1904, p. 1.
137 *NZPD*, 122 (1902), p. 933.
138 Attestation-Form, 17 March 1900, Robert Walter Gordon Collins, AABK 18805 W5515 0006267, ANZ; *SCNZ*, 1903, p. 356.
139 *Wanganui Chronicle*, 10 April 1902, p. 3.
140 *AJHR*, 1901, H-6A, p. 4; Attestation-Form, n.d., Ernest Barnett Lockett, AABK 18805 W5515 0003249, ANZ; Attestation-Form, 17 March 1900, Robert Walter Gordon Collins, AABK 18805 W5515 0006267, ANZ.
141 Attestation-Form, 8 April 1901; Medical History of an Invalid, 21 May 1902, Albert Rosanowski, AABK 18805 W5515 0004831, ANZ.
142 Memorandum from Defence Minister, 8 December 1903; Government Life Insurance, 5 May 1905; A. Rosanowski to the Chief Staff Officer, 29 August 1906, Albert Rosanowski, AABK 18805 W5515 0004831, ANZ; *SCNZ*, 1903, p. 356.
143 Albert Rosanowski to W. E. Butler, 5 January 1907; Medical report by Dr N. E. Herbert regarding Albert Rosanowski, 9 February 1907, Albert Rosanowski, AABK 18805 W5515 0004831, ANZ.
144 South African Contingents: (Deaths of Members of) In South Africa and Since Leaving South Africa, and Particulars as to Locality, etc., of Graves, *AJHR*, 1903, H-6A, p. 9; Medical History of Invalid, 21 May 1902, Albert Rosanowski, AABK 18805 W5515 0004831, ANZ.
145 *SCNZ*, 1902, p. 345.
146 *AJHR*, 1903, H-6, p. 3.
147 Proceedings of the Medical Board, 10 March 1901; Proceedings of the Medical Board, 10 March 1909, Albert Marr Beath, AABK 18805 W5515 0000308, ANZ.
148 *SCNZ*, 1903, p. 345.
149 Albert Beath to Alfred Robin, 5 January

1903, Albert Marr Beath, AABK 18805 W5515 0000308, ANZ.
150 Alfred Robin to Albert Beath, 14 May 1909, Albert Marr Beath, AABK 18805 W5515 0000308, ANZ.
151 *NZPD*, 113 (1900), p. 113.
152 *Otago Daily Times*, 23 April 1903, p. 6.
153 Albert Beath to Under-Secretary for Defence, 9 May 1910, AABK 18805 W5515 0000308, ANZ.
154 Penelope Beath to Under-Secretary for Defence, 19 August 1912, AABK 18805 W5515 0000308, ANZ.
155 Attestation-Form, 20 March 1902, Robert Hall Bakewell; Proceedings of Medical Board, 6 February 1904, AABK 18805 W5515 0000220, ANZ.
156 Re: Proposed Grant of Military Pension Report to R. H. Bakewell, n.d., AABK 18805 W5515 0000220, ANZ.
157 *Otago Witness*, 26 December 1900, p. 29.
158 *Otago Witness*, 26 December 1900, p. 29; Summary of data regarding Private M. Canavan, 4th Contingent, n.d., Michael Canavan, AABK 18805 W5515 0000921, ANZ; *SCNZ*, 1903 p. 356.
159 Michael Canavan to Richard J. Seddon, 19 July 1901, AABK 18805 W5515 0000921, ANZ.
160 J. H. Witheford to the Minister for Defence, 19 June 1903; Richard J. Seddon to J. H. Witheford, 9 July 1903, Michael Canavan, AABK 18805 W5515 0000921, ANZ.
161 *AJHR*, 1902, Session I, H-6, p. 12.
162 *New Zealand Herald*, 3 March 1902, p. 5; *AJHR*, 1903, Session I, H-6A, p. 3.
163 *New Zealand Herald*, 4 March 1902, p. 5; *Wanganui Herald*, 17 March 1902, p. 2; *AJHR*, 1903, Session I, H-6 p. 12.
164 Arthur Douglas to Samuel Finch, 25 June 1902, Harry Finch, AABK 18805 W5515 0001783, ANZ.
165 *AJHR*, 1903, Session I, H-6A, p. 3.
166 *Otago Daily Times*, 28 March 1900, pp. 2–3.
167 Arthur Penton to Secretary, Royal Patriotic Fund, 3 May 1900, John McIntosh Patterson; E. Dockrill to Colonel Pole-Penton, 13 Sept, 1900; Mary Paterson to Lieutenant-Colonel John Ellis, 5 November 1901, AABK 18805 W5515 0004394, ANZ; *Otago Daily Times*,

28 March 1900, pp. 2–3. The *Otago Daily Times* claimed Patterson had seven children.
168 Mary Paterson to Lieutenant-Colonel John Ellis, 5 November 1901, AABK 18805 W5515 0004394, ANZ.
169 *AJHR*, 1901, Session I, H-6E, p. 3.
170 Attestation-Form, 4 April 1901; Widows' Pension and Allowances declaration by Georgina Love, 6 March 1903, Daniel Joseph Love, AABK 18805 W5515 0003272, ANZ.
171 J. Kennedy Elliott to Defence Department, 20 September 1901, Daniel Joseph Love, AABK 18805 W5515 0003272, ANZ.
172 Under-Secretary of Defence to Georgina Love, 4 October 1901, Daniel Joseph Love, AABK 18805 W5515 0003272, ANZ; *AJHR*, 1903, H-6, p. 3; *SCNZ*, 1902, p. 345.
173 Georgina Love to the Minister of Lands and Agriculture, 22 July 1908, Daniel Joseph Love, AABK 18805 W5515 0003272, ANZ.
174 Lord Onslow (for the Secretary of State) to Lord Ranfurly, 12 December 1902, Daniel Joseph Love, AABK 18805 W5515 0003272, ANZ.
175 B. Hart to James Bennet, 30 July 1902, Robert McKeich, AABK 18805 W5515 0003519, ANZ.
176 *SCNZ*, 1903, p. 356.
177 Military Pensions Amendment Act 1903 (3 EDW VII 1903 No 31).
178 H. Griffen to Georgina Love, 7 July 1905, Daniel Joseph Love, AABK 18805 W5515 0003272, ANZ.
179 Antonia Betcke to Richard J. Seddon, 26 July 1901, Frederick Betcke, AABK 18805 W5515 0000374, ANZ.
180 Attestation-Form, 17 April 1901, Frederick Betcke, AABK 18805 W5515 0000374, ANZ.
181 *SCNZ*, 1903, p. 356.
182 *Taranaki Daily News*, 4 March 1903, p. 4.
183 *New Zealand Truth*, 24 November 1917, p. 5; *Sun*, 9 November 1917, p. 9.
184 *Wanganui Chronicle*, 21 November 1917, p. 4.
185 *NZPD*, 110 (1899) p. 94.
186 A. D. Aitken-Connell to Richard J. Seddon, 8 February 1900, John Vaille Aitken-Connell, AABK 18805 W5515 0000034, ANZ.
187 *SCNZ*, 1901, p. 330; Attestation-Form, 3 April 1901, Arnold Douglas Aitken-Connell; Form acknowledging receipt of clasps for South

Africa 1901 and South Africa 1902 medals, 19 January 1909; George Inglis to Captain Seddon, 5 December 1908, AABK 18805 W5515 0000033, ANZ.
188 Joseph Wylie to the Under-Secretary for Defence, 3 October 1901, Frederick William Wylie, AABK 18805 W5515 0006917, ANZ.
189 *London Gazette*, 9 July 1901, p. 4556.
190 Captain Barron to Captain Reid, 21 July 1901, Frederick William Wylie, AABK 18805 W5515 0006917, ANZ.
191 Walter Kennaway to Guy Fleetwood Wilson, 30 October 1903, Frederick William Wylie, AABK 18805 W5515 0006917, ANZ.
192 *AJHR*, 1903, H-6A, pp. 10, 12.
193 Patrick Lee to Richard J. Seddon, 1 November 1905; Patrick Lee to Joseph Ward, 26 July 1903; J. Grey for the Under Secretary of Defence to the Secretary of the Southland Patriotic Relief Fund, 13 June 1904, Patrick William Lee, AABK 18805 W5515 0003165, ANZ.
194 Patrick Lee to Joseph Ward, 26 July 1903, Patrick William Lee, AABK 18805 W5515 0003165, ANZ.
195 *NZPD*, 110 (1899), p. 92.
196 *New Zealand Tablet*, 19 October 1899, p. 17.
197 *Auckland Star*, 15 March 1902, p. 5.
198 Preliminary Medical Examination, n.d., Percy Charles Leary, AABK 18805 W5515 0003156, ANZ.
199 *New Zealand Herald*, 17 March 1902, p. 5.
200 Attestation-Form, 26 February 1902, Percy Charles Leary, AABK 18805 W5515 0003156, ANZ.
201 *New Zealand Herald*, 17 March 1902, p. 5.
202 *New Zealand Herald*, 17 March 1902, p. 5.
203 *Auckland Star*, 15 March 1902, p. 5.
204 *New Zealand Herald*, 18 March 1902, p. 5.
205 *New Zealand Herald*, 17 March 1902, p. 5; *Auckland Star*, 17 March 1902, p. 5.
206 J. M. Poynton to Under-Secretary for Defence, 10 June 1902, Percy Charles Leary, AABK 18805 W5515 0003156, ANZ.
207 J. M. Poynton to Under-Secretary for Defence, 18 July 1902, Percy Charles Leary, AABK 18805 W5515 0003156, ANZ.
208 Arthur Douglas to Officer Commanding Auckland District, 14 May 1902, Percy Charles Leary, AABK 18805 W5515 0003156, ANZ.
209 *Manawatu Evening Standard*, 26 March 1902, p. 3.
210 *New Zealand Herald*, 19 March 1902, p. 5.
211 *New Zealand Herald*, supplement, 16 November 1901, p. 5.
212 *Mataura Ensign*, 17 May 1900, p. 2.
213 *New Zealand Herald*, supplement, 16 November 1901, p. 5; *Mataura Ensign*, 17 May 1900, p. 2.
214 *Otago Daily Times*, 4 May 1900, p. 5; *Evening Post*, 3 July 1900, p. 5.
215 *Evening Post*, 3 July 1900, p. 5.
216 *Otago Witness*, 12 September 1900, p. 63.
217 H. V. Drew to Defence Department, 21 December 1899, AAYS 8638 AD1 344 / ax D1899/4548, ANZ.
218 *New Zealand Herald* (supplement), 16 November 1901, p. 5; Roll of individuals entitled to the South African Medal and clasps. Edinburgh & East of Scotland Hospital, n.d., National Archives, Kew, UK, WO 100/225, pp. 216, 217.
219 Roll of individuals entitled to the South African Medal and clasps. Edinburgh & East of Scotland Hospital, n.d., National Archives, Kew, UK, WO 100/225, p. 215.
220 *Evening Post*, 27 August 1900, p. 6; Roll of individuals entitled to the South African Medal and clasps. Civil Surgeons, 31 October 1903, National Archives, Kew, UK, WO 100/226, p. 76.
221 *Colonist*, 14 November 1914, p. 5.
222 *NZPD*, 111 (1900), p. 612.
223 *Timaru Herald*, 17 July 1900, p. 3.
224 *Star*, 16 June 1900, p. 7.
225 *Star*, 27 July 1900, p. 1.
226 *NZPD*, 111 (1900), p. 159.
227 William Henry Hosking to Richard J. Seddon, 8 March 1900, AD1 347 0 D1900/476, ANZ.
228 Norman Smith to William Henry Hosking, 9 February 1903, AD1 347 0 D1900/476, ANZ.
229 William Henry Hosking to Richard J. Seddon, telegram, 17 December 1901, AD1 347 0 D1900/476, ANZ.
230 C. A. Durie to Richard J. Seddon, 30 July 1900, Appointment of Nurses, March 1900 –

February 1957, AD34 2/2 8008, ANZ.
231 *Outlook*, 4 August 1900, p. 10.
232 Marien Hegglin to Richard J. Seddon, 15 August 1900, AD1 360 co D1900/3340, ANZ.
233 Penelope Farquharson to William Farquharson, July 1900; Alex Farquharson to William Farquharson, 17 September 1900, NAM NZ 1998.11.
234 *AJHR*, 1901, H-6E, p. 2; Penelope Farquharson to William Farquharson, 11 February 1900, NAM NZ 1998.11; Attestation-Form, 22 March 1900; Death Report, 25 February 1901, Douglas Malcolm Corson, AABK 18805 W5515 0001198, ANZ.
235 T. F. Grey to Helen Corson, 27 April 1904, AD1 439 0 D1904/972, ANZ.
236 T. F. Grey to Officer Commanding Volunteer District, 9 January 1904, AD1 439 0 D1904/972, ANZ.
237 John Cooney to Chief Detective, 14 January 1904, AD1 439 0 D1904/972, ANZ.
238 Proceedings of Pension Board, 11 April 1904, AD1 439 0 D1904/972, ANZ.
239 T. F. Grey to Helen Corson, 27 April 1904, AD1 439 0 D1904/972, ANZ.
240 Attestation-Form, n.d., Frederick Saville Broome, AABK 18805 W5515 0000596, ANZ; *Nelson Evening Mail*, 24 July 1900, p. 2.
241 Richard J. Seddon to C. S. Broome, 1900, AABK 18805 W5515 0000596, ANZ.
242 Medical report by Dr H. Potter, 9 April 1901, Frederick Saville Broome, AABK 18805 W5515 0000596, ANZ.
243 Colonel A. P. Penton to His Worship the Mayor of Wellington, 9 August 1900, Frederick Saville Broome, AABK 18805 W5515 0000596, ANZ.
244 Richard J. Seddon to James Colvin, telegram, n.d., AABK 18805 W5515 0001104, ANZ.
245 *Auckland Star*, 8 July 1902, p. 3.
246 Orders by Colonel R. H. Davis, 6 July 1902, 'P' Brigade Orders of 1st New Zealand Brigade. (Eighth N.Z. Contingent), AD34 5 5 8021/2, ANZ; *AJHR*, 1902, H-6C, pp. xi, 70.
247 *AJHR*, 1902, Session I, H-6C, pp. 70, 74.
248 *AJHR*, 1903, Session I, H-6A, p. 12; *AJHR*, 1902, H-6, p. 12; *Western Mail* (Aus.), 26 July, 1902, p. 44.
249 *AJHR*, 1903, Session I, H-6A, p. 14; *Marlborough Express*, 1 August 1902, p. 2.
250 *Star*, 1 August 1902, p. 3.
251 *AJHR*, 1903, Session I, H-31, pp. 38, 49.
252 Attestation-Form, 9 April 1902, James Edward Nicholson, AABK 18805 W5515 0004198, ANZ.
253 *AJHR*, 1902, Session I, H-6C, pp. 47–48.
254 *AJHR*, 1902, Session I, H-6C, p. 47; Attestation Form, 7 January 1902, Wallace Nicholson, AABK 18805 W5515 0004201, ANZ.
255 *AJHR*, 1903, Session I, H-6A, p. 14.
256 *Star*, 23 August 1902, p. 4.
257 *AJHR*, 1902, Session I, H-6C, p. xiii.
258 *AJHR*, 1902, Session I, H-6C, p. xiii.
259 *AJHR*, 1902, Session I, H-6C, p. 74.
260 *AJHR*, 1902, Session I, H-6C, p. 74.
261 *AJHR*, 1902, Session I, H-6C, p. xi.
262 *AJHR*, 1902, Session I, H-6C, p. 184.
263 *AJHR*, 1902, Session I, H-6C, p. 48.
264 *AJHR*, 1902, Session I, H-6C, p. 198.
265 *AJHR*, 1902, Session I, H-6C, p. 12.
266 *Mercury* (Aus.), 5 August 1902, p. 3.
267 *AJHR*, 1902, Session I, H-6C, p. x.
268 *AJHR*, 1902, Session I, H-6C, pp. xi–xii.
269 *AJHR*, 1903, Session I, H-31, p. 38.
270 *NZPD*, 121 (1902), p. 379.
271 *AJHR*, 1903, Session I, H-6A, pp. 13, 14.
272 *AJHR*, 1902, Session I, H-6C, p. xi.
273 *AJHR*, 1903, Session I, H-6C, pp. 129–30.
274 *Marlborough Express*, 15 July 1902, p. 2; *Otago Daily Times*, 5 August 1902, p. 5.
275 *West Coast Times*, 6 August 1902, p. 2.
276 *Evening Post*, 12 August 1902, p. 5; *Press*, 7 August 1902, p. 6.
277 *Wanganui Herald*, 11 August 1902, p. 2.
278 *Star*, 8 August 1902, p. 3.
279 *Evening Post*, 14 August 1902, p. 6; *Feilding Star*, 26 August 1902, p. 2.
280 *AJHR*, 1902, Session I, H-6C, p. 252.
281 *Taranaki Herald*, 25 August 1902, p. 2; *NZPD*, 122 (1902), p. 2. In *NZPD* Wray is described as 'D. A. Wray'.
282 Proceedings of Medical Board, 26 September 1902, AABK 18805 W5515 0006173, ANZ.
283 *NZPD*, 122 (1902), p. 2.
284 *Otago Daily Times*, supplement, 25 August 1902, p. 1.

285 *Wanganui Chronicle*, 13 August 1902, p. 5; *Press*, 13 August 1902, p. 8; *AJHR*, 1902, Session I, H-6C, p. 236; *Otago Witness*, 20 August 1902, p. 44; *NZPD*, 125 (1903), pp. 437–38; *Wanganui Chronicle*, 13 August 1902, p. 5; *Press*, 13 August 1902, p. 8; *Waikato Argus*, 13 October 1902, p. 2.
286 *Evening Post*, 16 August 1902, p. 5; *NZPD*, 125 (1903), pp. 437–38.
287 *NZPD*, 125 (1903), pp. 437–38.
288 *NZPD*, 125 (1903), p. 438.
289 *Press*, 13 August 1902, p. 8.
290 *NZPD*, 121 (1902), p. 371.
291 *NZPD*, 121 (1902), p. 371.
292 *Star*, 14 August 1902, p. 2; *Press*, 13 August 1902, p. 8.
293 *Press*, 13 August 1902, p. 8; *Auckland Star*, 14 August 1902, p. 5.
294 *Wanganui Chronicle*, 12 August 1902, p. 5; *Wanganui Chronicle*, 13 August 1902, p. 5.
295 *Star*, 23 August 1902, p. 4.
296 Proceedings of Medical Board, 27 August 1902, AABK 18805 W5515 0001490, ANZ.
297 W. H. Symes, 26 August 1902, AABK W5515 0001490, ANZ.
298 W. H. Symes to Under Secretary for Defence, 27 October 1902, AABK 18805 W5515 0001490, ANZ.
299 *Ashburton Guardian*, 15 July 1902, p. 3.
300 *AJHR*, 1903, Session I, H-31, p. 18.
301 *AJHR*, 1903, Session I, H-31, p. 18.
302 *AJHR*, 1903, Session I, H-6A, pp. 13, 14; *Colonist*, 21 August 1902, p. 3; 10th Contingent, casualty record, AD34 17 17, ANZ pp. 1–2.
303 Charles E. Browne, 11 September 1902, AABK 18805 W5515 0003714, ANZ; 10th Contingent, casualty record, AD34 17 17, ANZ, pp. 1–2.
304 *New Zealand Herald*, 19 August 1902, p. 5.
305 *New Zealand Herald*, 20 August 1902, p. 5.
306 *New Zealand Herald*, 19 August 1902, p. 5.
307 Margaret Lunn to Joseph Ward, 20 August 1902, AABK 18805 W5515 0003301, ANZ.
308 Margaret Lunn to Joseph Ward, 20 August 1902; Carl A. Schauer to Under-Secretary of Defence, 12 September 1902, AABK 18805 W5515 0003301, ANZ.
309 *New Zealand Herald*, 19 August 1902, p. 5; *New Zealand Herald*, 20 August 1902, p. 5.
310 *AJHR*, 1903, Session I, H-31, p. 18.
311 *AJHR*, 1903, Session I, H-6A, p. 13.
312 *Taranaki Herald*, 30 August 1902, p. 2.
313 L. Joyce to Officer Commanding Auckland District, 24 August 1902, AABK 18805 W5515 0000444, ANZ; *Taranaki Herald*, 30 August 1902, p. 2.
314 *AJHR*, 1903, Session I, H-31, p.18.
315 *AJHR*, 1903, Session I, H-31, p. 19.
316 *AJHR*, 1902, Session I, H-31, p. 8.
317 *AJHR*, 1903, Session I, H-6A, p. 12.
318 *AJHR*, 1903, Session I, H-6A, p. 12; *Thames Star*, 18 July 1901, p. 4.
319 *NZPD*, 116 (1901), p. 363.
320 *AJHR*, 1903, Session I, H-31, pp. 18–19.
321 *Argus*, (Aus.), 25 July 1901, p. 5.
322 Staff Diary, Natal Command, May 1902, UK National Archives, Kew, WO 32-8110, p. 25.
323 *AJHR*, 1902, Session I, H-31, p. 8.
324 *AJHR*, 1902, Session I, H-31, p. 8.
325 *AJHR*, 1903, Session I, H-31, p. 38.
326 *AJHR*, 1903, Session I, H-31, p. 8.
327 *AJHR*, 1903, Session I, H-31, p. 4.
328 *AJHR*, 1903, Session I, H-31, p. 43.
329 *AJHR*, 1903, Session I, H-31, p. 43.
330 *AJHR*, 1903, Session I, H-31, pp. 5, 28.
331 *AJHR*, 1903, Session I, H-31, p. 8.
332 *AJHR*, 1903, Session I, H-31, pp. 37–38, 42.
333 *AJHR*, 1903, Session I, H-31, p. 42.
334 *AJHR*, 1903, Session I, H-31, p. 42.
335 *AJHR*, 1903, Session I, H-31, p. 42.
336 *AJHR*, 1903, Session I, H-31, p. 42.
337 *Hastings Standard*, 5 July 1900, p. 4.
338 Proceedings of Medical Board, 19 August 1901, Robert Anderson Drinnan, AABK 18805 W5515 0001561, ANZ.
339 *NZPD*, 125 (1903), p. 695.
340 *Wanganui Collegian*, No. 58, April 1902, p. 6; *New Zealand Times*, 28 December 1901, p. 5.
341 D. Colquhoun, M.D., 22 December 1915, Robert Anderson Drinnan, AABK 18805 W5515 0001561, ANZ.
342 Robert Anderson Drinnan to James Allen, March 1916, AABK 18805 W5515 0001561, ANZ.
343 M. W. Fleming to James Allen, 6 January

1916, AABK 18805 W5515 0001561, ANZ.
344 Alfred William Robin to Frederick John Rockstrow, Minute Sheet, 12 January 1916, AABK 18805 W5515 0001561, ANZ.
345 Frederick John Rockstrow to Alfred William Robin, Minute Sheet, 14 January 1916, AABK 18805 W5515 0001561, ANZ.
346 A. J. Cross to District Headquarters, Military Forces, Dunedin, 8 April 1918, AABK 18805 W5515 0001561, ANZ.
347 M. W. Fleming to James Allen, 9 April 1918, AABK 18805 W5515 0001561, ANZ.
348 Daniel Aloysius Hickey to Headquarters, N.Z. Military Forces, Wellington, 3 May 1918, AABK 18805 W5515 0001561, ANZ.
349 Frederick Lund, New Zealand Military Forces, South African War, n.d., Frederick Lund, AABK 18805 W5515 0003296, ANZ; Attestation-Form, 14 April 1902, John Lund, AABK 18805 W5515 0003297, ANZ.
350 Report of Constable T. Barrett concerning W. Lund, 20 November 1904; Medical Report, 16 January 1903, John Lund, AABK 18805 W5515 0003297, ANZ.
351 Walter Lund to District Office, N.Z. Defence Forces, 15 May 1905, John Lund, AABK 18805 W5515 0003297, ANZ.
352 Frederick Lund, New Zealand Military Forces, South African War, n.d., Frederick Lund, AABK 18805 W5515 0003296, ANZ; Walter Lund to Mr. Platman, 3 November 1904, John Lund, AABK 18805 W5515 0003297, ANZ.
353 Acting Under-Secretary for Defence to Inspector of Police, 14 November 1904; Acting Under-Secretary of Defence Memorandum, 28 November 1904, John Lund, AABK 18805 W5515 0003297, ANZ.
354 *Star*, 7 October 1902, p. 2; *AJHR*, 1901, H-6E, p. 2.
355 *Thames Star*, 2 January 1901, p. 3; *Southland Times*, 3 January 1901, p. 2; *Auckland Star*, 2 January 1901, p. 5; *Wanganui Herald*, 31 December 1900, p. 2; *Otago Daily Times*, 31 December 1900, p. 7. *Evening Post*, 2 January 1901, p. 6.
356 James Arthur Shand, 'O'er Veldt and Kopje – The Official Account of the Operations of the New Zealand Contingents in the Boer War' [ca 1931], qMS-1790-1793, p. 545.
357 Evidence of Major Alfred Edwards Perkins at Board of Inquiry into Salter's death, 17 December 1900; Diet Sheet of Private Salter, John Salter, AABK 18805 W5515 0004932, ANZ.
358 *AJHR*, 1901, H-6E, p. 2.
359 *Evening Post*, 2 January 1901, p. 6; *AJHR*, 1903, H-6A, p. 14.
360 *Ohinemuri Gazette*, 21 March 1900, p. 3.
361 Trooper named Arthur to his mother and father, 8 June 1900, NAM NZ 1999.2002.
362 *New Zealand Herald*, 19 December 1901, p. 5.
363 *Wanganui Collegian*, No. 52, Apr 1900, p.13.
364 Attestation-Form, n.d., David Hannibal Waldie, AABK 18805 W5515 0005753, ANZ; *Otago Witness*, 17 April 1901, p. 54; Lonnie (nee) Farquharson to William Farquharson, 8 March 1901, NAM NZ 1998.11.
365 *Otago Daily Times*, 17 April 1901, p. 54.
366 *Wanganui Collegian*, No. 53, August 1900, p. 12; *Wanganui Collegian*, No. 55, April 1901, p. 19.
367 *Otago Witness*, 26 September 1900, p. 30; *New Zealand Times*, 22 September 1900, p. 7.
368 *New Zealand Times*, 22 September 1900, p. 7.
369 Roll of individuals entitled to the South African Medal and clasps, Protectorate Regiment Frontier Force, 18 February 1902, National Archives, Kew, UK, WO 100/263, p. 252; Roll of individuals entitled to the South African Medal and clasps, British South African Police, 2 November 1901, National Archives, Kew, UK, WO 100/238, p. 9.
370 *Wanganui Collegian*, No. 53, August 1900, p. 19.
371 Supplement to the *Wanganui Collegian*, August 1901, n.p.
372 *Wanganui Collegian*, No. 55, April 1901, p. 19; *Evening Post*, 4 August 1900, p. 5.
373 *Evening Post*, 23 July 1900, p. 5.
374 *Evening Post*, 22 September 1900, p. 5.
375 *Wanganui Collegian*, No. 54, December 1900, p. 8.
376 *Wanganui Collegian*, No. 54, December 1900, p. 4.
377 Certificate of Discharge, 1902, Alfred Davitt, AABK 18805 W5515 0001407, ANZ; *Auckland Star*, 27 August 1902, p. 2.

378 *Auckland Star*, 27 August 1902, p. 2.
379 *Bay of Plenty Times*, 3 October 1910, p. 2; Statement of Service of Alfred Davitt in the South African War 1899–1902 and the war of 1914–1919; Adjutant-General to Alfred Davitt, 26 April 1937, Alfred Davitt, AABK 18805 W5515 0001407, ANZ.
380 Defence Office, Wellington to the Officer Commanding Otago Military District, 27 February 1904; Report of Mounted Constable Emerson, 12 March 1904, Patrick William Lee, AABK 18805 W5515 0003165, ANZ.
381 John C. Adams to Agent-General for New Zealand, 23 October 1900; G. E. Way to J. F. Grey, 2 May 1901, Hugh McDonagh, AABK 18805 W5515 0003382, ANZ.
382 *NZPD*, 111 (1900), p. 604.
383 *Tuapeka Times*, 29 September 1900, p. 3.
384 *NZPD*, 111 (1900), p. 604.
385 *Akaroa Mail and Banks Peninsula Advertiser*, 24 January 1902, p. 2.
386 Evidence of Sir Arthur Percy Douglas, Minutes of Evidence taken before the Royal Commission on the War in South Africa, Vol. I, London: His Majesty's Stationery Office, 1903, p. 426.
387 *Otago Witness*, 12 February 1902, p. 29; Certificate of Discharge, 6 August 1901, Henry Godfrey Heywood, AABK 18805 W5515 0002541, ANZ.
388 George Leece to his mother, 15 July 1900, ATL MS-Papers-8464-05.
389 *Ashburton Guardian*, 25 February 1901, p. 2.
390 *Marlborough Express*, 3 May 1900, p. 4; *Feilding Star*, 1 May 1901, p. 4.
391 *Feilding Star*, 1 May 1901, p. 4; Roll of individuals entitled to the South African Medal and clasps. Kimberley Light Horse, 16 July 1901, National Archives, Kew, UK, WO 100/255, p. 18.
392 Denis Hickey to Stuart Newall, 1 September 1901, AABK 18805 W5515 0002544, ANZ.
393 *SCNZ*, 1903, p. 345; Joseph Culling to the Commandant N.Z. Forces, 16 September 1901, AABK 18805 W5515 0001308, ANZ.
394 Record of Military Service, Arthur Stuckey, AABK 18805 W5515 0005399, ANZ.
395 Herbert Ernest Hart diary, 2, 28 July 1902, NAM NZ 1990.1024.
396 Herbert Ernest Hart diary, 24 October 1902, NAM NZ 1990.1024; Ian McGibbon (ed): *The Oxford Companion to New Zealand Military History*, Auckland: Oxford University Press, 2000, p. 215.
397 *Otago Witness*, 6 February 1901, p. 40; *Wanganui Collegian*, No. 53, August 1900, p. 17.
398 *Wanganui Collegian*, No. 60, December 1902, p. 13.
399 *Wanganui Collegian*, No. 60, December 1902, p. 13; N.Z. High Commissioner to William Massey, 27 March 1923, Walter John Borlase, AABK 18805 W5515 0000474, ANZ.
400 Attestation-Form, 15 March 1902; Percy Arthur Cohen to the N.Z. Secretary of Defence, 24 November 1903, Percy Arthur Cohen, AABK 18805 W5515 0001064, ANZ; *Evening Post*, 29 August 1930, p. 11.
401 *Evening Post*, 29 August 1930, p. 11.
402 Attestation-Form, William James Dunnet, 27 May 1902; M. Dunnet to E. Chaytor, 17 October 1904, William James Dunnet, AABK 18805 W5515 0001617.
403 Attestation-Form, John Sanderson, 25 May 1900; Reference Papers, John Sanderson, AABK 18805 W5515 0001617, ANZ; *Otago Daily Times*, 11 September 1902, p. 2.
404 Attestation-Form, Robert Simmers, 22 March 1900, AABK 18805 W5515 0005086, ANZ.
405 *Wanganui Collegian*, No. 61, April 1903, pp. 18–19.
406 M. Dunnet to E. Chaytor, 17 October 1904, William James Dunnet, AABK 18805 W5515 0001617, ANZ.
407 *AJHR*, 1906, B-1, p. 62.
408 Sheryl Kendall and David Corbett, *New Zealand Military Nursing: A History of the Royal New Zealand Nursing Corps — Boer War to Present Day*, Auckland: Sheryl Kendall & David Corbett, 1990, p. 6; Ellen Ellis, *Teachers for South Africa: New Zealand Women at the South African War Concentration Camps*, Paekakariki: Hanorah Books, 2010, pp. 166–67.
409 John R. Hurrey to Richard J. Seddon, 28 May 1901, AABK 18805 W5515 0002723, ANZ.
410 Richard J. Seddon to Under-Secretary of Defence, 4 June 1901, AABK 18805 W5515

0002723, ANZ.
411 John R. Hurrey to Richard J. Seddon, 27 May 1901, AABK 18805 W5515 0002723, ANZ.
412 *Lake Country Press*, 30 May 1901, p. 5; *Otago Daily Times*, 20 July 1901, p. 2; *Queensland Times*, 28 May 1901, supplement, p. 1.
413 Attestation-Form, 13 January 1901, Ernest Charles Hurrey, AABK 18805 W5515 0002722, ANZ; Attestation-Form, 14 February 1900, John Alexander Hurrey, AABK 18805 W5515 0002724, ANZ.
414 J. W. Hazledine to Commandant of Forces, 30 May 1901, AABK 18805 W5515 0002723, ANZ.
415 Staff Officer to J. W. Hazledine, 1 June 1901, AABK 18805 W5515 0002723, ANZ.
416 Albert Andrew to Under-Secretary for Defence, 1 August 1901, Irving Stanley Hurrey, AABK 18805 W5515 0002723, ANZ.
417 *Evening Post*, 15 July 1901, p. 5; Albert Andrew to Under-Secretary for Defence, 1 August 1901; J. S. Purdy to Major Sherman, 21 May 1901, Irving Stanley Hurrey, AABK 18805 W5515 0002723, ANZ.
418 Casualty Return, 2 October 1901, William Mathews, AABK 18805 W5515 0003801, ANZ; *AJHR*, 1903, H-6A, p. 9.
419 *AJHR*, 1903, Session I, H-6A, pp. 5, 11.
420 *Otago Witness*, 30 November 1904, p. 41.
421 *Wanganui Collegian*, No. 54, December 1900, p. 16; *AJHR*, 1900, H-6, p. 6; *AJHR*, 1901, H-6E, p. 1; *AJHR*, 1903, H-6A, p. 3.
422 *AJHR*, 1903, H-6A, p. 3.
423 Henry Smith to Richard J. Seddon, 25 September 1900, Lionel Eric Smith, AABK 18805 W5515 0005194, ANZ.
424 Extract, New Zealand Government Agency, 5 February 1901, Lionel Eric Smith, AABK 18805 W5515 0005194, ANZ.
425 *Wairarapa Daily Times*, 26 March 1902, p. 2.
426 Frances Davis to Richard John Seddon, 4 April 1904, John Wyllie, AABK 18805 W5515 0006200, ANZ.
427 *Poverty Bay Herald*, 13 August 1902, p. 2.
428 *New Zealand Mail*, 8 October 1902, p. 57; *Wairarapa Daily Times*, 9 February 1903, p. 2; *Poverty Bay Herald*, 7 February 1903, p. 4; *Ashburton Guardian*, 10 February 1903, p. 2; *Bush Advocate*, 18 February 1903, p. 2.
429 *Nelson Evening Mail*, 6 February 1908, p. 4; *Taranaki Daily News*, 3 November 1910, p. 3.
430 Governor General to the Earl of Liverpool, 6 January 1913, IA1 1256 [15] 1913/1793, ANZ.
431 *Dominion*, 22 April 1910, p. 6.
432 *Observer*, 7 June 1902, p. 2.

Chapter Seven: 'These wars will always be popular'

1 *Otago Daily Times*, 17 August 1899, p. 5.
2 *Otago Witness*, 17 August 1899, p. 23.
3 *NZPD*, 109 (1899), p. 570.
4 *NZPD*, 109 (1899), p. 570; *SCNZ*, 1898, p. 201.
5 *Hawke's Bay Herald*, 30 September 1899, p. 3.
6 *Feilding Star*, 9 October 1899, p. 2.
7 *New Zealand Tablet*, 5 October 1899, p. 31.
8 *NZPD*, 111 (1900), p. 20.
9 *Star*, 30 September 1899, p. 6.
10 *New Zealand Tablet*, 5 October 1899, p. 31.
11 *Evening Post*, 15 December 1899, p. 2.
12 *New Zealand Tablet*, 5 October 1899, p. 31.
13 *Hawke's Bay Herald*, 30 September 1899, p. 3.
14 *Hawke's Bay Herald*, 30 September 1899, p. 3.
15 *Ashburton Guardian*, 12 October 1899, p. 2.
16 *AJHR*, 1901, H-27A, p. 2.
17 *AJHR*, 1901, H-27A, p. 2.
18 *Thames Star*, 28 July 1902, p. 4.
19 *AJHR*, 1901, H-27, p. 1.
20 *Otago Witness*, 12 March 1902, p. 7.
21 *Star*, 6 September 1900, p. 3; *Marlborough Express*, 12 September 1900, p. 4; *Otago Daily Times*, 8 November 1900, p. 4.
22 *NZPD*, 116 (1901), p. 230; *Marlborough Express*, 12 September 1900, p. 4.
23 *SCNZ*, 1898, pp. 188, 192, 201, 209.
24 *SCNZ*, 1898, p. 201.
25 *SCNZ*, 1898, p. 201; *SCNZ*, 1900, pp. 209–10.
26 *SCNZ*, 1898, p. 201; *SCNZ*, 1900, p. 210.
27 *Star*, 21 March 1900, p. 3.
28 *Colonist*, 6 February 1902, p. 4.
29 *Evening Post*, 17 November 1900, p. 4.
30 *SCNZ*, 1899, p. 206; *SCNZ*, 1900, p. 210; *SCNZ*, 1901, p. 216.
31 *SCNZ*, 1900, p. 210; *SCNZ*, 1901, p. 216; *SCNZ*, 1902, p. 220.
32 *AJHR*, 1900, H-11, p. xix.
33 *Otago Witness*, 1 March 1900, p. 38.

34 *SCNZ*, 1899, p. 206; *SCNZ*, 1900, p. 210; *SCNZ*, 1901, p. 216; *SCNZ*, 1902, p. 220; *SCNZ*, 1903, p. 229.
35 *SCNZ*, 1898, p. 234.
36 *SCNZ*, 1901, p. 251; *SCNZ*, 1902, p. 256.
37 *AJHR*, 1900, H-6I, p. 3.
38 *Otago Witness*, 17 September 1902, p. 29.
39 *NZPD*, 139 (1907), p. 624.
40 *NZPD*, 139 (1907), p. 623.
41 *NZPD*, 139 (1907), pp. 568–69.
42 *NZPD*, 139 (1907), p. 568.
43 *NZPD*, 139 (1907), p. 624.
44 *NZPD*, 139 (1907), p. 624.
45 *NZPD*, 139 (1907), pp. 624–25.
46 *NZPD*, 139 (1907), p. 625.
47 *SCNZ*, 1905, p. 280; *SCNZ*, 1910, pp. 202–3.
48 *North Otago Times*, 29 August 1908, p. 1.
49 *AJHR*, 1901, H-29, p. 1.
50 *Star*, 10 October 1901, p. 2.
51 *SCNZ*, 1901, p. 253; *SCNZ*, 1902, p. 258; *SCNZ*, 1903, p. 269; *SCNZ*, 1904, p. 282.
52 *Star*, 12 October 1901, p. 2.
53 *SCNZ*, 1901, pp. 208–9; *SCNZ*, 1902, pp. 211–12.
54 *Otago Witness*, 22 January 1902, p. 9.
55 *AJHR*, 1900, Session I, B-6, p. ix.
56 *Evening Post*, 6 December 1901, p. 6.
57 *Evening Post*, 6 December 1901, p. 6.
58 *Evening Post*, 8 December 1899, p. 4.
59 *Poverty Bay Herald*, 15 December 1899, p. 4.
60 *SCNZ*, 1899, p. 207; *SCNZ*, 1900, p. 210; *SCNZ*, 1901, p.217; *SCNZ*, 1902, p. 220.
61 *SCNZ*, 1899, p. 198; *SCNZ*, 1902, p. 211; *SCNZ*, 1903, p. 219.
62 *Evening Post*, 3 June 1901, p. 4; *Ashburton Guardian*, 3 June 1901, p. 2.
63 *Evening Post*, 3 June 1901, p. 4.
64 *SCNZ*, 1899, p. 199; *SCNZ*, 1900, p. 202; *SCNZ*, 1901, p. 209; *SCNZ*, 1902, p. 212; *SCNZ*, 1903, pp. 220–21.
65 *Otago Daily Times*, 14 March 1900, p. 2.
66 *AJHR*, 1902, H-39A, pp. 1–3.
67 *NZPD*, 110 (1899), p. 185.
68 *Evening Post*, 13 June 1901, p. 5.
69 *Auckland Star*, 10 June 1901, p. 4.
70 *Auckland Star*, 10 June 1901, p. 4; *New Zealand Herald*, 11 June 1901, p. 4.
71 *Manawatu Evening Standard*, 11 May 1901, p. 4; *Press*, 30 May 1901, p. 5.
72 Evidence of Lieutenant-Colonel A. P. Penton, Minutes of Evidence taken before the Royal Commission on the War in South Africa, Vol. I, London: His Majesty's Stationery Office, 1903, p. 359.
73 *AJHR*, 1902, H-39A, p. 3.
74 *SCNZ*, 1899, p. 193.
75 Evidence of Lieutenant-Colonel A. P. Penton, Minutes of Evidence taken before the Royal Commission on the War in South Africa, Vol. I, London: His Majesty's Stationery Office, 1903, p. 359; *Evening Post*, 13 October 1899, p. 5.
76 *SCNZ*, 1900, p. 196.
77 *SCNZ*, 1901, p. 203.
78 *SCNZ*, 1902, p. 206.
79 *SCNZ*, 1900, p. 238.
80 *Evening Post*, 12 October 1899, p. 5; *Otago Witness*, 1 Mar 1900, p. 38.
81 *Otago Witness*, 1 March 1900, p. 38.
82 *AJHR*, 1902, H-28, p. 2.
83 *AJHR*, 1902, H-28, p. 2.
84 *AJHR*, 1902, H-28, p. 2.
85 Accounts and Papers – War in South Africa, Horses Purchased for Shipment to, 1900/168 LE1 375, ANZ.
86 *Otago Witness*, 1 March 1900, p. 4.
87 *NZPD*, 112 (1900), pp. 302, 462.
88 *NZPD*, 112 (1900), pp. 302, 462.
89 *Wanganui Herald*, 1 November 1901, p. 3.
90 *Thames Star*, 3 December 1901, p. 4.
91 *Wanganui Herald*, 23 December 1901, p. 2.
92 *AJHR*, 1907, H-5, p. 47.
93 *Evening Post*, 14 June 1901, p. 4.
94 *Otago Witness*, 5 March 1902, p. 43.
95 Secretary of State for the Colonies to Lord Ranfurly, received 11 February 1902, ACHK 16561 G5 Box 6, ANZ.
96 *Akaroa Mail and Banks Peninsula Advertiser*, 13 February 1903, p. 2; *Star*, 24 February 1903, p. 2; *Otago Witness*, 25 February 1903, p. 6.
97 *NZ Truth*, 4 March 1916, p. 5; *Maoriland Worker*, 29 September 1915, p. 6.
98 *Star*, 14 April 1900, p. 1.
99 Walter Kennaway to Under-Secretary

of State for the Colonies, 21 April 1900, CO209/261.
100 *NZPD*, 114 (1900), pp. 103–4.
101 *Oamaru Mail*, 6 June 1901, p. 4.
102 *North Otago Times*, 23 October 1899, p. 1.
103 *Oamaru Mail*, 5 March 1900, p. 1.
104 *Oamaru Mail*, 26 November 1900, p. 1.
105 *Hawera & Normanby Star*, 23 October 1901, p. 2.
106 *Southland Times*, 2 October 1901, p. 2.
107 *Evening Post*, 23 November 1900, p. 4.
108 *Press*, 26 October 1901, p. 5.
109 *AJHR*, 1902, H-17, p. 10.
110 *Evening Post*, 12 September 1902, p. 2.
111 *Evening Post*, 23 November 1900, p. 4.
112 Lord Ranfurly to Secretary of State for the Colonies, received 15 February 1902, ACHK 16561 G5/6, ANZ.
113 Secretary of State for the Colonies to Lord Ranfurly, received 13 February 1902, ACHK 16561 G5/6, ANZ.
114 Richard J. Seddon to Joseph Chamberlain, 28 May 1902, SEDDON1 2 4/16, ANZ.
115 *AJHR*, 1903, H-27, pp. 2–3.
116 *AJHR*, 1902, I-10, p. 38.
117 *AJHR*, 1902, I-10, p. 83.
118 *AJHR*, 1902, H-39A, p. 2.
119 *AJHR*, 1902, H-39A, p. 2.
120 *AJHR*, 1902, H-39A, p. 2.
121 Secretary of State for the Colonies to Lord Ranfurly, received 23 February 1902, ACHK 16561 G5/6, ANZ.
122 *SCNZ*, 1901, p. 222; *SCNZ*, 1902, p. 228; *SCNZ*, 1903, p. 238.
123 *SCNZ*, 1899, p. 215; *SCNZ*, 1900, p. 218; *SCNZ*, 1901, pp. 224–25; *SCNZ*, 1902, p. 229.
124 *Colonist*, 6 February 1902, pp. 3–4.
125 *Sydney Morning Herald*, 25 January 1902, p. 12.
126 *Evening Post*, 25 January 1902, p. 5.
127 *New Zealand Tablet*, 5 October 1899, p. 31; *Sydney Morning Herald*, 25 January 1902, p. 12.
128 *AJHR*, 1902, A-1, p. 31.
129 *AJHR*, 1902, A-1, p. 31.
130 *NZPD*, 122 (1902), p. 438.
131 *AJHR*, 1902, H-27A, p. 1.
132 *AJHR*, 1902, H-27A, pp. 1–4.
133 *AJHR*, 1902, H-27A, p. 3.
134 *AJHR*, 1901, H-27D, p. 2.
135 *Evening Post*, 19 January 1901, p. 5.
136 *Wanganui Herald*, 12 October 1899, p. 3; *Wanganui Chronicle*, 21 November 1899, p. 4.
137 *AJHR*, 1900, H-6A, p. 3.
138 *Otago Witness*, 26 March 1902, pp. 27–28.
139 Richard J. Seddon to William Hall-Jones, telegram, 6 February 1902 pp.1–2, ATL MS-Papers-5755-57.
140 *Otago Witness*, 3 September 1896, p. 14.
141 *Otago Witness*, 3 September 1896, p. 14.
142 *NZPD*, 109 (1899), p. 570.
143 *Otago Daily Times*, 17 January 1900, p. 4.
144 *Otago Daily Times*, 17 January 1900, p. 4.
145 *Ashburton Guardian*, 23 November 1900, p. 2.
146 *Evening Post*, 6 December 1900, p. 7.
147 *NZPD*, 121 (1902), p. 397.
148 *NZPD*, 119 (1901), p. 329; *NZPD*, 121 (1902), p. 384; *Otago Witness*, 15 February 1900, p. 46.
149 *NZPD*, 119 (1901), p. 329; *NZPD*, 121 (1902), p. 384; *NZPD*, 121 (1902), p. 397; *AJHR*, 1901, B-6, p. xxi.
150 *NZPD*, 121 (1902), p. 397.
151 *NZPD*, 121 (1902), p. 397.
152 *NZPD*, 119 (1901), p. 329. This was presumably a reference to the S.S. *Indradevi*.
153 *NZPD*, 119 (1901), p. 329.
154 *AJHR*, 1901, H-27D, p. 1.
155 *NZPD*, 121 (1902), p. 384.
156 *NZPD*, 119 (1901), p. 329.
157 *Press*, 18 May 1900, p. 4.
158 *AJHR*, 1901, A-1, p. 5; Richard J. Seddon to Sir John Hall, 27 March 1900, Hall Papers, MS-Papers-1784-237, ATL.
159 *AJHR*, 1901, A-2, p. 16.
160 *AJHR*, 1901, A-1, p. 16.
161 *AJHR*, 1902, H-28, p. 1.
162 *AJHR*, 1902, H-28, p. 2.
163 *Otago Witness*, 22 January 1902, p. 9.
164 *Sydney Morning Herald*, 25 January 1902, p. 20; *AJHR*, 1902, H-27, p. 1.
165 *AJHR*, 1902, H-27, p. 7.
166 *AJHR*, 1902, H-27, p. 11.

167 *Sydney Morning Herald*, 25 January 1902, p. 20; *AJHR*, 1902, H-27, p. 1.
168 *NZPD*, 121 (1902), p. 384; *AJHR*, 1902, H-27, p. 2.
169 *NZPD*, 121 (1902), p. 384; *AJHR*, 1902, H-27, p. 2.
170 *NZPD*, 121 (1902), p. 386.
171 *NZPD*, 121 (1902), p. 398.
172 *NZPD*, 121 (1902), p. 384.
173 *Evening Post*, 8 October 1902, p. 4.
174 *NZPD*, 131 (1904), pp. 151–52.
175 *NZPD*, 131 (1904), p. 152.
176 *NZPD*, 131 (1904), p. 153.
177 *NZPD*, 131 (1904), p. 152; *AJHR*, 1903, H-17, p. 3.
178 *NZPD*, 131 (1904), p. 152.
179 *AJHR*, 1903, H-17, pp. 3–4.
180 *Motueka Star*, 16 June 1903, p. 3.
181 *Evening Post*, 9 July 1906, p. 5.
182 *AJHR*, 1900, H-11, p. xvii.
183 *Ohinemuri Gazette*, 6 March 1901, p. 2.
184 *Hawke's Bay Herald*, 13 March 1901, p. 4.
185 *New Zealand Herald*, 7 January 1902, p. 5.
186 *Wairarapa Daily Times*, 26 March 1902, p. 2.
187 *Evening Post*, 19 January 1901, p. 2.
188 *Timaru Herald*, 6 December 1900, p. 2.
189 *Auckland Star*, 30 April 1908, p. 6.
190 *Evening Post*, 23 December 1899, p. 5.
191 *NZPD*, 117 (1901), p. 89.
192 Hugh Pollen to Colonial Secretary, 5 April 1906, AAYS 8638 AD1 Box 474 ae D1906/1016, ANZ.
193 Charles Baré to Thomas Kay Sidey, 2 August 1903, Charles M. Baré, AABK 18805 W5515 0000234, ANZ.
194 F. Shaw to Edward Walter Clervaux Chaytor, 12 June 1903, AABK 18805 W5515 0003117, ANZ.
195 Frederick Gordon Armstrong to R. J. Seddon, 30 October 1902, AABK 18805 W5515 0000134, ANZ.
196 Frederick Gordon Armstrong, J46 Box 276 COR 1904/354, ANZ.
197 *NZPD*, 116 (1901), p. 379.
198 *Press*, 30 September 1902, p. 5.
199 *Hawera & Normanby Star*, 13 December 1902, p. 2.
200 *New Zealand Gazette*, No. 3, 8 January 1903, p. 89; *New Zealand Gazette*, No. 62, 6 August 1903, p. 1727.
201 Attestation Form, 17 February 1902, John Bowden, AABK 18805 W5515 0000498, ANZ.
202 *Ohinemuri Gazette*, 18 May 1903, p. 2.
203 *Otago Witness*, 12 September 1900, pp. 48, 63; *Clutha Leader*, 7 June 1901, p. 5; *Auckland Star*, 4 January 1901, p. 5.
204 *Timaru Herald*, 10 May 1901, p. 3.
205 *Press*, 11 February 1902, p. 3.
206 *Press*, 11 February 1902, p. 3.
207 *Press*, 11 February 1902, p. 3; *Press*, 2 September 1921, p. 8.
208 *Press*, 21 October 1902, p. 2.
209 *NZPD*, 117 (1901), p. 483.
210 *Press*, 21 August 1901, p. 4.
211 *Press*, 21 October 1902, p. 2.
212 *Evening Star*, 17 October 1902, p. 5; *Otago Daily Times*, 18 October 1902, p. 4.
213 *Evening Post*, 9 September 1901, p. 6.
214 *Otago Daily Times*, 21 October 1902, p. 3.
215 *Evening Post*, 9 September 1901, p. 6.
216 *Evening Post*, 9 September 1901, p. 6.
217 Evidence of Sir Arthur Percy Douglas, Minutes of Evidence taken before the Royal Commission on the War in South Africa, Vol. I, London: His Majesty's Stationery Office, 1903, p. 425.
218 F. J. Knight to Lieutenant-Colonel Newall, 1 January 1899 [1900], Charles Oscar Hagenson, AABK 18805 W5515 0002276, ANZ.
219 *Press*, 30 September 1902, p. 5; *Press*, 9 October 1901, p. 6; *Press*, 16 October 1901, p. 4.
220 *NZPD*, 116 (1901), p. 189.
221 *AJHR*, 1901, D-2, p. 15.
222 *AJHR*, 1903, D-2, Appendix A.
223 *AJHR*, 1903, D-2, p. 11.
224 *NZPD*, 111 (1900), p. 375.
225 *AJHR*, 1901, D-1, p. 133.
226 *AJHR*, 1902, D-1, p. 61.
227 *AJHR*, 1902, C-1, p. xxii.
228 *AJHR*, 1901, C-1, p. xxv.
229 *AJHR*, 1902, C-1, pp. xxii, xxv.
230 *AJHR*, 1901, C-1, n.p., p. xxv.
231 *AJHR*, 1901, C-1, p. xxv.
232 *AJHR*, 1901, C-1, p. xxv.

233 *AJHR*, 1902, C-1, pp. xxii, xxv.
234 *New Zealand Times*, 2 April 1900, p. 3.
235 *New Zealand Herald*, 19 March 1902, p. 5.
236 *New Zealand Herald*, 19 March 1902, p. 5.
237 *AJHR*, 1901, Session I, H-16, p. 3.
238 South African War 1899–1902, AABK 18805 W5515 0004519, ANZ.
239 *AJHR*, 1897, H-38, p. 3.
240 Herbert Pilcher to Richard J. Seddon, AABK 18805 W5515 0004519, ANZ.
241 Relative to Transvaal map, Patriotic Fund, and Public Service presentation to Major General Baden Powell, LS-W1 383 19480, ANZ.
242 *Evening Post*, 28 May 1902, p. 6.
243 *New Zealand Herald*, 20 January 1900, p. 5.
244 *Otago Witness*, 28 June 1900, p. 30; *AJHR*, 1899, D-3, p. 23.
245 *Otago Witness*, 28 June 1900, p. 30.
246 *Evening Post*, 23 June 1900, p. 5.
247 Norman Mitchell Keane, 17 May 1902, Permit to Land in South Africa, ACGO 8333 IA1 Box 851/[76], 1902/1743, ANZ .
248 Attestation Form, n.d., Frederick Harcourt, AABK 18805 W5515 0002336, ANZ; *Wanganui Collegian*, No. 55, April 1901, p. 22.
249 *AJHR*, 1901, F-1, pp. iii; Attestation-Form, 25 January 1901, Charles Borland Tasker, AABK 18805 W5515 0005468, ANZ; Attestation-Form, 25 January 1901, Manson Waterson Chant, AABK 18805 W5515 0006256, ANZ; Attestation-Form, 3 April 1901, Charles Federick Normanby Minifie, AABK 18805 W5515 0003917, ANZ; Attestation-Form, 17 February 1902, Edgar Spiers, AABK 18805 W5515 0005274, ANZ.
250 *AJHR*, 1900, F-1, p. iii.
251 *AJHR*, 1900, H-6, p. 2; *AJHR*, 1901, H-6, pp. 11–12; *AJHR*, 1901, H-6A, pp. 3, 5, 13; *AJHR*, 1902, H-6A, pp. 14, 20, 23, 27; *AJHR*, 1902, pp. 2, 5, 9, 14, 18, 22, 24. Others who gave their occupation on enlistment as 'telegraphist' were John Martin of the Sixth Contingent; George Chetwyn Parker and Private Harry Bateman of the Seventh Contingent; Private William Iorns, Private Kenneth Rennie, Signaller Robert Crump and Corporal Otto List of the Eighth Contingent; Reginald Darrow, Clyde McGilp and James Jordan of the Ninth Contingent; and Private Arthur Card, Private John Dobson, Private James Bateman, Lance-Corporal Henry Lean, and Private Hugh Menzies and Regimental Quarter-Master Sergeant John Montgomery of the Tenth Contingent.
252 *AJHR*, 1902, H-6A, p. 26.
253 *AJHR*, 1902, F-1, p. iii.
254 *AJHR*, 1903, B-1, p. 60.
255 Post and Telegraph Rifle Volunteers, capitation roll, 13 March 1902, ARM41 142 1911/73d, ANZ.
256 Attestation-Form, 26 February 1902, Clyde McGilp, AABK 18805 W5515 0003446, ANZ.
257 *Poverty Bay Herald*, 23 March 1901, p. 2; *Poverty Bay Herald*, 10 September 1901, p. 2; *Manawatu Evening Standard*, 26 March 1901, p. 2; *Wanganui Chronicle*, 21 June 1900, p. 2; *New Zealand Herald*, 1 April 1901, p. 6; *Free Lance*, 29 December 1900, p. 3; Attestation-Form, 21 April 1902, Harry Holford, AABK 18805 W5515 0002611, ANZ.
258 *Manawatu Evening Standard*, 12 December 1900, p. 2.
259 *AJHR*, 1899, Session I, F-5, pp. 27, 30.
260 *Poverty Bay Herald*, 10 September 1901, p. 2; *Manawatu Evening Standard*, 26 March 1901, p. 2.
261 *Manawatu Evening Standard*, 28 August 1902, p. 2.
262 *AJHR*, 1902, Session I, F-1, p. vii.
263 *AJHR*, 1906, B-20C, p. 5.
264 *AJHR*, 1901, Session I, F-1, p. xii.
265 *NZPD*, 111 (1900), p. 364.
266 *AJHR*, 1901, F-1, p. xii; *AJHR*, 1902, F-1, p. xi; *AJHR*,1903, F-1, p. xvi; *AJHR*, 1904, F-1, p. xiii.
267 *AJHR*, 1901, F-1, p. xii; *AJHR*, 1902, F-1, p. xi; *AJHR*, 1903, F-1, p. xvi. Three hundred and nine parcels were sent to New Zealand from South Africa in 1900, 886 in 1901 and 570 in 1902.
268 *AJHR*, 1902, H-27A, p. 1.
269 Penelope Farquharson to William Farquharson, 25 August 1900, NAM NZ 1998.11.
270 John Edward Thomas Burnett, diary, 28 July 1900, NAM NZ 2004.543.
271 *Otago Witness*, 12 July 1900, p. 51.
272 *AJHR*, 1901, F-1, p. vi.

273 *AJHR*, 1900, F-1, p. v.
274 *AJHR*, 1900, H-6, pp. 6, 9, 16, 31; *AJHR*, 1901, H-6A, p. 5; *AJHR*, 1902, H-6B, p. 14; *AJHR*, 1902, H-6, p. 2. These included Trooper John O'Reilly and Lieutenant George Crawshaw of the Second Contingent, Trooper Joseph Orford of the Fourth Contingent, Corporal Wolsey Kain of the Fifth Contingent, Trooper Leo Espagne of the Seventh Contingent, Sergeant-Major Alexander Charters of the Eighth Contingent, and Captain David Cosgrove of the Sixth and Tenth contingents.
275 Attestation-Form, 11 April 1902, David Cosgrove, AABK 18805 W5515 0001204, ANZ; *New Zealand Herald*, 10 September 1920, p. 6; *Manawatu Evening Standard*, 22 May 1901, p. 2.
276 *Manawatu Evening Standard*, 22 May 1901, p. 2; Attestation-Form, n.d., John Joseph O'Reilly, AABK 18805 W5515 0004283, ANZ.
277 George Dutton to Elizabeth Dutton, 20 May 1900, p. 3, ANZ, Ref. 86-037.
278 *Colonist*, 1 March 1901, p. 4.
279 *SCNZ*, 1899, p.198; *SCNZ*, 1900, p. 203; *SCNZ*, 1901, p. 240.
280 *Evening Post*, 18 March 1901, p. 5.
281 Thomas William Porter to Commandant, telegram, 27 August 1901, C337 593 AD1 385 av D1901/3686, ANZ.
282 F. D. Morrison to Captain Hughes, memo, 26 August 1901, C337 593 AD1 385 av D1901/3686, ANZ.
283 James O'Sullivan to Under-Secretary of Defence, memo, 11 October 1901, C337 593 AD1 385 av D1901/3686, ANZ.
284 Herbert Pilcher to James O'Sullivan, 23 January 1902, C337 593 AD1 385 av D1901/3686, ANZ.
285 *Evening Post*, 12 October 1899, p. 2.
286 *Evening Post*, 23 November 1900, p. 4; *Evening Post*, 19 January 1901, p. 5.
287 Vincenzo Almao to Under-Secretary for Defence, 8 December 1884, AD1 181 es M&V 1885/2020, ANZ; Dunedin Cavalry Volunteers Corps, Form of Enrolment of Volunteers, Vincenzo Almao, 23 January 1883, AD1 181 es M&V 1885/2020, ANZ.
288 Dunedin Cavalry Volunteers Corps, Form of Enrolment of Volunteers, Vincenzo Almao, 23 January 1883, AD1 181 es M&V 1885/2020, ANZ; *Evening Star*, 22 July 1885, p. 2; *Evening Star*, 24 July 1885, p. 4.
289 *Evening Post*, 8 September 1899, p. 1.
290 *Otago Daily Times*, 1 February 1900, p. 2.
291 John Ballance to Sir William Jervois, 18 June 1885, AD1 181 es M&V 1885/2020. ANZ; James O'Sullivan to Under-Secretary of Defence, 18 April 1901, AD1 376 br D1901/1285. ANZ.
292 *Evening Post*, 3 February 1900, p. 5.
293 *Evening Post*, 20 January 1902, p. 7.
294 *Wairarapa Daily Times*, 23 January 1900, p. 3.
295 *Star*, 28 October 1899, p. 5; *Star*, 23 December 1899, p. 5.
296 *Bruce Herald*, 20 March 1900, p. 3; *Wanganui Chronicle*, 5 April 1900, p. 2; *Southland Times*, 19 May 1900, p. 3.
297 *Daily Telegraph*, 9 March 1900, p. 1; *Otago Daily Times*, 27 October 1900, p. 6; *Ashburton Guardian*, 15 February 1900, p. 3; *Evening Post*, 22 May 1900, p. 3; *Taranaki Herald*, 20 June 1900, p. 2, *Feilding Star*, 30 June 1900, p. 3.
298 *Daily Telegraph*, 16 December 1899, p. 2.
299 *Poverty Bay Herald*, 2 April 1900, p. 4.
300 *Free Lance*, 18 May 1901, p. 6.
301 *Otago Daily Times*, 16 February 1900, p. 8.
302 *New Zealand Herald*, 25 July 1901, p. 7.
303 *Southland Times*, 24 March 1902, p. 1.
304 *Taranaki Herald*, 1 March 1901, p. 1; *Ashburton Guardian*, 12 Oct 1899, p. 3.
305 *Hastings Standard*, 9 September 1896, p. 2; *Taranaki Herald*, 23 November 1899, p. 2.
306 *Wanganui Herald*, 4 November 1899, p. 3.
307 *Evening Post*, 18 December 1899, p. 6.
308 *Southland Times*, 19 May 1900, p. 3; *Evening Post*, 20 December 1899, p. 6.
309 *Evening Post*, 20 December 1899, p. 6; Penelope Farquharson to William Farquharson, transcript of letter, 25 June 1900, NAM NZ 1998.11.
310 *Wairarapa Daily Times*, 23 April 1902, p. 1.
311 *Evening Post*, 18 December 1899, p. 6.
312 *Poverty Bay Herald*, 13 June 1900, p. 2.
313 *Northern Advocate*, 10 February 1900, p. 2; *Northern Advocate*, 24 February 1900, p. 2.
314 *Northern Advocate*, 24 February 1900, p. 2;

Northern Advocate, 3 March 1900, p. 2.
315 *Otago Daily Times*, 19 January 1900, p. 1.
316 *Otago Daily Times*, 19 January 1900, p. 1.
317 *Evening Post*, 26 March 1900, p. 1.
318 AJHR, 1900, B-7, p. 15. A further £600 was allocated for celebrations in connection with the war, including activities commemorating peace, though the conflict would continue for another year.
319 *Press*, 10 August 1901, p. 5.
320 Frank Perham, diary, 8 May 1901, NAM NZ 2000.736.
321 Frank Perham, diary, 7 February 1901, 16 March 1901, 25 March 1901, NAM NZ 2000.736. Perham confuses the day and date in his diary. It is unclear which is correct.
322 Penelope Farquharson to William Farquharson, transcript of letter, n.d., July 1900; 2 December 1900, NAM NZ 1998.11.
323 George Leece to Edwin, 1 April 1901, ATL MS-Papers-8464-06.
324 James Farquharson to William Farquharson, transcript of letter, 10 March 1901, NAM NZ 1998.11; Herbert Pilcher to Under Secretary for Defence, 29 October 1901, AABK 18805 W5515 0001638, ANZ.
325 James Farquharson to William Farquharson, transcript of letter, 6 July 1900, 15 August 1900, 10 March 1901, NAM NZ 1998.11.
326 Herbert Ernest Hart diary, 1 July 1902, NAM NZ 1990.1024; *Wanganui Collegian*, No. 58, April 1902, p. 15.
327 *Wanganui Collegian*, No. 55, April 1901, p. 19.
328 *Wanganui Collegian*, No. 54, December 1900, p. 14.
329 *Wanganui Collegian*, No. 53, April 1900, p. 18.
330 *Tuapeka Times*, 13 June 1900, p. 2.
331 *Wanganui Herald*, 2 March 1901, p. 1.
332 *Wanganui Collegian*, No. 54, December 1900, p. 16.
333 *Colonist*, 1 March 1901, p. 4; *Evening Post*, 21 June 1900, p. 3.
334 *Wairarapa Daily Times*, 23 January 1900, p. 3.
335 *Tuapeka Times*, 3 March 1900, p. 2.
336 James H. Birch, *History of the War in South Africa: Containing a Thrilling Account of the Great Struggle Between the British and the Boers*, Wellington: Milton Porter [1900], p. 596, Appendix pp. 12–15.
337 J. P. Fitzpatrick, *The Transvaal from Within: A Private Record of Public Affairs*, New York: Frederick A. Stokes, 1900; Pakenham, *The Boer War*, London: Abacus, 1992, p. 88.
338 *NZPD*, 119 (1901), p. 314; *Grey River Argus*, 25 August 1900, p. 2.
339 *Thames Star*, 8 June 1900, p. 2; *Ashburton Guardian*, 25 September 1900, p. 4; *NZPD* 119 (1901), pp. 312–13.
340 *NZPD*, 110 (1899), p. 75.
341 *AJHR*, 1901, A-1 p. 10.
342 Evidence of Lieutenant-Colonel A. P. Penton, Minutes of Evidence taken before the Royal Commission on the War in South Africa, Vol. I, London: His Majesty's Stationery Office, 1903, p. 360.
343 *AJHR*, 1901, B-6, p. xxi.
344 Evidence of Lieutenant-Colonel A. P. Penton, Minutes of Evidence taken before the Royal Commission on the War in South Africa, Vol. I, London: His Majesty's Stationery Office, 1903, p. 360; *Otago Daily Times*, 11 January 1900, p. 4; *AJHR*, 1901, A-1, p. 4.
345 *Star*, 6 May 1901, p. 3.
346 Evidence of Sir Arthur Percy Douglas, Minutes of Evidence taken before the Royal Commission on the War in South Africa, Vol. I, London: His Majesty's Stationery Office, 1903, p. 425.
347 *Otago Daily Times*, 2 February 1900, p. 3.
348 *Evening Post*, 30 January 1900, p. 5.
349 *AJHR*, 1900, H-6c, pp. 1–2.
350 *AJHR*, 1900, H-6c, pp. 1–2.
351 *AJHR*, 1901, B-6, p. xxi.
352 *AJHR*, 1902, B-20, p. 1.
353 *AJHR*, 1910, B-5, p. 6; *AJHR*, 1911, B-5, p. 7.
354 *AJHR*, 1900, B-5, pp. 4–5; *AJHR*, 1901, B-5, pp. 4–5; *AJHR*, 1902, B-5, pp. 4–5; *AJHR*, 1903, B-5, pp. 6–7; *AJHR*, 1904, B-5, pp. 6–7; *AJHR*, 1905, B-5, pp. 6–7; *AJHR*, 1906, B-5, pp. 6–7; *AJHR*, 1907, B-5, pp. 6–7; *AJHR*, 1908, B-5, pp. 6–7; *AJHR*, 1909, B-5, pp. 6–7; *AJHR*, 1910, B-5, pp. 6–7; *AJHR*, 1911, B-5, pp. 6–7.
355 *AJHR*, 1901, A-1, p. 4; *AJHR*, 1901, B-1, p. 64.
356 *AJHR*, 1900, B-19, pp. 1–2, 7.
357 Paymaster General to Major H. D. Jackson, 30 September 1904, Henry Drummond

358 Jackson, AABK 18805 W5515 0006366, ANZ. Paymaster General to Major H. D. Jackson, 30 September 1904, AABK 18805 W5515 0006366, ANZ; *AJHR*, 1903, H-6A, p. 12.
359 Major H. D. Jackson to the Under Secretary for Defence, letter received 3 April 1903, AABK 18805 W5515 0006366, ANZ.
360 *NZPD*, 119 (1901), p. 413.
361 *AJHR*, 1900, B-1, p. 42. The 1900 financial year was from 1 April 1899 to 31 March 1900.
362 *AJHR*, 1901, B-1, p. 42. The 1901 financial year was from 1 April 1900 to 31 March 1901.
363 *New Zealand Official Year-Book*, Wellington: Government Printer, 1902, p. 456.
364 *AJHR*, 1901, B-5, p. 5.
365 *AJHR*, 1904, B-1, p. 46. The 1904 financial year was from 1 April 1903 to 31 March 1904.
366 *AJHR*, 1905, B-1, p. 48. The 1905 financial year was from 1 April 1904 to 31 March 1905.
367 *Marlborough Express*, 22 December 1899, p. 2; *Otago Daily Times*, 22 December 1899, p. 5.
368 *Otago Witness*, 1 March 1900, p. 26.
369 *Star*, 12 February 1900, p. 3; *AJHR*, 1902, B-1, p. 52.
370 *New Zealand Herald*, 28 October 1899, p. 5.
371 *Otago Daily Times*, 22 December 1899, p. 5.
372 *Evening Post*, 30 January 1900, p. 5.
373 *AJHR*, 1903, B-7, p. 101.
374 *NZPD*, 118 (1901), p. 615.
375 Paymaster General to Manager, Bank of New Zealand, Wellington, 15 March 1904, AABK 18805 W5515 0004601, ANZ.
376 John Pringle to Paymaster General, 19 April 1904, John Pringle, AABK 18805 W5515 0004601, ANZ.
377 Manager, Bank of New Zealand, Wellington to Paymaster General 29 March 1904, AABK 18805 W5515 0004601, ANZ.
378 *AJHR*, 1902, B-1, p. 52; *Manawatu Evening Standard*, 20 June 1900, p. 4. While the Palmerston South memorial gives Sim's unit as Kitchener's Horse, the 1902 *AJHR* claims that Sim served in the South African Light Horse, and this was repeated in the *Manawatu Evening Standard*.
379 *Otago Daily Times*, 25 January 1901, p. 3; *Otago Witness*, 30 January 1901, p. 71; *Ashburton Guardian*, 31 January 1901, p. 4.
380 *AJHR*, 1902, B-1, p. 52; Extract, n.d., Frederick Saville Broome, AABK 18805 W5515 0000596, ANZ.
381 *AJHR*, 1901, H-6F.
382 *AJHR*, 1904, B-1, p. 56; *Thames Star*, 27 November 1901, p. 4; *AJHR*, 1902, B-20B, p. 2.
383 *AJHR*, 1907, B-1, p. 62.
384 Joseph Culling to Under-Secretary for Defence, 6 December 1900; Joseph Culling to the Defence Department, 15 November 1906, AABK 18805 W5515 0001308, ANZ; Receipt from the City Buffet Hotel Wellington for Sergeant-Major Lockett's board and residence, 12 February 1904, AABK 18805 W5515 0003249, ANZ.
385 *AJHR*, 1904, B-1, p. 56; *AJHR*, 1906, B-1, p. 62.
386 A. A. Marks to Patrick W. Lee, 6 November 1906; Under-Secretary Defence to the Minister of Defence, 13 November 1905, AABK 18805 W5515 0003165, ANZ.
387 Norman Smith to unknown, 24 August 1901; W. Hall-Jones to Commandant of Forces, 7 April 1902; J. R. MacDonald to R. J. Seddon, 8 September 1902; R. J. Seddon to Commandant, 10 October 1903; AAYS 8663 AD34 Box 4/4 8015, ANZ; E.W.C. Chaytor to Minister of Defence, 28 October 1920, AAYS 8663 AD34 Box 4/4 8015, ANZ.
388 J. W. Heenan to Government Printer, 19 August 1935; H. K. Kippenburger [Kippenberger] to Government Printer, 25 June 1947, GP18 5/141/48, ANZ.
389 Evidence of Sir Arthur Percy Douglas, Minutes of Evidence taken before the Royal Commission on the War in South Africa, Vol. I, London: His Majesty's Stationery Office, 1903, p. 428; Otago newspapers put the figure at £194,189. *Otago Daily Times*, 19 August 1905, p. 4; *Otago Witness*, 23 August 1905, p. 5.
390 *Thames Star*, 14 November 1903, p. 4.
391 *Otago Daily Times*, 19 August 1905, p. 4; *Otago Witness*, 23 August 1905, p. 5.
392 D. O. W. Hall, *The New Zealanders in South Africa 1899–1902*, Wellington: War History Branch, Department of Internal Affairs, 1949, p. 9.
393 *AJHR*, 1900, B-1, p. 42; *AJHR*, 1901, B-1, p. 42; *AJHR*, 1902, B-1, p. 42; *AJHR*, 1903, B-1, p. 44; *AJHR*, 1904, B-1, p. 46; *AJHR*, 1905, B-1, p. 48.
394 *AJHR*, 1903, A-7, p. 32.

395 *SCNZ*, 1900, p. 210; *AJHR*, 1900, B-1, p. 42; *AJHR*, 1901, B-1, p. 42; *AJHR*, 1902, B-1, p. 42; *AJHR*, 1903, B-1, p. 44; *AJHR*, 1904, B-1, p. 46.

396 *SCNZ*, 1899, p. 206; *SCNZ*, 1900, p. 210; *SCNZ*, 1901, p. 216; *SCNZ*, 1902, p. 220.

397 *SCNZ*, 1899, p. 241; *SCNZ*, 1900, p. 245; *SCNZ*, 1901, p. 251; *SCNZ*, 1902, p. 256; The 1902 statistics include exports to Orange River Colony and Transvaal Colony.

398 *SCNZ*, 1899, p. 243; *SCNZ*, 1900, p. 245; *SCNZ*, 1901, p. 253; *SCNZ*, 1902, p. 258. This percentage includes £1150 of parcel post to Cape Colony and £84 parcel post to Natal.

399 *Sydney Morning Herald*, 25 January 1902, p. 12.

400 *Colonist*, 6 February 1902, p. 4.

Epilogue

1 *NZPD*, 110 (1899), p. 90.
2 *Evening Post*, 14 July 1938, p. 22.
3 *New Zealand Herald*, 16 July 1938, p. 15.
4 Louisa Hallam, 21 April 1902, Permit to Land in South Africa, ACGO 8333 IA1 849/[9] 1902/1288, ANZ; Isabella Hunter, 1 May 1902, Permit to Land in South Africa, ACGO 8333 IA1 849/[84] 1902/1421, ANZ; Emily Hagenson, 11 June 1902, Permit to Land in South Africa, ACGO 8333 IA1/854/[34] 1902/2147, ANZ; Maggie Hagenson to Colonial Under-Secretary, 1 May 1902, ACGO 8333 IA1/854/[14] 1902/2117, ANZ; Amelia Nicol Fleming Tennent, 27 March 1902, ACGO 8333 IA1 846/[5], 1902/955, ANZ; Jane Anne Nielsen, 5 June 1902, Permit to Land in South Africa, ACGO 8333 IA1/853/[15] 1902/1979, ANZ.
5 *AJHR*, 1903, H-19B, p. 2.
6 *Feilding Star*, 5 November 1913, p. 2; *Evening Star*, 6 November 1913, p. 8; *AJHR*, 1881, H-2, p. 69; *AJHR*, 1900, H-6, p. 25.
7 *Observer*, 15 November 1913, p. 4.
8 Attestation form, James Thorn, 10 February 1900, AABK 18805 W5515 0005589, ANZ.
9 Jim McAloon. 'Thorn, James', *Dictionary of New Zealand Biography, Te Ara — the Encyclopedia of New Zealand*, https://teara.govt.nz/en/biographies/4t15/thorn-james (accessed 23 September 2020).
10 *NZ Gazette*, No. 54, 4 May 1916, p. 1625; Vercoe, Henry Ray, WW1 16/161 – Army, AABK 18805 W5557 27/ 0117200, ANZ.
11 Evidence of Sir Arthur Percy Douglas, Minutes of Evidence taken before the Royal Commission on the War in South Africa, Vol. I, London: His Majesty's Stationery Office, 1903, pp. 425–26.
12 Evidence of Lieutenant-Colonel A. P. Penton, Minutes of Evidence taken before the Royal Commission on the War in South Africa, Vol. I, London: His Majesty's Stationery Office, 1903, p. 362.
13 *Waihi Daily Telegraph*, 22 December 1904, p. 3.
14 *Wairarapa Daily Times*, 19 November 1904, p. 5; T. W. Porter to Richard J. Seddon, 15 February 1905, AD1 452 e D1905/155, ANZ; *Observer*, 26 November 1904, p. 11.
15 William Hobbs to Officer Commanding 1st Wellington Battalion, 31 August 1904; James O'Sullivan to Colonel W. Webb, 21 September 1904; R. S. Lewis to Colonel W. Webb, 29 November 1904; James O'Sullivan to Colonel Webb, 6 December 1904; James O'Sullivan to Under-Secretary for Defence, 15 December 1904; F. Harrison to Lieutenant-Colonel Drummond, 7 February 1906; Under-Secretary for Defence to Commandant of New Zealand Forces, 19 March 1906; Lieutenant-Colonel Drummond to Colonel Webb, 27 January 1906, AD1 472 a D1906/504, ANZ; *Wairarapa Age*, 22 January 1906, p. 4.
16 *AJHR*, 1906, H-19, p. 1.
17 *Auckland Star*, 10 October 1907, p. 3.
18 *NZPD*, 148 (1909), p. 1418.
19 *Wanganui Collegian*, No. 59, August 1902, p. 2.

GLOSSARY

Te reo Māori

Māori words can have a number of meanings. The definitions given are those considered most appropriate to the context in which the words appear in this text.

haka	a ceremonial dance or challenge with actions and rhythmically shouted words
Kīngitanga	Māori King movement
mana	prestige and authority
mere	short, flat weapon of stone, often of greenstone
pā	fortified village, fort or stockade
poi	light ball on a string which is rhythmically twirled during a singing performance
pouri	sad, sorrowful
rangatira	Māori of chiefly status
tā moko	traditional Māori tattooing
taiaha	long weapon made of wood
tiki	carved ornament, usually made of greenstone, worn around the neck
toi moko	preserved, tattooed, sacred Māori head
utu	retribution, reciprocity

Afrikaans

assegai	short African stabbing spear
burgher	Boer citizen of a town or city
drift	river ford
kaffir	derogatory term for a black African
kloof	steep-sided, wooded valley
kommando	standard Boer military unit
kraal	traditional African village of huts, often enclosed by a fence
knobkerrie	type of African club
kopje	small hill
sangar	defensive position usually constructed of rocks
sjambok	short Boer whip
spruit	creek
Vierkleur	Boer flag of the South African Republic

ACKNOWLEDGEMENTS

I wish to acknowledge Sally Hatcher, without whom I would never have gone to university, James Watson for his advice and encouragement, David Retter for his constant assistance, and my brother Brian for reading the draft text. Although I am deeply indebted to many, I especially wish to thank Richard Bourne of Whanganui Collegiate School, Jim McAloon, Tom Brooking, Natalie McConnell, Murray and Carol Hemphill, Alastair Hosking, the Jollie family, Jon and Peter Judson, Jocelyn and Kate MacIntyre, Dolores Ho, Zabeth Botha and Judith Pryor.

ABOUT THE AUTHOR

Nigel Robson is a senior historian at the Office of Māori Crown Relations — Te Arawhiti. This, his first book, comes out of his master's thesis (2013) supervised by Associate Professor James Watson, Massey University. Robson has contributed articles to journals, including 'A Warrior Chief: Tuta Nihoniho, Porourangi and Māori in the Second Boer War 1899–1902', *The Volunteers: The Journal of the New Zealand Military Historical Society* (2012), and presented conference papers, including 'Chinaman and Trooper', a paper examining the impact of the South African War on New Zealand's Chinese community (2019 Dragon Tails 'Transformation and Transformation' conference held at Victoria University) and 'What a Trophy for one Christian to loot from another!', at the New Zealand Historians Association conference (2019), a paper which examined the looting of religious items by New Zealand soldiers in South Africa during the South African War.

INDEX

Page numbers in **bold** refer to images

Abbott, Captain Fred 110
Abd Allah, Muhammad Ahmad 12
Acland, Hugh 249
A'Court, Sydney 112, 114
Adams, John 268
Afrikaans 11
Agriculture Department 126, 287, 289, 290, 291, 292
Aislabie, Corporal Allan 135–36
Aitken, W. Francis, *Baden-Powell, the Hero of Mafeking* 26, 29, **193**
Aitken-Connell, Arnold and Mary 245
Aitken-Connell, Private John 90, 91, 245
alcohol: drunk and disorderly behaviour 170–72, **174**, 175–78, 179, 192, 215; heavy drinking after return from South Africa 268
Aldworth, Corporal Robert 98, 99
Allen, James 124, 125, 213, 264–65, 283, 284
Allen, Lance-Corporal Walter 108
Almao, Vincent 318–19
Anderson, Lance-Corporal Duncan 96
Andrew, Major Albert 192, 217, 272
Andrews, Henry 126
Anglican Church 71
Anstey, John 341
Aotea Maori Council 156
Arden, Trooper Leolin 46, 48
Argentina, competition for South African trade 294–95, 296, 333, 339
Arkell, Charles 221, 222, 230
Armstrong, Louisa 222
Armstrong, Private Leonard 184, 187
Armstrong, Trooper 234
Armstrong, Walter Douglas 203, 207, 221, 222, 230, **231**
Arnold, George 107
Arnold, James 298
Arnott, Fred 31
Ashburton Woollen Factory 317

Ashton, Corporal John 93–94
Asiatic Restriction Act 1896 172
Auckland, exports to South Africa 297
Auckland Chamber of Commerce 296
Auckland Ladies Benevolent Society 60
Auckland Peace Association (APA) 48, 78
Auckland Star 37–38
Audit Office 327
Australia: competition for South African trade 292, 293, 294, 295, 296; New Zealand military relationship 125–29, 134; New Zealand trade 278, 281, 284–85; participation in the war 147
Australian Bushmen's Contingent 71
Australian Press Association (APA) 37

Babington, Major-General 134, 235, 265
Baden-Powell, Colonel Robert 20, 22, 24, 26, 64, 68, 126, 321; gifts from New Zealanders 25, 156, 161, **194**, 313; namesakes 25
Baden-Powell, the Hero of Mafeking (Aitken) 26, 29
Baigent, Trooper Ivanhoe 132–33, 135, 136
Bain, Wilhelmina 56, 58–59, 338
Baker, Private Horace 111
Bakers' Union 74
Bakewell, Surgeon-Captain Robert 62–63, 114–15, 129, 241, 248, 250
Banks, Lieutenant-Colonel 129, 179
Barclay, Alfred 41, 74, 75, 77, 83–84, 86, 186
Baré, Charles 107, 305
Barratt, Albert 210
'Bastard's Drift' 133, 137, 203
Batchelor, Trooper Arthur 184
Bates, Daniel 68, 72, 76
Bauchop, Major Arthur 134, 316–17
Bayley, Walter 126

Bayne, John 304
Beamish, Captain Francis 331
Beath, Trooper Albert 93, 241
Beira, Portuguese East Africa 117, 124, 129, 227, 315, 324
Bell, Sergeant Allen 30
Bell, Francis 79
Benjamin, Ethel 70
Bergl and Co, 293, 294
Berland, Quarter-Master Sergeant Prosper 91–92, 270
Berry, Charlotte 237–38
Berry, Lieutenant William John 237
Betcke, Frederick 244–45
Betcke, Otto and Antonia 244
Bewicke, Charlotte 83, 86, 338
Bezar, Albert 127
Bezar, Frederick 114
Bibles 188–90, **198**
Bing, Harris, and Company 329
Birch, James, *History of the War in South Africa* 324
black African population 11, 66, 116, 119, 145–46, 148, 184, 218; siege of Mafeking 20, 26, 29; *see also* Matabele; Zulu
Blair, Sergeant Duncan 50, 98–99, 99, 106, 186, 221
Blue Spur medallion **194**
Blue Star Line 302–03
Bluff 278, 280, 282, 283, 284
Blyde, Trooper Albert 260
Boase, William 282
Bodle, Jessie 55
Boer forces 14, 20–21; age range 56; Boer guns brought to New Zealand 330–31; capture of Sixth Contingent members 45; cattle used to conceal movements 94, 100–01; guerrilla warfare 183–84, 186; knowledge of terrain 98; New Zealand media reports 22, 40, 53; proposals to inter prisoners in New Zealand 86, 149–50, 307–08; reports of treatment of British women and children 53; summary executions by British and New Zealanders 39, 45; white flag abuses 186

402

Index

Boer War *see* South African War
Boers 11, 12; Bibles 188–90; British destruction of homes and livestock 41, 42, 43; farmers' visit to New Zealand, 1903 291; German reports of British soldiers raping Boer women and girls 39; New Zealand destruction and pillaging of homes and farms 184, **185**, 186–90, 192, 218, 340 (*see also* looting and theft); New Zealand pro-Boer sentiments 24, 48, 56, 58–59, 69, 73, 74, 76–86, 144, 155, 287, 338; women and children in concentration camps 41–42, 63, 69, 74, 336
Bokfontein Farm 99
Bolton, Corporal 225–26
Bonar, Archibald 120
Bonar, Lieutenant-Colonel James 120
Booth, Harold 112
Border Mounted Rifles 29
Borlase, Walter 270
Bothasberg, Battle of *see* Langverwacht
Bottle Lake tent camp 257, 258–59
Bottom, Trooper Oscar 92, 226
Boxer Uprising, China 65, 79, 130, 171, 175, 285
Brabant's Horse regiment 103, 112, 127, 128, 130–31, 179, 237, 313
Bradburne, Lieutenant Henry 99, 133, 134
Bradford, Trooper George 223, 225, 307
Brebner, Lieutenant Charles 97–98
Brewer, Florence 58
Britain: New Zealand support of British actions in South Africa 14, 21–22, 31–32, 50, 72, 338 (*see also under* South African War); New Zealand system of government and laws 338; trade 12, 72
Britannic 107, 146, 216–17, 253–57, 258, 260, 261, 262, 263, 264
British Army 19, 20–21, 64, 69, 70, 109, 132, 137, 138, 169, 182, 184, 215, 217; accusations of atrocities 39, 163–64; dragoons 213, 215; New Zealand attitudes 100, 138; Royal Munster Fusiliers 213, 215;

South African Light Horse regiment 70; Yorkshire Regiment 91, 138
British Empire 9, 12, 14, 19, 22, 31–32, 36–37; New Zealand's loyalty 14, 15, 35–36, 38, 49, 50, 53, 59–60, 63, 65, 66, 68, 96, 154–55, 225, 338, 342
British South Africa Company 30, 66; police 12, 26, 119 (*see also* South African Police)
Broadwood, General Robert 91
'Broncho George's Wild West Show', Whangārei 321
Bronkhorstspruit, Battle of (1880) 21, 323
Brooke-Smith, Mabel 62, 136
Broome, Private Frederick 252–53, 330
Broughton, Edward Renata 146, 152, 340
Brown, Henry 246
Brown, Trooper James 189
Brown, Private Robert 233
Browne, Lieutenant Henry 116, 119
Bruce, Catherine 136
Buchanan, Walter 303
Buck, Sergeant Peter 158
Bucknall Line 294
Buddo, David 268, 269, 311, 316
Buller, General 59
bulletproof shields 248
Bullock, Margaret 56, 58, 59
Bunten, Trooper William 107, 190
Burleigh, Bennet, *Natal Campaign* 324
Burnett, Trooper John 71, 184, 316
Burnett, Trooper Richard 323
Bushveldt Carbineers 39, 136
businesses, impact of the war 317–25
Butler, Trooper John 179
Byrne, Thomas 136
Byrne, Corporal William 110, 111, 136, 137–38, 233, 341

cable communications 36–37, 84, 225, 315–16; *see also* postal services
cadets: Parnell Lady Cadets 50, **51**; school cadet system 50, 52
Calkin, Sergeant A. **157**
Calkin, Sergeant-Major C. **157**
Callaway, Sergeant-Major Walter 76, 90, 112, 133, 152,

154–55, 163, 165, 257, 258
Cameron, Robert 164
Campbell, George 108
Campbell Street School, Palmerston North 194
Campbell, W. D. 37–38
Canada 147, 291; competition for South African trade 280, 293
Canavan, Lieutenant Michael 38–39, 92, 241
Cape Colony (Kaapkolonie) 11, 12, 31, 72, 184, 248, 250, 313, 316; New Zealand direct shipping connection 295; New Zealand exports **208**, 277, 281, 282, 289, 297, 332, 333
capitalism, as motivation for war 58, 69, 73, 76–77
Cardale, Colour-Sergeant 37
Carmichael, Private Peter 94
Carncross, Walter 335, 336
Carpenters' Union 74
Carr, Corporal Robert 94
Carroll, James 127, 147, 160, 161, 164
Carson, Gilbert 58, 75, 76, 152
Carter, Fred 127
Casey, Theodore 172
casualties: Australian 125, 126; Boer 90, 95; British 21, 39; casualty reports 225; *see also* diseases
casualties, New Zealand 221–26; first New Zealander killed 223, 225; injuries sustained just before and after the war 234–35; *see also* pensions
casualties, New Zealand: accidents: fall from horse 213; firearms 110–11, 228, 341; train accident near Machavie 97, 246, 331; Worcester incident 176, 217
casualties, New Zealand: engagements: 'Bastard's Drift' 133; Bronkhorstspruit 323; Doornbutt 125; Elandslaagte 230; Klipfontein 134, 245; Laing's Nek 314, 330; Langverwacht Hill 10, 93, 94–97, 110, 136, 228, 240, 242, 247; Ottoshoop 92, 222–23, 224, 226; Paadeplaats 125, 133; Patriotsfontein 133; Rensburg 112, 245;

403

Rhenosterkop 92–93, 241;
Roode Kopjes 98–99, 133;
Slingersfontein 90; Vet
River 273; Waterval 226
Catholic Church 69–70, 71
Caulton, Lieutenant Stapylton 96
censorship 36, 44–46, 48
Chamberlain, Joseph 36, 49, 68,
  79, 91, 97, 147, 148–49, 150, 182,
  294, 295, 301, 320, 325, 326
Chambers, Reverend 65
Chant, Manson 314, 315
chaplains 71–72, 215, 235, 259
Chapman, Trooper Charles 105
Chatham Islands 307–08
Chaytor, Major-General
  Edward 189, 256, 331–32,
  338, 339
children and young people,
  New Zealand: Church Lads'
  Brigades 66; commemorative
  souvenirs 321; fundraising 64–
  65; interest in contingents and
  war-related activities 63–65, 97;
  military and physical drill 50,
  82; school cadet system 50,
  52, 341–42; selling contingent
  badges 321
Chinese in New Zealand 171–72,
  174, 175, 217–18
Chisholm, Robert 19, 21–22, 56,
  58, 69
Chodowski, Adolph 70
Christie, Trooper James 39, 136
Christ's College Register 40, 52
Chudleigh, Edward 308
Church Lads' Brigades 66
churches: opposition to British
  actions in South Africa 66–
  68; Protestant–Catholic
  tensions 69; reconciliation
  of spiritual beliefs with
  warfare 65–66, 68–72; soldiers'
  access to spiritual guidance 70–
  71; *see also* Anglican Church;
  Catholic Church; Presbyterian
  Church; Salvation Army
Chute, Major-General Trevor 127
Clark, Captain James 136
Clarke, James 188
Cleary, Henry 246
Cleland, Andrew 294, 295
climate 232–33; *see also* weather
  extremes
clothing and equipment
  contracts 298; clothing supplied

to returned soldiers 296
Clough, Trooper Edgar 308
Clutha Mounted Rifles 107, 223
Coates, Reverend 65
Cohen, Isodore 70, 178
Cohen, Mark 77
Cohen, Percy 70, 270
Cole, John 74
Coleman, Arthur 234, 238
Colenso, Battle of (1899) 20,
  107, 158
Colledge, Lieutenant James 96
Collins, A. H. 78
Collins, Alfred 126
Collins, Lieutenant (Robert's
  son) 240
Collins, Lieutenant Robert 92,
  239, 240, 312–13
Collins, William 85
Collison, Trooper Hugh 213, 234
Colonial Office, Britain 149, 150,
  291, 292, 301
Colvin, James 120, 253
Colvin, Maria 31
commemorations *see* memorials
  and commemorations
Compton, Arthur 72
concentration camps 41–42, 63,
  69, 74, 291, 336
contingents, New Zealand 35; age
  range of members 56, 129–32;
  British praise for soldiers 89, 97,
  100, 138; censorship of soldiers'
  letters 45–46; character
  assessments on discharge
  papers 137; concerns about the
  number of contingents 48–49;
  defection of men to South
  African Police 101–03;
  departures 319–20; diversity
  of members 89, 104–05,
  107–08; fundraising for troop
  support 53–54, 55, 60,
  73; gifts for members 319;
  horsemanship 105,
  107–08; marksmanship
  and weapons 108–12, 114,
  137; operation as imperial
  troops 101, 117, 183, 325, 326,
  327–28, 338; pay rates 325, 326;
  phonographic recordings sent
  to Wellington 37; politicians'
  sons 90, 91, 101, 119–25; racist
  attitudes 145–46; recognition
  for valour 99, 120, 126, 132–38,
  245, 264, 340 (*see also* medals);

recruitment activities 103–04,
  129–30; reluctance of soldiers to
  talk about experiences 267–68;
  souvenir badges, medals
  and mementoes 319, 321,
  323; unrealistic view of
  capabilities 100; veterans of
  Britain's earlier wars 114–19;
  war cries 163; *see also* Eighth
  Contingent; Fifth Contingent;
  First Contingent; Fourth
  Contingent; Ninth Contingent;
  Second Contingent; Seventh
  Contingent; Sixth Contingent;
  Tenth Contingent; Third
  Contingent
Contingents Transport
  Commission, 1902 report 217,
  255–56
Cook, Otto 127
Cooper, Private 270
Coronation Contingent, 1902 164
Corson, Trooper Douglas 250, 252
Corson, Helen 252
Corson, William 252
Cossgrove, Captain David 317
cost of New Zealand's
  participation in the war 325–33;
  imperial funds 325, 326, 327–28;
  public contributions 326, 329
Coutier, Charles 314
Coutts, Captain Henry 172, 215
Coutts, Trooper Henry
  Donald 135
Cox, Lieutenant-Colonel
  Charles 93, 95
Cox, Fred and Alfred 315
Cradock, Major 44, 99, 102, 252
Crawford, Percy 111, 341
Crawford, Thomas 111
Crawshaw, Lieutenant George 323
Crespin, Staff Sergeant
  Frederick 94–96
Cronjé, General Piet 41, 64, 93, 321
Culling, Trooper Joseph 90, 270,
  331
Curzon-Siggers, Reverend 66

Dannevirke Volunteer Rifles
  311, 341
Davidson, Reverend 68
Davies, Captain/Major 37, 102
Davies, Colonel Richard 212, 256
Davis, Colonel 181
Davis, Frances 273
Davis, Sergeant-Major

Index

Nathaniel 179–80
De Beers Cold Storage 293, 294
De la Rey, Koos 100
De Labrosse, Corporal Tyrell 254
De Latour, Lieutenant-Colonel 223, 261
De Wet, General Christiaan 42, 46, 91, 93, 97, 98, 100, 321
Defence Department 9, 114, 143, 158, 229, 248, 260, 272, 289, 296, 318, 338; contracts 318–19, 340; Māori service 149; medals 136; nurses 60, 62; pay rates 245, 309, 327, 330; pensions and payments 115, 238, 241, 242, 247, 252, 258, 265, 268, 340; recruitment and enrolment of troops 103–04, 109
defence force, New Zealand: Imperial Reserve proposal 48; Permanent Force 35, 65, 328; *see also* contingents, New Zealand; Volunteer defence force
Delaney, Trooper Fred 92
Denniston, George 69, 277
Department of Industries and Commerce 283, 303
Department of Lands and Survey 312–13
Department of Public Health 259, 260, 261, 262; graph showing increase in measles 262, **263**, 264
Devereux, Corporal Rodney 92
Dewar, Lieutenant Arthur 116–17, 119
Dewar, Private Peter 314
diamond mines 29, 58
Dickinson, Trooper Harold 187
discipline and behaviour 169–70, 217–18; crime and punishment 178–84; drunk and disorderly behaviour 170–72, **174**, 175–78, 179, 192, 215; embezzlement 210–11; officers 192, 209–11; Tasker transported to England for imprisonment 46, 181–82; theft, looting and destruction of Boer homes 178, 179, 183, 184, **185**, 186–90, 192, 209, 210–11, 212, 218, 340, 341
diseases 221, 222, 226, 227, 230, 246–68; brought back to New Zealand by returning soldiers 256–57, 261–64;

deaths 114, 249, 252, 253, 254, 255, 256, 257, 260, 261, 262, 264, 265, 313; *see also* dysentery; malaria; measles; pneumonia; smallpox; typhoid (enteric fever)
Distinguished Conduct Medal (DCM) 95, 133, 134, 136, 179, 306
Distinguished Service Order 210
Ditely, Lytton 96
Dixie, Trooper Herbert 176, 217
Dobson, Sergeant William 94
Dockrill, Edward 242
doctors 114–15, 136, 248–50, 256, 324, 339
dog tax 142, 143
dogs dispatched to head off Boer cattle 100–01
Donald, Elizabeth 31
Donald, Lieutenant 66
Donnelly, Airini 156, **159**
Donnelly, Ann 228, 229
Donnelly, Robert, William and Peter 228, 229
Doornbutt 125
Doornkop 30
Douglas, Sir Arthur 309, 326, 332, 340
Douglas, Lady Mary 60, 197
Douglas, W. R. 103
Doyle, Arthur Conan 272–73; *The Great Boer War* 324
Dresden Company 329
Drew, H. V. 248
Drinnan, Corporal Robert 134, 264–65
drunk and disorderly behaviour 170–72, **174**, 175–78, 179, 192, 215
Duigan, Henry 233
Duigan, John Evelyn 130–32, **131**, 270
Duirs, Alexander 30
Duke of Edinburgh's Own Volunteer Rifles 115
Duncan, Thomas 306, 307
Dunedin 74, 297, 298, 336; celebration of relief of Mafeking 19, 21–22, **23**, 24, 58; fundraising for troops 53, 55
Dunedin Chamber of Commerce 73, 277
Dunedin Presbytery 41–42, 70–71
Dunnet, William 270, 271
Duthie, John 144
Dutton, Captain-Chaplain Daniel 71, 72, 198, 199

Dutton, George 317
dysentery 117, 222, 227, 228, 232, 246–47, 264

Earle, William 323
Easterfield, Thomas 321
economic impact of the war 15, 49, 268–69, 274, 292; *see also* cost of New Zealand's participation in the war; labour availability during the war; trade with South Africa
Education Department 63, 317
Edward VII, King 134, 164, 182, 212
Egan, Farrier James 71
Eighth Contingent 43, 99, 100, 114, 127, 128, 129, 145–46, 338; behaviour and discipline 55, 170–71, 177, 178, 210, 211, 212; chaplain 71, 72; departure 56, 311–12; diseases 253–57, 260; Machavie rail accident 97; Māori membership 164, 165, 166; opposition to dispatch 73–74; returned soldiers 308; Seddon's son 101, 120; sheep and cattle dogs 101; watches 319
Elandslaagte 230
Ellice Street Rifles 65
Elliott, James 248, 249, 264
Ellis, Lieutenant-Colonel 242
Eloff, Field Cornet 267
Emery, Alfred 126
employment *see* labour availability during the war
Empson, Walter 50, 190
Evening Post 37, 46
Evening Star 53, 56, 64, 77
exports to South Africa *see* trade with South Africa

Fahey, Trooper James 227
farming expertise, New Zealand 291
Farquhar, Private Peter 234
Farquharson, Leonie 55
Farquharson, Penelope 48, 55, 241, 250, 323
Farquharson, Trooper William 44, 241, 250, 323
Farrow, Trooper Pressney 180
Feasley, Private John 99
Feldwick, Henry 262
Fenwick, Surgeon-Captain Percival 130

405

Field Punishment No. 1 180
Field, William 9, 308
Fifth Contingent 71, 114, 116, 117, 125, 127, 129, 161, 163, 166, 176, 180–81, 205, 228, 320, 325, 326, 329, 338
Fiji 49
Finch, Harry 241–42
Finch, Martha and Samuel 241–42
Finch, William and Samuel 'Frederick' 241–42
Findlay, George 320
Finn, Bertram 126
First Anglo-Boer War (1880–81) 11, 12, 14, 21
First Contingent 37, 40, 49, 50, 64, 65, 73, 107, 121–22, 250; behaviour and discipline 190; casualties 223; costs 325, 326; debate about participation in the war 15, 31–32, 35, 75, 76, 77, 78, 80, 85, 109, 146–47, 335; fundraising 38; gifts from home 55; kinematograph scenes of departure 320; Slingersfontein 89–91; soldier with Māori heritage 152; 'Soldiers of the Queen' souvenir booklet **202**; uniform 233, 298, 318–18; war cry 163; weapons 109–10, 121–22
First New Zealand Mounted Rifles Association 44
First World War 15, 72, 73, 125, 158, 232, 233, 249, 292, 336, 340, 342; South African War veterans 98, 121, 131, 136, 166, 268, 270, 339
Fisher, Captain Francis Marion 120, 209, 226, 312
Fisher, George jnr 226
Fisher, George snr 40, 120, 226, 325, 329–30
Fisher, Lance-Corporal George 226
Fitzgerald, Sergeant Edward 106
Fitzherbert, Patrick 40–41, **231**, 232, 266
Fitzpatrick, J. P., *The Transvaal From Within* 324–25
Flatman, Frederick 112
Fleming, D. 80, 217
Fleming, M. W. 265
Foley, Dean 71
Forbes, Trooper Frederick 261
Forsythe, Lieutenant William 96

Foster, Corporal Edmund 95
Fourth 'Rough Riders' Contingent 42, 49, 69–71, 73, 76, 102, 108, 145, 146; boots 329; casualties 92, 222–23, 225–26, 227, 228, 229, 234, 239, 240; consulting surgeon 250; costs 326, 327; departure 311, 319–20; fundraising 54, 171; mail 39, 42; Māori members 108, **113**; piano 329; Seddon's son 119–20, 253; souvenir badges and medals 319; William Hutchison 122–24
Fowlds, George 21
Francis, Catherine 238
Francis, Lieutenant-Colonel Frederick Wyatt 238, 240
Francis, Major Frederick 124
Fraser, Alfred 163, 238, 278
Fraser, Farrier-Sergeant Patrick 108
Fraser, Simon 250
Free Lance 40
Freeth, Pierce 49
French, Major-General John 30, 54, 59, 89, 107
Friedlander Brothers 305
Friedlander, Hugo 304–05
Fritz, Joseph Johannes 188
Fullers Royal Waxworks and Big Vaudeville Company 320
Fulton, Captain Harry 124
fundraising for troop support 53–54, 55, 60, 64–65, 73, 321, 335; by Māori 155, 160–61; public contributions 326, 329; *see also* names of individual organisations, e.g. More Men Fund

Galpin, Edgar 323
Gannon, Arthur Te Wawata 113
Gatland, Edwin 126
George, James 286
Germans in New Zealand 83
Germany: expansion of interests in Africa 11; manufacture of Boer weapons 64; New Zealand antipathy, 1902 39–40; in Sāmoa 12, 49–50
Gibb, Reverend James 68
Gilfedder, Michael 75, 85
Gillespie, Sergeant-Major John 29, 30, 269

Gisborne Women's Political Association (GWPA) 58
Gladstone, William 226
Glass, Walter 126, 127
Glenie, Trooper Herbert 97
Godfray, Surgeon-Captain Sidney 93, 134
Goedgevonden Farm camp 110
Goffe, N. E. **157**
gold mines 58, 68, 77
Goldsmith, Farrier Frank 108
Goldsmiths' and Silversmiths' Depot, Wellington 319
Goldstein, Rabbi 70
Gordon, Major-General Charles 12, 21
Gore 25, 194
Gourley, Hugh 90–91, 120
Gourley, Sergeant Samuel 90–91
government departments and the war 44, 311–17
Gow, J. Graham 285–86, 293–94, 303
Graham, John 36, 245
Grahame, Trooper Charles 233
Grahame, Trooper Hugh 133–34
Grand Maori Carnival, Gisborne, 1900 160
Grant, Corporal 30
graves and burials 99, 271–74, 336
Gray family 228
Great Barrier Pigeongram Agency 101
Grey, Charles 304
Grey, James Grattan 84–86, 338
Grey River Argus 41
Greymouth Khaki Corps 60, **61**
Greymouth Oriental Bazaar Committee 60
Gribben, St Leger Hugh 248, 249
Grieve, Robert and Elsie 30–31
Griffin and Sons Ltd 54
grooms 126–27
Groot Rietspruit Farm 94
Gudgeon, Walter 142
Guild of Loyal Women of South Africa 273, 336
Guinness, Arthur 152, 154
Gunn, Mafeking Baden Powell 25

H. W. Lloyd, Wellington 319
Hagensen, Charles 311
Hall, Charles 137
Hall, David, *The New Zealanders in South Africa 1899–1902* 332
Hall, Sir John 301

Index

Hall-Jones, William 89, 100, 112, 135, 234, 258
Hallam, Louisa 31
Hamer, Lieutenant James 116, 119
Hamilton, General Ian 89, 340
Hanan, Josiah 50, 52, 282
Handcock, Peter 39
Hanmer Springs Sanatorium 259
Harcourt, Trooper Frederick 102, 314
Hardham, Farrier-Sergeant William 132, 178
Harper, Archdeacon 169
Harper, 'Lance-Sergeant' Reginald 118
Harris, Private Arnold 187
Harrison Brothers, Kaitoke 72
Harrison-Lee, Bessie 258
Hart, Ernest 175–76
Hart, Trooper Herbert 212, 213, 216, 235, 270, 323, 339
Harvey, Trooper Henry 180
Harvey, Captain John 92, 222–23, **224**, 253, 272; 'Captain Harvey' daffodil 223
Haslett, Farrier-Sergeant Samuel 180
Hattaway, Vincent 234
Hauraki Band Bazaar 60
Hawke's Bay Freezing Company 294
Hawthorne, Sydney 72
Hay, Bessie 273
Hazeldine, J. W. 272
Hazlett, Sergeant 30
Health Department 254, 261–62
Heasley, Private John 99
heat, effect on health 232–33
Heckler, Captain Henry 253–54, 255–56
Hegglun, Marilen 250
Helm, Trooper John 133
Hemphill, Corporal Frank 133, 137, 203
Hemphill, James 228
Hemphill family 228, 229
Henderson, Trooper Donald 186
Heretaunga Mounted Rifles 49, 129
Herries, William 44, 135
Hewitson, Reverend 68
Hewitt, Dudley 92
Hewson, William 179, 184
Heywood, Lieutenant Henry 45, 269
Hiatt, Annie 62

Hickey, Major Daniel 96, 265
Hickey Private Denis 270
Hill and Son 318
Hillside Railway Workshops, Dunedin 74; employees who served in South Africa **200**
Hipango, Wāta Wiremu 156, **157**
Hoban, William 308
Hodge, Corporal Francis 314
Hogg, Alexander 188
Hōhepa, Āhere Te Koari (Arthur Joseph) 112, **113**, 154–55
Hokianga Mounted Rifle Volunteers 155
Holden, George 154
Holden, Grey 154
Holden, Richard 154
Holford, Fred 315
Holford, Harry 315
Holloway, Frank 30
Hongi Hika 155
horses 105–08, 127, 156, 223, **288**, 305, 311, 312, 329; demand for horse feed 280, 281; New Zealand exports 287, 289–90, 297, 317
Hosking, Sergeant Rupert 25, **28**, 38, 234, 251, 267–68
Hosking, William 234, 249–50, 251, 267–68
hospitals 222, 227, 228, 230, 248, 249, 250
Houchen, Henry 111
Houghton, Trooper Roland 227–28
Howe, Trooper Onesimus 180
Hudson, Frank 293
Hughes, Blanche 104
Hughes, Lieutenant Frederick 103–04, 126, 136
Huia Ladies Khaki Contingent, Dannevirke 60
Hunter, Provost Sergeant Edwin 98
Hurrey, Ernest Charles 271–72
Hurrey, Trooper Irving 271, 272
Hurrey, John 271
Hurrey, John Alexander 272
Hutcheson, John 50, 75, 76, 119, 324–25
Hutchison, George 109, 121–22, 278
Hutchison, William (*Outlook* editor) 41–42, 69, 86
Hutchison, William jnr 122, **123**, 124–25, 270–71

Hutchison, William snr (father of George Hutchison) 122
Hutton, Farrier-Sergeant (Trooper) Archibald 108, 114, 127
Hutton, Major-General Edwin 89
Hyde, D. D. 291
Hyde, Private George 92
Hynes, Trooper Thomas 232–33

Imperial Cold Storage and Supply Company 293, 295
Imperial Light Horse 107, 222, 232
Imperial Reserve proposal 48
Imperial South African Association 35; *The British Case Against the Boer Republics* 325
Inglis, Alice St Clair 55
Invercargill 50, 52, 282–83, 284, 297
Irish: nationalism 76, 84; pro-Boer sympathies 83; in the South African war 69–70
Isandlwana, Battle of 12

Jack, Trooper James 110
Jack, Robert 30
Jackson, Frank 127
Jackson, Major Henry 328
Jacobs, Bugler Leonard Noly 172, **174**, 175
Jameson, Leander Starr 11–12, 30, 320
Jameson Raid (1895–96) 12, **13**, 30, 64, 78, 79, 117, 152, 320, 324
Janse Van Vuuren, B. M. 188, 189
Jansen, Captain 77–78
Jeffreys, Constance Geraldine 62, 134
Jellicoe, Edwin 182
Jenkins, Graeme 189–90
Jennings, William 210, 258
Jensen, Jack 131
Jensen, Vilhelm 79–80
Jewell, Trooper Claude 39, 42
Jewish community 70, 77, 172
Jickell, Sergeant-Major William 92
jingoism, New Zealand 53–55, 74–75, 85, 338; *see also* patriotism, New Zealand
Johannesburg Reform Committee 324
Johnsonville, street lamp memorial 9, 10, 336
Johnston, Trooper Daniel 30, 107
Joint Defence Committee 48, 49
Jollie, Sergeant-Major Edward

'Teddie' 25–26, **26**, **27**, 267, 323
Jollie, Sarah 27, **27**
Jones, Melita 62
Joubert, General Piet 64
Jowsey, Major Thomas 44, 105–06, 177–78, 179, 183, 198
Julius, Bishop, prayer books given to Third Contingent **198**

'kaffir' 145–46
Kaffrarian Mounted Rifles 163, 186, 226
Kaiapoi Woollen Manufacturing Company 329
Kaihau, Hēnare 146, 156
Kaire, Bugler M. **157**
Kamau Taurua Quarantine Island 257, 261
Karaitiana, Kuku 160
Karori Camp 36, 53, 65
Kawakawa Rifle Volunteers 155
Keane, Norman 314
Keddell, Walter 126
Kelly, James 75
Kelly, Thomas 36
Kendall, Trooper Arthur 261
Kennaway, Walter 292
Kerei, Hori **157**
'khaki fever' 48, 49, 60, 155; *see also* militarism
Khartoum 12, 21
Kimberley: diamond mines 29, 269; siege 20, 29, 30, 31, 70, 161
Kimberley Light Horse 29, 30
kinematograph 320
King, Surgeon-Captain Frederick 247, 258
Kingi, Captain **157**
Kingi, Takarangi Mete 156, **157**
Kingi, Weraroa **157**
Kīngitanga (Māori King movement) 141, 165
King's South Africa Medal 136, 137
Kipling, Rudyard 35; 'The Absent-Minded Beggar' 53
Kirikiri School 80–81
Kirk, Charles 30
Kirkwood, James 93–94, 111
Kitchener, Herbert Horatio, 1st Earl 20, 45, 89, 93, 96, 97, 104, 180, 186, 190, 320, 321
Kitchener's Fighting Scouts 131, 179
Kitchener's Horse regiment 127
Klee, Emily 229
Klee, Trooper George 94, **224**, 228, 229

Klee, Lance-Corporal Louis 94, **224**, 228, 229
Klee, Trooper Victor 94, **224**, 228, 229
Klipfontein 134, 245
Knight Templar 103, 105–06
Koorn Spruit 91–92, 100, 120, 135
Kruger, Paul 12, 22, 35, 64, 65, 226, 320, 321; caricatures with toi moko 161, **162**
Kūmara 59

labour availability during the war 304–11
labour movement 72–74
Ladysmith, siege of 20, 29, 38, 64, 70, 73, 79, 118, 146, 161, 222, 230
Laing's Nek 314, 330
Lang, Frederick 312
Langham, James 306
Langverwacht Hill 10, 93–97, 100, 110, 112, 125, 136, 228, 240, 242, 247, 336
Lascelles, Sergeant Edward 117
Laurenson, George 150
Laurie, Lance-Corporal William 175
Law, Trooper John 133
Law, Trooper Leonard 133
Lawrence, Private Robert 45
Lawrence, Trooper William 254
Leary, Trooper Percy 115, 247–48
Lee, Trooper Patrick 246, 268, 331
Lee, Robert 50, 64
Leece, Lieutenant George 45, 133, 269, 323
Len Shing 172, **174**, 175
Leslie, Trooper William 102–03
Letts, Trooper Ralph 133
lightning strikes 233–34, 238
Lindsay, Lieutenant Michael 268
Little Barrier Purchase Act 1894 143
Littlecott, Gertrude 62, 271
Lloyd George, David 86
Lloyd's Patriotic Fund 115, 234, 237, 246, 268
Lockett, Sergeant-Major Ernest 132–33, 179, 238, 240, 331
Lockett, Hughes 30
looting and theft 178, 179, 183, 184, 186–90, 192, 209, 210–11, 212, 340, 341
Lorigan, Lieutenant William 306–07
Louw family 189

Love, Sergeant-Major Daniel 133, 242
Love, Georgina 242, 244
Loveday, Major 129
Luck, Corporal Arthur 176, 177
Lund, John 265
Lund, Walter 265
Lunn, Trooper John 260
Lunn, Margaret 260
Lusk, Trooper Charles 133
Luxford, Captain Chaplain John 72, 215, 235
Lynch, T. 73
Lyttelton 278, 280, 281, 283, 293, 297, 300

MacDonald, Lieutenant John 42, 312
MacFarlane, Arthur 111
Machavie train accident 54, 97–98, 100, 246, 331
Mackenzie, Thomas 300
Macpherson, Trooper William 90
Madden, Patrick 29
Madocks, Captain William 37, 90, 91, 102
Mafeking, relief of (1900) 20, 26, 38, 49, 64, 65, 70, 267, 323; Aitken, W. Francis, *Baden-Powell, the Hero of Mafeking* 26, 29, **193**; commemorative ribbon and patriotic badge **193**; New Zealand celebrations 19, 21–22, **23**, 24–25, 58, 73, 77, 156, 161, 335
Magersfontein, Battle of (1899) 20
Mahon, Brigadier-General Bryan 89, 103
Mahood, Sergeant William 116, 119
Mahupuku, Hamuera Tamahau 156, 160
Majuba Hill, Battle of (1881) 12, 21, 226, 242
malaria 222, 226, 227, 261, 264, 270, 306
Malcolm, Sergeant Kenneth 95, 96
Manawatu Racing Club 329
Mann, George 314
Manning, Wilfred 54
manufacturing, New Zealand 296, 297–98, 329, 340
Māori: Coronation Contingent, 1902 164; Jubilee Contingent, 1897 147, 152, 154, 156, 160; land 141, 142–43, 144, 149, 151;

Index

language 161, 163; post-1860 Māori–Pākehā conflict 142–44; racist attitudes towards Māori 142, 144–46, 151, 155; regulation of weapon sales 143–44; rights 141, 151; supporters of the Crown 141–42, 160; threat of diseases spread by returning troops 262
Maori Carnival, Wellington, 1990 160–61
Māori responses to South African War: contingent members 108, 112, **113**, 130, 133, 146, 152–55, 165–66; fundraising 155, 160; Māori support for the war 146–47, 155–56, 158, 160–61, 163–66, 340; opposition to Māori participation 145, 147, 148–49, 150, 158, 160, 164–65, 166, 339–40; Pākehā use of cultural imagery and language 161–63; pro-Boer allegations 144, 155, 161, 164; proposal for history of Māori participation 331; support of Māori participation 147, 148, 149, 150–52, 154; Taranaki Māori views 152
maps 319, 324
Markham, Captain Nigel 192
Marsh, Fanny 31
Marshall, John Taylor 127, 237
Mason, James 257–58
Massey, William 75, 101, 277–78, 303
Matabele 30, 146
Matabele Mounted Police 30, 117
Matabele War, First (1893) 117, 118
Mathews, Trooper William 272
Matiu Somes Island 254, 255, 257, 258, 260, 296
Matson, Sergeant 267
Matthews, Leo 108
Maxwell, Corporal Kerr 127
McCartney, Corporal Angus 125
McConway, Trooper 30
McDonagh, Hugh 115, 268
McDonald, Peter 112
McDonald, Lance-Corporal William 179
McDonnell, Captain **157**
McDougall, Trooper Septimus 92, 224, 272
McFarlane, Trooper William 235
McGilp, Clyde 315
McGowan, Cyril 312

McKegg, Corporal 30
McKegg, Trooper Amos 190
McKeich, Emily 238, 244
McKeich, Lieutenant Robert 234–35, **243**, 244, 328
McKeich, Walter 235
McLaren, Trooper Duncan 133, 134
McLean, Donald 144–45
McNab, Angus 248
McNab, Robert 54, 75, 77, 248–49
McNicol, Reverend 65
McRae, Jack 132
measles 253–57, 259, 260, 261, 262; Department of Public Health graph showing increase after *Britannic* and *Orient* arrival 262, **263**, 264
medals 29, 76, 95, 115, 124, 132–34, 135–37, 170, **203**, 244, 245
media, New Zealand: cable news 36–37, 45, 84; contributions to the war effort 38; depictions and accounts of warfare 40–41; diversifying to profit from the war 324; lessening of support for war 44; number of newspapers 38; opinion of Boer forces 22, 40, 53; responses to pro-Boer views 77, 78, 79, 80, 83, 84; servicemen working as civilian reporters 39; shaping of New Zealanders' attitudes 36, 38; soldiers' complaints about access to news 42, 44; South African War coverage 36–42, 44; use of letters from men in South Africa 38–39
media, South Africa 44
medical costs 331
Mellish, Lieutenant 249
Melville, Alex 229
Melville, David 229
Melville, Hamilton 229, 230
Melville, Hugh 229, 230
Melville, Trooper James 189
Melville, Leslie Seton 207, 221, 222, 229, 230, **231**
Melville, Louisa 229–30
Melville brothers 29, 31, 229
memorials and commemorations 223, 225, 272, 307, 336, **343**
mental health issues 266, 268
Meredith, Richard 81–82, 278

Messenger, William 164
Methuen, Lieutenant-General Lord 100
Mfecane 11
Miles, William Lancelot 118–19
militarism 48–52, 56, 58, 66, 78, 82
Military Pensions Act 1866 141, 236
Military Pensions Amendment Act 1903 244, 252
Military Pensions Bill 1902 240
Military Pensions Extension to Contingents Act 1900 236, 237
Millar, John 82, 255–56, 296, 298, 305
Miller, George 91
Miller, Henry 120
Miller, Sergeant Walter 95
Milner, Sir Arthur 115
Minifie, Sergeant Charles 95
Mitchell, Trooper George 30, 107
Mitchell, Robert 312
Moetara, David 155
Monaghan, Trooper 30
Monk, Ernest 30
Monk, Richard 36, 75, 170, 235–36, 280
Montgomerie, Hew 186
Montgomerie, Lieutenant John 42, **43**, 188
Montgomery's Kinematograph Company 320
Montrose 114, 217, 259, 260, 261, 265
Moolman family 190
Moore, Private Edward 101
Moore, James 186
Moore, Trooper Robert 265–66
Morant, Harry 'Breaker' 39
More Horses Fund, Otago 289
More Men Fund 38, 156, 299, 319, 321, 326; programme of event organised by Wellington More Men Fund **206**; Wanganui Native Committee 156, **157**
Morris, Private Charles 179
Morrison, Trooper Francis 42
Morton, Alf C. 39, 227–28
Morton, W. 73
Motuihe Island Quarantine Station 260
Moultray, John 37, 39
Mowlem, James 131
Mowlem, Percy 175
Mulhern, Michael 257

409

Mulhern, Patrick 188
Murray, Captain-Surgeon
　A. L. 115, 247–48
Murray, Donald 248, 249
Murray, James 80–81, 82
Murray, John 137
myriorama 321

Napier 24–25, 80
Napier, William 49, 109
Napier Guards Rifle
　Volunteers 154
Natal 11, 31, 45, 72, 184, 229, 248, 250, 261; New Zealand direct shipping connection 295; New Zealand exports **208**, 281, 282, 289, 297, 332, 333
Natal Border Mounted Rifles 222, 229–30
Nathan, Arthur 287
Nation, Lance-Corporal Percy 96, 233–34
National Council of Women (NCW) 24, 56–59
national identity, New Zealand 15
Naughton, William 74
Neave, Lieutenant Arthur 52
Nekitini, Te Aohau 156, **157**
New South Wales Citizens' Bushmen's Contingent 126
New South Wales Mounted Infantry 126, 232
New South Wales Mounted Rifles, 3rd 93
New Zealand and African Steamship Company 302–03
New Zealand Clothing Factory 329
New Zealand Crown Mines, Karangahake 54–55
New Zealand Ensign Act 1901 80–81
New Zealand Hill 89–91, 138
New Zealand Mounted Rifles 90, 339
New Zealand Patriotic Fund 156, 245, 252
New Zealand Press Association (NZPA) 37
New Zealand Rifle Association 52
New Zealand Shipping Company 299, 316
New Zealand Soldiers' Graves Upkeep Fund 273
New Zealand Tablet 38, 41, 45, 69, 70, 75, 246

New Zealand Wars 141, 147, 165, 221, 232
New Zealand Young Ladies' Contingent 60, **197**
Newall, Lieutenant Stuart 270
Newcastle, Natal 209, 211–13, **214**, 215–18, 339, 340, 341
Newcombe, Katie 59
Newell, Lieutenant-Colonel Stuart 36, 146, 338–39
Newlyn, John 150
newspapers *see* media
Newtown Park military camp: costs 329–30; protest 169–70
Ngāi Te Rangi 164
Ngāi Tūmapuhiārangi 156
Ngāpua, Hōne Heke 143, 146
'Ngapuhi Nursing Sisters' 155, 156, **157**
Ngātai, Hōri 164
Ngāti Apa 156
Ngāti Kahungunu 156, 160, 163, 164
Ngāti Kauwhata 160
Ngāti Maniapoto 164
Ngāti Porou 158, 160, 163, 165
Ngāti Porou Rifles 149, 158, **159**, 160
Ngāti Raukawa 160
Ngāti Toa 160
Nichol, Trooper George 110
Nicholls, Flora 83
Nicholson, James 254–55, 256
Nicholson, Wallace 254–55
Nielsen, Jane 31
Nightingale, Florence 62, 63
Nihoniho, Tuta (Matutaera) 158, **159**, 160, 170, 204
Ninth Contingent 49, 100, 114–15, 130, **243**; behaviour and discipline 171, 175, 183, 209, 211, 212, 217; boots 298; casualties 235; costs 238; departure 76, 311–12; disease 247–48, 253–57; Māori members 146, 152; medals 136; occupations of members 105, 108; opposition to dispatch 75; programme for church service **199**; returned soldiers 308; sheep and cattle dogs 100–01; soldiers' faith 71, 72; Trooper Leolin Arden's death 46–47; weapons accident 111
Noot, William 74

Norfolk Island 281
Norris, Farrier George 180
Nunneley, Harold 135–36
Nurse, Corporal Charles 114
nurses 29, 59, 60, 62–63, 134, 155, 236, 249, 250, 258, 271, 273, 336, 339

Ōamaru 280, 281, 282, 293, 297; unveiling of Troopers Memorial **343**
O'Brien, Major Lucius 216–17
O'Dowd, Corporal 134
O'Hagan, Hugh 'Smoke' 107
O'Keefe, Trooper Alexander 110
Olivier, Commandant J. H. 126
Olivier, J. S. 189
Omdurman, Battle of (1898) 14, 20, 21
O'Neill, Father 69
Operative Sausage Case and Skinmakers Union 73
Opium Prohibition Act 1901 172
Opunake Mounted Rifles 71–72
Orange Free State (Oranje Vrystaat) 10, 11, 83, 227
Orcana (hospital ship) 62
O'Regan, Patrick 69
O'Reilly, Trooper John 114, 317
Orford, Private Joseph 42, 106, 107, 112
Orient 255, 257–59, 262, 263, 266
ostrich feathers 316, 323
Otago and Southland Nurses' Fund 62
Otago Daily Times 42, 44, 56
Otago Patriotic Committee 240
Otago Patriotic Fund 246
Otago Presbytery 71
Otago Witness 40, 41, 54, 58–59
Ōtanake soldier settlement, King Country 307
Ottoshoop 92, 100, 125, 222–23, 224, 226, 240, 312
Outlook, The 40, 41–42, 45, 66, 67, 69, 70
Owen, Alan 248, 249

Paardeberg, Battle of 40–41, 52, 115, 126, 266
Paardeplaats 125, 133
pacifism 76, 339
Page, Corporal John 133
Page, Sergeant Thomas 175
Palmerston, Norman 163, 186, 324
Palmerston North 25

Index

Pāpāwai: fundraising festival 161; Māori gathering, 1902 151, 156
Papawai (Native) Rifles 149
Parata, Katherine Te Rongokahira 161
Parata, Tame 146, 156
Parihaka 142, 164, 339
Park, James 39
Parkinson, Campbell 207
Parliamentary Union debates 64
Parnell Lady Cadets 50, **51**
Parsons, Sergeant William 313
Paterson, 'Banjo' 190, 192
patriotism, New Zealand 14, 19, 21–25, 31–32, 35–36, 39–40, 52, 54, 68, 76, 77, 81, 193, 266; Chinese 171; civil servants 313; decision to participate and early stages of the war 335; Māori 155–61, 163–64; schoolchildren 64–65, 194; souvenir badges, medals and mementoes 319; trade unionists 72–73; women 55, 58, 59, 60; *see also* fundraising for troop support; jingoism, New Zealand
Patriotsfontein 133
Patterson, Private James 242
Patterson, Mary 242
Payne Family of Bellringers 25
Pearless, Surgeon-Major Walter 256
Pearson's Weekly 292
Peat, Trooper Louis 29, 130
Peat, Willis 29
Peckover, Joseph 78
Peddie, Alexander 'Sandy' 112, 131
Pennycook, Major 209
pensions and ancillary payments: amount affected by rank 238, 240–41; death by disease 247; estimated bill per year 332; fathers 245–46, 340; mothers and sisters 240, 244, 245, 330, 340; soldiers disabled in South Africa 235–46, 264–65, 340; widows and orphans 236, 237–38, 240, 242, 244, 252, 340
Penton, Lieutenant-Colonel Arthur 21, 44, 52, 72, 90–91, 98, 102–03, 104, 152, 169, 170, 253, 325, 340
Pere, Wī 146
Perham, Trooper Frank 71, 107, 114, 175, 176, 180–81, 187, 323

Perham, Trooper Luke 42, 46, 99, 102, 133, 187, 221
Peter, Jane 62
Petersen, Julius 72
Peterson, Sergeant Andrew 40
Phair, Lieutenant Charles 95, 96
Philips, Philip 70
Physical Drill in Public and Native Schools Act 1901 82
Pickett, Private Mark 134
Pierson, Charles 62
Piet Retief 190
Pilcher, Major Herbert 190, 228, 249
Pinkerton, David 114
Pirani, Frederick 120, 240, 328
Pirani, Percy 120
Pitt, Lieutenant-Colonel 147
Pitt, Sergeant-Major William Tutepuaki 165, 210
Plimmer, John 326
Plumer, General 133, 192
pneumonia 114, 222, 227, 254, 255, 256, 257, 259, 260, 261, 313
Poland, Constable James 313
Pollen, Henry 256
Pollen, Hugh 305
Porourangi greenstone mere 158, **204**
Porter, Lieutenant-Colonel 111, 134, 318
Post and Telegraph Department 314–16; postcards **195**, 317
Post and Telegraph Rifles 315
postage stamps **195**, 317, 323
postal services 39, 42, 46, 228, 316–17; *see also* cable communications
Potchefstroom 54, 97, 98
Potter, W. S. 78
Powell, Arthur William Baden 25
Powell, Trooper George 188
Power, Sergeant Henry 313
Presbyterian Church 45, 68–69, 71; Dunedin Presbytery 41–42, 70–71; Otago Presbytery 71; programme for service at Dunedin's First Church for Ninth Contingent **199**
Press 38, 42, 80
Prince of Wales Light Horse regiment 103, 104, 135
Pringle, Captain John 330
prisoners of war: Boer soldiers 86, 149–50, 307–08; New Zealand soldiers 42, 91–92, 179
Protectorate Regiment 267
Protectorate Regiment Frontier Force 28
Public Works Department 312
Pukehika, Hōri **157**
Purdy, Sergeant-Captain John 190

'Queen's Scarf' 135
Queen's South Africa Medal 29, 124, 134, 136, 172, 179, 183, **203**, 247
Queensland Mounted Rifles 126

R. Hannah and Company 298
racism, New Zealand 142, 144–46, 151, 155, 171–72, **174**, 175, 217–18
Raikes, Trooper Francis 175
Railways Department 44, 118, 311–12, 313–14
Rakiura Stewart Island 86, 149, 307
Ralph, Julian, *Towards Pretoria* 324
Ranfurly, Uchter Knox, 5th Earl 21, 46, 63, 104, 182, 183, 307; casualties 136, 225, 235; contingents 49, 70, 90, 97; costs of New Zealand's partition in the war 326, 327; looting 190, 212; Māori 144, 148, 150; trade issues 291, 294, 295, 296, 301
Rangatira 278, 280
Rangirihau, E. **157**
Rash and Gooder 318
Rayne, Lieutenant Henry 235, **243**
Raynes, Private William 111, 137, 185
Ready, Reverend 71
Red Cross Brigade 60, 95–96
Redstone, Nellie 134
Reed, Bernard 165–66
Reeves, William Pember 46, 84, 182, 287, 289, 290, 294–95, 327
Reid, Robert Samuel 176, 177, 205
Reid, Trooper T. J. 126
religion *see* churches; Jewish community
Rensburg 112, 245
Reporting Debates and Printing Committee 85
Retter, Darcy 228
Retter, Hector 228, 336
Retter, Leonard 9, 10, 228, 336
returned soldiers: clothing 296; diseases spread by returning

411

troops 262; employment and land 305–07, 308–09, 311; heavy drinking after return 268; payment 327; *see also* veterans
Reuters 36
Rhenosterkop, Battle of 92–93, 125, 134, 233, 241, 323
Rhodes, Cecil 11, 29, 64, 67, 68, 78, 79, 117, 118, 119
Rhodes, Captain Robert Heaton jnr 120–21, 132
Rhodesian Charter Company 269
Rhodesian Field Force 124
Richards, Lieutenant Robert 108
rifle clubs 52, 108, 341
Rigg, John 77
Roberts, Frederick, 1st Earl 90, 182, 184, 189, 320, 321; jug featuring image of Lord Roberts **202**; Porourangi greenstone mere gifted by Tuta Nihoniho 158, **204**
Robert's Horse regiment 106, 178, 232
Robin, Major Alfred 37, 44, 64, 92, 96, 101, 102–03, 109–10, 125, 138, 156, 232, 233, 235, 241, 265, 273, 298, 318, 321
Robley, Major-General Horatio 161, 162
Rockstrow, Captain 265
Roddick, Helena 240
Roddick, Lance-Corporal William 240
Rolleston, Lieutenant George 120
Rolleston, William 120
Ronalds, Guy 234
Roode Kopjes 98–99
Roodewal 46
Roosevelt, President Franklin D. 145
Rosanowski, Lance-Corporal Albert 240
Rose, Captain John 177–78, 192
Ross, Trooper Hugh 145
Rough Riders 73, 112, 155, 171, 194, 203, 240, 319; *see also* Fourth 'Rough Riders' Contingent; Third 'Rough Riders' Contingent
Rouse, Farrier-Sergeant William 125
Rowley, Emily 249
Rowley's Waxworks and London Company 320–21
Royal Army Medical Corps 203, 222, 231, 249, 264

Royal Canadian Regiment 126
Royal Commission on the War in South Africa, 1903 309, 325, 326, 332, 340
Royal Commission on War Stores, South Africa 283, 284
Royal Munster Fusiliers 213, 215
Royal Red Cross 134
Rumble, Private 134
Russell, Sergeant Frederick 93
Russell, George 290
Russell, Lieutenant Lionel 264
Russell, Sir William 35, 222, 236, 264
Rustenburg 42, 43, 188
Ryan, Father 69–70
Ryan, Lance-Corporal (later Lieutenant) Francis 93, 125, 132–34

Salih, Abdol 176–77, 217
Salter, Trooper John 266
Salvation Army 77–78
Sāmoa 12, 49–50
Samuel, Private Albert 70, 234
Sanderson, John 270
Sanna's Post 270
Sargood and Sons 296
Saunders, Alan 130
Saunders, Samuel 130
Saunders, Trooper William 39, 44
Saxby, Captain Conrad 92, 107, 108, 216
Saxby, Ronald 101, 232, 270
Saxon, Trooper John 227–28
Scheepers, Commandant Gideon 135
Schnackenberg, Annie 58
school cadet system 50, 52, 341–42
Scotland, Henry 48, 74–75, 77, 119, 169, 278, 292
Scott, Agnes 58
Scott, Stanley 127
scouts 98, 100, 111, 118, 120, 125, 131, 135, 137, 179, 184
Searell, Edith 175
Second Contingent 37, 43, 55, 90, 128, 130, 177, 188; commemorative medallion **203**; costs 326; horses 305; Rhenosterkop 92–93; Roode Kopjes 99–99; support for dispatch 120–21, 160, 282; uniforms 329
Seddon, Louisa 122, 156, 273

Seddon, Mary 59–60, 156
Seddon, Richard 9, 36, 38, 39, 49, 72–73, 75, 79, 100, 104, 105, 115, 119–20, 143, 211, 242; assistance with contingent enrolments 121, 129; behaviour and discipline of troops 46, 170, 180, 182, 190, 210–11; Boer prisoners 307, 308; Boxer uprising 130; burial and graves of soldiers 271, 272, 273; casualties 92, 222–23, 224, 228; chaplains to New Zealand contingents 71–72; compulsory drill for schoolchildren 50; contingent dispatches 44; costs of New Zealand's participation in the war 325, 326, 327, 328, 329–30; diseases suffered by soldiers 249, 250, 252–53, 258, 261; First Contingent debate and dispatch 15, 31–32, 35, 75, 76, 77, 78, 80, 85, 109, 335; Imperial Reserve proposal 48; Mafeking 21; Māori involvement in the war 147, 149, 150–52, 153–54, 155, 166, 339–40; Māori language 163; medals 132, 133; memorials 225; military contribution in support of Empire 14, 21, 31–32, 35, 50, 339; nurses 62; objection to contingent members becoming police recruits 101–02; Pacific territories 49–50; pensions and payments for disabled soldiers, widows and orphans 29, 235–37, 238, 241, 245; post and cable services 316; responses to pro-Boer views 84, 85, 86; returning soldiers 305, 306, 308–09, 312; Sāmoa 49–50; school cadet system 50; Second Contingent dispatch 120–21, 160, 282; sheep and cattle dogs to move Boer livestock 100–01; Tasker 46; trade issues 277–78, 280, 285, 290–91, 294, 295–96, 298, 300, 304, 333, 339; transport ships for New Zealand troops 301; Waikato visit, 1900 163; weapons for troops 109, 114; and William Hutchison jnr 122, 124–25; women's patriotic groups 60
Seddon, Richard John Spotswood 119–20, 136

Index

Sellers, John 111
Seventh Contingent 107, 116, 121, 130, 211, 229; behaviour and discipline 79, 172, 180, 188, 189; casualties 228, 229, 240, 242; costs 325; Langverwacht 10, 93–97, 228, 247; mail 38, 39; Māori member 152; media 38, 40; returning soldiers 305–06; uniforms 317; weapons 110, 111, 112
Sew Hoy, Henry 171, **173**
Shand, James 37, 39, 45, 91, 266; 'O'er Veldt and Kopje' 332
Shappere, Harry 29
Shappere, Rose 29, 230, 249
Shaw, Savill and Albion 299
Shaw, Trooper Frederick 91, 266
Sheppard, Kate 56
Shields, Harvey 178–79
shipping to South Africa 39, 278, 280, 287, 289–90, 316; freight charges 285–86, 300, 302, 303, 330; New Zealand direct shipping connection 286, 291, 294, 295, 299–300, 301–04; via Australia 280–81, 302; wartime demands 300–01
Sieberhagen, Elizabeth 210, 211
Sievwright, Margaret 58, 59, 77
Siggins, Arthur James 119
Signal, Trooper Edward 105
Sim, George 314, 330
Simmers, Robert 272
Sinclair, Duncan 125–26
Sing, Nehora 212
Sixth Contingent 45, 46, 56, 107, 130, 230; behaviour and discipline 71, 171, 175–76, 178–79, 180, 181, **185**, 189–90, 192; burials 271, 272; Christmas card **201**; costs 325, 327; Māori inclusion 148–49; uniform 318; weapons accident 111
Skerman, Surgeon-Major Sidney 190
Sleigh, Harold C. 302–03
Slingersfontein 89–91, 100
smallpox 257–58
Smiers, Father 144
Smith, Gunner Cecil 228
Smith, Private Ernest 232
Smith, Hugh 107
Smith, Lionel 273
Smith, Norman 37
Smith, William 110–11, 118

Snyman, General J. P. 22
Soldiers' Graves Guild 273–74
Sommerville, Lieutenant-Colonel 56
South Africa: applications for permits for New Zealanders to visit 31; New Zealand civilians in 25–31; New Zealand links 31–32; New Zealand soldiers remaining after the war 101–02, 269–71, 306, 312, 315, 324, 336
South African Light Horse regiment 70, 107, 178, 314, 330
South African Police 101–03; *see also under* British South Africa Company
South African Republic (Suid-Afrikaanse Republiek) 9, 11, 25, 183, 277
South African War: catalysts 12; debate about New Zealand's participation 15, 31–32, 35–36, 55–56, 58–60, 66–68, 72–86, 109, 169, 335; domestic impact 14–15, 274, 338–40, 342; initial defeats and inconclusive victories 20–21; literature 324–25, 331–32; Māori support 146–47, 155–56, 158, 160–61, 163–66, 340; names 10–11, 53; New Zealanders in forces of other countries 125–26; public awareness 335–36; public entertainment material 320–21; trophies 321–22, 331; war-weariness 75–76, 89, 101
South African War Veterans' Association of New Zealand 225, 336
South Island Regiment 97, 338
South Rhodesian Volunteers 186
souvenir badges, medals and mementoes 319, 321, 323
Spiers, Edgar 314
Spion Kop, Battle of 40
Springburn 55
Squirrell, George 63
St Hill, George 118
St John Ambulance Association 29
Stafford and Collins 318
State-School Children Compulsory Drill Bill 1900 82
Statham, Edith 59
Stead, George 104, 305

Stead, William T. 86
Stephens, Trooper Henry Alexander 234, 238, **239**
Stevens, Trooper Bert 46, **47**, 120; postcard sent to his father **196**
Stevens, Delia 47
Stevens, Howard Waldo 47
Stevens, John (Manawatū MHR) 292
Stevens, Trooper John 62, 98, 99
Stevens, Nora 62
Stevenson, Walter 95
Stewart Dawson and Company 54
Stewart, Sir John 136
Stewart, Sir William 258, 307–08
Steyn, Marthinus 93
Stock, Sergeant Arthur 211–12
Stormberg, Battle of (1899) 20
Stout, Sir Robert 48–49
stowaways 127–29, 244
Strange-Mure, Trooper Horace 133
Strawbridge, Trooper Henry 133
Stuckey, Arthur 270
Sue Har 172, 173
suicide 266, 306
Sullivan, Michael 218
Sussex Imperial Yeomanry 176
Swan, George 238
Swanwick, Frank 269
Symes, W. H. 259

Tabakplaats 111
Tagus 261, 331
Tailoresses' Union 73
Tanner, Captain Thomas 117
Tansley, Hettie and Florence 25
Taplin, Burton 230, 232
Taplin, Jane and Samuel snr 230, 232
Taplin, Samuel jnr (Frederick) 230, 232
Taplin, Thomas 232
Taplin, William 230
Taranaki Māori Council 152
Tasker, Trooper Charles 42, 46, 56, 181–83, 314, 315, 339
Tasker, Marianne 56, **57**, 59, 181
Tatton, Trooper Harry 95
Tauranga Mounted Rifle Volunteers 152
Taylor, Thomas 75, 76–77, 84, 119, 189
Te Arawa 152
Te Aute College 156, 158, 163
Te Aute Native College Rifle Cadets 156

413

Te Hauturu-o-Toi Little Barrier Island 142–43
Te Karetu, Hēnare 154
Te Kerei 157
Te Kooti 150, 160
Te Kopa, Rea 154
Te Owai, Māta 165
Te Papapa Camp 171, 246–47
Te Puni, Private Hohiana 147
Te Tau, Taiāwhio 156
Te Ua, Captain Taranaki 147, 164
Te Waharoa, Tupu Atanatiu Taingākawa 165
Te Whiti-o-Rongomai 142, 152
teachers 63, 80–81, 271, 273, 317, 336
Teape, Bessie 187, 249, 250
Tenetahi 143
Tennent, Robert 71
Tennent, Trooper Hobart 38, 41
Tenth Contingent 71, 72, 120, 121, 129, 131, 136, 235; behaviour and discipline 209, 211–13, 216–17; diseases 114, 253–57, 259–60; dispatch 75, 97, 100; Māori member 164; returned soldiers 308; saddlery and clothing 298
Third Contingent Marsden Rough Riders' Fund 321
Third 'Rough Riders' Contingent 73, 108, 122, 127, 171, 233, 325; behaviour and discipline 177, 179, 187; commemorative medallion 203; costs 325, 326, 327; diseases 250; horses 105–06; Māori member 113, 154, 155; prayer books given by Bishop Julius 198; Rhenosterkop 92–93, 125; Roode Kopjes 98–99; war cry 163; weapons 114
Thom, Alfred 111
Thomas, Colonel Owen 103
Thomas, John 222
Thompson, John 111
Thompson, Robert 30
Thomson, James 54, 75, 77, 85
Thomson, Farrier-Sergeant William 107–08, 152, 153, 154–55
Thorn, James 339
Thornton, John 158
Thorpe, James 152
Tīmaru 278, 280, 281, 293, 297
Timaru Agricultural and Pastoral Society 294

Timber Yards Workers' Union 74
Tohu Kākahi 142, 152
Tōia, Hōne 143
Tomoana, Hēnare 160
Totman, Private Clement 55
Totman, Rebecca 55
Towers, Veterinary-Surgeon Captain 105
Townsend, Trooper John 54
trade 285–86; Britain 12, 72; China 285; tariffs 284–85
trade unions *see* labour movement; unions
trade with South Africa 72, **208**, 277, 285–87, 290–91, 297, 300, 303, 311, 317, 318, 332; butter and cheese 286–87, 291, 297, 300; cold storage facilities 286, 293–94, 295; competition and regionalism 292–98, 299; foodstuffs 278, 291; forage 277–78, 292, 297; hay 290; livestock 303; meat 291, 293–95, 296, 297, 300; oats 278, **279**, 280–84, 285, 287, 291, 292–93, 297, 300, 303, 305, 332; poultry and eggs 291, 300, 303; profits 278, 292, 297; revenue exceeded expenditure on the war 332–33, 339; uniforms and equipment 297–98; wheat 286, 303; *see also* shipping to South Africa
Trades and Labour Council 73
train accident near Machavie 54, 97–98, 100, 246, 331
Transvaal Contingent Benefit Fund 60
Transvaal Refugees' Fund Committee 59
Transvaal Relief Fund 329
Transvaal State *see* South African Republic (Suid-Afrikaanse Republiek)
Transvaal War Fund 160, 236
Trask, Captain Arthur 101
Treaty of Waitangi 141, 146, 147, 151
Trentham Camp 79, 165, 170
Tuapeka Mounted Rifles 244
Tucker, Lieutenant Frederick 93
Tuckey, Lieutenant Henry 313
Tudor, Lieutenant (later Captain) Piers 92–93, 101, 271
Tukiri, Tirua 164
Tunuiarangi, Captain 147

Tūrei, Titiripa 166
typhoid (enteric fever) 221, 222, 226, 230, 232, 242, 249, 250, 252, 253, 261, 262, 271, 306
Tyser Line 299, 300

Ulundi, Battle of (1879) 14
unemployment 304–06, 308–09, 311
uniforms 232–33, 298, 317, 318–19, 329, 340
Union Steam Ship Company 301, 316
unions 72–74
United Furniture Trades Union 74
United Press Association (UPA) 36, 37

Vaal Bank 134, 245
Van Rensburg, Mrs J. 189
Vanderberg, Peter 189
Veld Cross of Honour 137, **203**
Vercoe, Lance-Corporal Henry Reiwhati 130, 133, 152, 340
Vereeniging Treaty 235, 336
veterans: 1913 general strike special constables 339; Anzac Day 1930 **337**; graves 273–74; medals 132; public sentiment 178; retaining weapons 112, 114, 341; South African service by veterans of other wars 114, 115; *see also* pensions and ancillary payments; returned soldiers; South African War Veterans' Association of New Zealand
veterinary surgeons 105, 128, 339
Victoria Cross (VC) 132–33, 134, 136, 178
Victoria, Queen 26, 52, 73, 135, 136, 223, 224, 323; Diamond Jubilee, 1897 147, 152, 154, 156, 160
Victorian Mounted Rifles, 5th 126
Vinsen, Trooper William 92
Vintiner, James 108
Vlakfontein 184
Vogan, Sergeant Arthur 135, 136–37
Voight, Farrier-Sergeant John 108
Volunteer defence force 35, 48, 49, 50, 52, 111, 148, 149, 154, 169–70, 223, 234, 244, 315, 328, 329, 330, 338, 341

Index

Waddell, Reverend Rutherford 66, 67, 68–69, 338
Waetford, Lieutenant G. **157**
Wahanui, Hare 164
Waikato Mounted Rifle Volunteers 154
Waimana Native School 156, 158
Waimate Advertiser 39
Waitara 297, 300
Wakapuaka Cable Station 37, 316
Wakatu Mounted Rifle Volunteers 101
Waldie, Trooper David 267
Walker, Captain George 210–11
Walker, Trooper John (Hōne Wāka) **113**
Walker, Second-Lieutenant 179–80
Wallis, Frederic 65
Wanganui Amazon Carbineers 60
Wanganui Chronicle 36, 80
Wanganui Collegian 15, 25, 40–41, 42, 46, 48, 64, 221, 268
Wanganui Collegiate School 15, 21, 25, 26, 29, 41, 42, 50, 64, 92, 117, 126, 163, 190, 229, 266, 314, 323, 342; South African War memorial window in chapel **207**
Wanganui General Committee for raising troops 59
Wanganui Ladies' Club 55
Wanganui Women's Political League 59
War Office, British 46, 116, 151, 182, 222, 237, 246, 289; contracts with New Zealand 277, 281, 282, 283, 284, 287, 289, 290–91, 292–93, 294–95, 296; costs of shipping New Zealand contingents 327
War Relief Fund 155
Ward, Sir Joseph 9, 59, 73, 122, 135, 257, 260, 269, 304, 313, 314; casualties, pensions and payments 226, 235, 237, 238, 241, 246; contingent behaviour and discipline 213, 217; trade issues 283, 284, 294, 296, 299, 300, 302, 303
Warmington, Mary 134, 250
Warner, Trooper William 95, 314
Waterval 226
Watt, David 313
Wattam, Annie 31
waxworks 320–21

Way, G. E. 268
weapons and marksmanship 108–12, 114, 137; firearm accidents 110–11, 228, 341
weather extremes 233–34; *see also* climate
Webb, Dora 271
Webster, Grace 62
Weekly Press 38
Wellington, exports to South Africa 297, 299
Wellington Amazon Contingent 60
Wellington Chamber of Commerce 299
Wellington Mayor's Patriotic Fund Committee 330
Wellington Peace and Humanity Society 40, 76, 78, 189
Wellington Typographical Union 74
Wellington Woollen Company 296
Wepener 131
West Yorkshire Regiment 264
Westport Harbour Board 80
Westropp, Trooper Roland 95, 96
Wharekauri Chatham Islands 142, 149–50
Wharf Lumpers Union 73
White, Lieutenant-General Sir George 230
White, W. 294, 295
White, William Henry 127–28
Whitehead, Jessie 54
Whitmore, George 91
Whitney, Trooper Alfred 110
widows 236, 237–38, 240, 242, 244, 252; 'unworthy conduct' 244
Wilkie, Trooper Alex 46
Wilkie, Veterinary-Surgeon Henry 105
Wilkinson, Sergeant-Major 172
Williamson, Janet 134, 250
Williamson, Jessie 59
Williamson, John 72
Willis, Alexander 121
Willis, Archibald 101, 120
Willis, Private Bertie 101
Willis, Frank 121
Wilson, Corporal George 39
Wilson, Trooper Wilfred 108, 186
Witheford, Joseph 121, 132, 150, 234, 241, 258, 305, 306, 307
Witheford, Robert 121

Withers, Trooper Frank 110
Witkop 133
women, New Zealand: anti-war sentiments 55–56, 58–59; career opportunities in South Africa 336; financial hardship of soldiers' wives and families 55–56, 178, 210–11; fundraising for nurses 62; Parnell Lady Cadets 50, **51**; patriotism 55, 58, 59, 60; return to South Africa after the war 336, 338; stowaway to South Africa 128–29; support of sending troops to South Africa 56, 58, 59–60; teachers 63; unions 73; *see also* nurses
Women's Christian Temperance Union 81
Women's Democratic Union 57
Women's Political League 81
Wood, Trooper Frederick 102
Wood, Trooper James 189–90, **191**
Wookey, Archibald 125
Woollcombe, Laura 62
Worcester, Cape Colony 176–77, 180–81, **205**, 217, 218, 340–41
Worthington, Leonard 128
Wray, Trooper Ernest 258
Wright, Susie 59
Wrigley, Trooper Ellis 110, 266
Wylie, Private Frederick 134, 245
Wylie, Joseph 245–46

Yorkshire Regiment 91, 138
Young Ladies' Contingent 60, **197**
young people, New Zealand *see* children and young people, New Zealand
Young She Sue 172

Zeerust 126
Zulu 11, 118, 146, 148, 339
Zulu War (1879) 12, 116

First published in 2021 by Massey University Press
Private Bag 102904, North Shore Mail Centre
Auckland 0745, New Zealand
www.masseypress.ac.nz

Text copyright © Nigel Robson, 2021
Images copyright © as credited, 2021

Design by Kate Barraclough
Front cover image (detail): 'A Tight Corner: A New Zealander', by George Montbard. The fanciful depiction of a New Zealand trooper in South Africa was reproduced in *Defenders of the Empire*, a collection of Montbard's paintings that the *Otago Witness* described as showing 'types of our army in South Africa'. *Defenders of the Empire* was published in London and sold in New Zealand by Whitcombe & Tombs in three parts each costing 3s. (Alexander Turnbull Library, A-245-002)
Back cover image: The New Zealand Young Ladies' Contingent, also known as the Wellington Amazons, with members of the Wellington Militia (Alexander Turnbull Library, 1/2-020186-F)

The moral right of the author has been asserted

All rights reserved. Except as provided by the Copyright Act 1994, no part of this book may be reproduced, stored in or introduced into a retrieval system or transmitted in any form or by any means (electronic, mechanical, photocopying, recording or otherwise) without the prior written permission of both the copyright owner(s) and the publisher.

A catalogue record for this book is available from the National Library of New Zealand

Printed and bound in China by Everbest Printing Investment Limited

ISBN: 978-0-9951407-0-7
eISBN: 978-0-9951229-1-8

The assistance of Creative New Zealand is gratefully acknowledged by the publisher